MAKING
CANCER HISTORY

MAKING CANCER HISTORY

Disease and Discovery at
the University of Texas
M. D. Anderson Cancer Center

JAMES S. OLSON

THE JOHNS HOPKINS UNIVERSITY PRESS
Baltimore

The Johns Hopkins University Press
2715 North Charles Street
Baltimore, Maryland 21218-4363
www.press.jhu.edu

Library of Congress Cataloging-in-Publication Data

Olson, James Stuart, 1946–
Making cancer history : disease and discovery at the University of Texas
M.D. Anderson Cancer Center / James S. Olson.
p. ; cm.
Includes bibliographical references and index.
ISBN-13: 978-0-8018-9056-7 (hardcover : alk. paper)
ISBN-10: 0-8018-9056-X (hardcover : alk. paper)
1. University of Texas M.D. Anderson Cancer Center—History.
2. Oncology—Texas—Houston—History.   3. Academic medical
centers—Texas—Houston—History.   I. Title.
[DNLM: 1. University of Texas M.D. Anderson Cancer Center.
2. Cancer Care Facilities—history.   3. Medical Oncology—history.
4. Neoplasms—history.   5. Neoplasms—therapy. QZ 24 O52m 2008]
RC277.T4O47 2008
362.196'994009764—dc22
2008010925

A catalog record for this book is available from the British Library.

All illustrations courtesy of the University of Texas
M. D. Anderson Cancer Center.

*Special discounts are available for bulk purchases of this book. For more
information, please contact Special Sales at 410-516-6936 or
specialsales@press.jhu.edu.*

The Johns Hopkins University Press uses environmentally friendly
book materials, including recycled text paper that is composed of at least
30 percent post-consumer waste, whenever possible. All of our book
papers are acid-free, and our jackets and covers are printed on paper
with recycled content.

To
Mary Jane Schier,
Steve Stuyck,
and Jim Bowen

CONTENTS

PREFACE

I had cancer. I have cancer. I will always have cancer. The reader should know that M. D. Anderson has been a part of my life since 1981 when I registered there as a patient and entrusted to them my future. Time and again, they redeemed that trust. Multiple primary tumors—epithelioid sarcomas, basal cell carcinomas, and an oligodendroglioma—accompanied by a string of recurrences have required multiple surgeries on my brain, multiple surgeries on and then the amputation of my left forearm, multiple rounds of chemotherapy, and multiple rounds of radiotherapy. Physicians treated me for primary tumors or for recurrences in 1981, 1983, 1984, 1985, 1987, 1991, 1999, 2002, and 2005.

In 1981, at the age of thirty-four, I entered the clinic there wondering whether to count my life in months or years instead of decades. In 2009, I will cash my first social security check, celebrate a forty-fourth wedding anniversary, and share in the lives of four children and sixteen grandchildren. Because I owe so much to M. D. Anderson, I cannot presume to tell its story dispassionately. I have been intimately involved with its history and the object of its tender mercies; the strands of my past and those of patients, physicians, and nurses there are now entwined tightly into a rope that tethers my life.

At the same time, however, I am by trade a historian, bound to cast a clear eye on the past. The history of M. D. Anderson requires neither apology nor embellishment, and I intend the book to be more than a psalm of grateful praise. Triumphal history is boring, uncritical, and dishonest, denying the vicissitudes of human nature, with its hubris and heroism, its flashes of miserable failure and soaring success.

In *Making Cancer History*, I set out to write a thoroughly researched monograph revealing the birth and rise of a remarkable institution, and through that lens to examine the course of modern oncology, with its torturous struggle against a bewildering complex of diseases, all played out within the larger context of United States history and the history of American medicine. Monographs rarely incorporate their authors as characters, but I take such liberty: if the goal of modern oncology is to transform cancer from an acute, deadly disease into a chronic disorder managed carefully during the quest for a normal life span, then I may very well be sitting, along with other cancer survivors, at its forefront.

In 1852, the physician Samuel Gross remarked, "Cancer—All we know, with any degree of certainty, is that we know nothing." That is no longer true. We know a great deal. In 1944, M. D. Anderson registered its first patient, a man suffering from an untreatable, incurable lymphoma that took his life swiftly. Since his death, M. D. Anderson has treated more than 700,000 individual patients. In 1944, cancer was a killer disease; patients with Hodgkin's disease, Wilms' tumor, Ewing's sarcoma, non-Hodgkin's lymphomas, choriocarcinoma, embryonal testicular cancer, rhabdomyosarcoma, osteosarcoma, and acute lymphoblastic leukemia had few options but to plan their funerals; today, most with early stages of these diseases can reasonably hope for cure or long-term control. Because of chemotherapy, targeted therapy, surgery, and radiotherapy, and combinations of them, most people today can expect to be cured of many early-stage genitourinary cancers and early-stage malignancies of the breast, head and neck, prostate, skin, colon, and rectum.

Treatments to produce cures today are far less radical and debilitating than their earlier counterparts. Because oncologists calibrate radiation and chemotherapy dosages more accurately, and because supportive care is more sophisticated, cancer patients are better able to tolerate the treatments; surgeons remove far less tissue to handle tumors; and cancer rehabilitation is no longer the stepchild of oncology. Pain is managed far more effectively, hospital stays are shorter, and, at M. D. Anderson, ambulatory care has been elevated to an art form. And because of the rise of cancer prevention, the knowledge is readily available to stave off or significantly postpone most malignancies of the stomach, lung, colon, cervix, uterus, esophagus, head and neck, anus, rectum, skin, and prostate. Whether as individuals and a society we will act on our knowledge remains to be seen, but the world of oncology in 2009 contrasts sharply with its forerunner of 1944, when M. D. Anderson opened its doors to patients.

In 1946, on a flight from New York to Houston, R. Lee Clark, about to assume the directorship of M. D. Anderson, sketched out plans to build a medical institution where patients could receive multidisciplinary treatment by day and return to their homes in the evening rather than endure extended hospital stays. I am a beneficiary of that vision; after twenty-seven years and hundreds of visits to M. D. Anderson for surgery, chemotherapy, radiotherapy, diagnostic tests, interviews, and physician consults, I have missed only twelve days of work, a remarkable tribute to ambulatory care. (I realize, of course, that my flexible work schedule as a university professor makes my case unlike that of many patients.)

Making Cancer History, I hope, does justice to my profession and to the University of Texas M. D. Anderson Cancer Center, where my future will unfold.

ACKNOWLEDGMENTS

Debts accumulate out of control in a project such as this, and I am grateful to a host of people. The members of the Historical Resources Committee at the University of Texas M. D. Anderson Cancer Center encouraged me at every turn and read the manuscript or portions of it carefully. They include Stephen Tomasovic, Mary Jane Schier, Kathy Hoffman, Walter Pagel, Steve Stuyck, Charles A. LeMaistre, Ralph Freedman, and Lesley Brunet. John Mendelsohn and James Bowen offered many helpful suggestions, as did the historian David Burner. My daughter Karin Williams, a board-certified internist, saved me from the errors of a nonphysician writing medical history. She has also been a devoted, loving daughter, helping me to negotiate cancer survivorship, as have my daughters Susan and Heather and my son, Brad. Marilyn Mehr, my sister-in-law, offered many helpful suggestions.

In presentations to the Rhetoric Group in the Department of History at Sam Houston State University, I received critical feedback from colleagues Ty Cashion, Robert Shadle, Brian Dimitrovic, Nancy E. Baker, Kersten Biehn, Kenneth Hendrickson, Kate Pierce, Carolina Castillo Crimm, Jeff Littlejohn, Rosanne Barker, Susannah Bruce, and Jeff Crane. In other departments at Sam Houston State University, I peppered colleagues with questions, and I am grateful for the patience and assistance of Rex Isham in physics and Harold F. Foerster and Anne R. Gaillard in biology. Frank Fair in philosophy helped me untie a knot in the organization of the book. Dick and Frances Cording read portions of the manuscript and graciously allowed me to share the story of their daughter

Susan. Ed Coffey helped tell the story of his wife, Joan, and Don and Marsha Stringer described their family's battle with leukemia.

Making Cancer History would never have found its way into print without Lesley Brunet, the archivist at M. D. Anderson, who guided me through the archival collections, gently kept me to the grindstone during my most recent illness, and read the manuscript carefully. She has the skills of an archivist and the instincts of a historian, a perfect combination for the career she pursues.

Two generations of M. D. Anderson physicians and nurses—too many to mention individually—have attended to my illnesses. I am grateful for their talent and their tenderness.

Jacqueline Wehmueller, my editor at the Johns Hopkins University Press, has seen me through two books—*Bathsheba's Breast: Women, Cancer, and History* (2002) and now this one. She has throughout been helpful, insightful, and encouraging.

Judy, my wife and best friend, has shared every moment of my cancer odyssey, leavening episodes of fear and despair with the yeast of optimism, and chanting, in mantra, monk-like fashion, "Get the book done." To her I am the most grateful.

MAKING
CANCER HISTORY

1

R. Lee Clark, History,
and the Dread Disease

*I am disposed to believe that there does not exist in the vegetable kingdom
an antidote to cancers. . . . It is not in my power to suggest a remedy
for the cure of the disorder in her breast.*
—BENJAMIN RUSH, 1789

Freddie Steinmark first noticed the pain early in the summer of 1969 while sliding into third base during a pickup baseball game. The dull ache above the left knee he attributed to a tendon pull. A scholarship athlete with sparkling eyes and the "never quit" stubbornness that coaches treasure, Freddie played safety on the University of Texas football team. Early in September, he showed up for drills worried about the leg. Defensive backs must be agile enough to backpedal faster than quick-as-gazelle wide receivers and tough enough to butt heads with fullbacks exploding through the line. "Every defensive back," Freddie later confessed, "lives with a fear of costing his team a ball game, of being beaten by a receiver for a game-winning touchdown." A few weeks into the season, teammates noticed a limp and nicknamed him "Ratso," after Dustin Hoffman's character in the film *Midnight Cowboy*.

Freddie played through the season, spending hours in the whirlpool. By December 6, the day of the national championship game against Arkansas, his leg "hurt, really hurt. . . . For the first time, I admitted to myself there might be something seriously wrong." Not playing, however, was unthinkable. The entire state crackled with nervous anticipation. Humanities professors ignorant of the difference between a punt and a point were reading the *Austin American-Statesman* sports pages, and foreign students could be found trying to play soccer with a football. Freddie was not about to surrender his shoulder pads to a sub. The opening kickoff in Fayetteville was delayed because inclement weather kept President Richard Nixon's helicopter grounded for a few minutes. Once he arrived, the game and the national television broadcast began. Freddie played, enduring pain so severe that he could not stand on the leg between plays and in the third quarter he had to be pulled. In what many consider the greatest game in the history of college football, Texas beat Arkansas 15 to 14. In the locker room after the game, Nixon congratulated the Longhorns personally.

The next day, an x-ray revealed a mass just above the knee. To Freddie, the "news was like a physical collision with something in the dark. A truck just rammed me in the stomach." In Texas, Christianity and football compete as state religions, devotion to goal posts rivaling reverence for the cross. Powerful football coaches move the heavens for their best players, and in the late 1960s, Darrell Royal, head coach of the Longhorns, was omnipotent, a man who collected fawning state legislators in his hip pocket and commanded adoring audiences everywhere in the state. Freddie would get the best treatment in the world. Charles A. "Mickey" LeMaistre, a physician, deputy chancellor of the University of Texas System, and Longhorn fan, would see to it. He suggested that Steinmark waste no time getting to M. D. Anderson Hospital and Tumor Institute in Houston. A component of the University of Texas System, M. D. Anderson had specialized in cancer since its founding in 1941. The UT athletic department chartered a single-engine plane for the flight from Austin. LeMaistre accompanied Steinmark. With warm eyes, a gentle smile, and a voice softer than the cotton bolls of his native Alabama, LeMaistre inspired confidence. He brought the x-ray film and showed Freddie the mass, indicating that it was probably a bone tumor. Freddie stoically peppered LeMaistre with questions, hoping more than anything to be told that he could play in the upcoming Cotton Bowl, when UT would take on Notre Dame. LeMaistre could not be reassuring. Toward the end of the flight, Freddie asked, "Tell me how long I have to live?" LeMaistre replied, "Freddie, this is a very bad tumor and I would rather you wait until we have a definite diagnosis before discussing prognosis." Freddie wanted more. "What is the average time that someone lives with this tumor?" LeMaistre urged patience. After arriving in Houston, they drove to the Anderson Mayfair, the hospital's hotel for outpatients. The next morning, Freddie limped into M. D. Anderson and the world of oncology.

For the next several days, he endured tests from men he called "superstars," not "second stringers." R. Lee Clark, head of M. D. Anderson since 1946, called in the first team. During his career, Clark had pioneered the idea of "multidisciplinary cancer care," making sure that patients received a collective assessment of their disease from a team of surgeons, oncologists, and radiotherapists, rather than isolated first and second opinions. Clark handed Freddie over to Drs. Robert C. Hickey and Robert D. Moreton, who quickly established a warm relationship with the football star and protected him from the media and from admiring, if sometimes obnoxious, fans. Clark caught up with Freddie during the day and welcomed him to the hospital. A talented athlete during his younger years, Clark followed UT football and knew of the scrappy player. At the University of Texas, administrators ignored football at their peril.

Technicians x-rayed Steinmark from "from toe to cowlick . . . three times," Freddie said. Clark suspected sarcoma, and because sarcoma cells often migrate to distant sites in a process known as metastasis, the staff also x-rayed lungs, liver, and bones. Moreton, a gifted radiologist acutely interested in patient welfare, checked and double-checked the films for "hot spots"—potentially lethal metastatic lesions beyond the leg—and found none. Except for the tumor in Freddie's leg, the films were negative.[1]

During the next several weeks, as the gravity of his situation settled in, Freddie's inquiries changed from "Can I play in the Cotton Bowl?" to "Can I go to the Cotton Bowl?" He informed his parents of the likelihood of amputation.

Richard G. Martin, a surgeon and sarcoma specialist, headed Steinmark's medical team. He examined lymph nodes throughout Freddie's body, pressing his fingers carefully into the groin, the underarms, and the neck, hunting for telltale lumps. He found none. A kind, gentle man, Martin had a knack for putting patients at ease. Some, he had learned in more than fifteen years at M. D. Anderson, want no more information than the next test they will undergo, while others want to know everything, including whether they will survive and exactly how long. To those patients, unless they stood at the very end of their lives, Martin counseled patience: "In medicine, or in life for that matter, we don't know the end from the beginning."[2]

Freddie Steinmark wanted to know. On the field, he competed with the mind of a coach. Now, in the biggest game of all, Freddie needed information. If the tumor was cancer—an osteosarcoma, a Ewing's sarcoma, or some other malignant aberration of bone cells—he faced amputation. But first, the possibility of radiation had to be explored. Under the leadership of Gilbert H. Fletcher,

M. D. Anderson's efforts in radiation oncology enjoyed global luster. Fletcher, tenacious and highly opinionated, had long crusaded to replace radical surgery with radiotherapy. Martin solicited his opinion and learned that radiation was not an option.[3]

In twenty years as a surgeon, Martin had severed a tangled mountain of malignant arms, legs, shoulders, breasts, hips, hands, and feet, a burden that weighed heavily on him, prompting notions of "limb salvage" surgery—treating bone and soft tissue sarcomas with excision rather than amputation—but only fools risk lives to save limbs. In performing conservative procedures, the twin threats of local recurrence and metastasis troubled Martin. He sought to remove bone tumors and enough surrounding bone to prevent local recurrences and then implant a permanent prosthesis to rebuild the limb, a way to leave patients like Freddie with a leg and a life. Eventually, Martin would help pioneer limb salvage surgery, in which chemotherapy before local surgery reduces tumor size while innovative chemotherapy and radiotherapy after the operation render metastasis less likely. Cadaver bone and stainless steel implants were to become the architecture for rebuilt arms and legs, but in 1969, limb salvage was still in the works.[4]

Over the course of several days, Martin explained Freddie's ironic encounter with bone cancer. Because the cancer had apparently not yet spread, Freddie would be blessed with the opportunity to have his leg amputated. If the sarcoma had already settled into the lungs, why put a doomed patient through an amputation? Since Steinmark's lungs, lymph nodes, and bones appeared clear, amputation offered the only course apart from doing nothing. The operation, to a surgeon at least, was routine. Once Freddie went under the anesthesia, Martin would carve out a piece of the tumor and forward it to pathologists. If they found the tissue malignant, he would remove the leg at the hip. Freddie wondered why the hip, since the tumor nested closer to the knee. Martin knew that sarcoma cells can migrate up the bone and along muscle bundles. The last thing he wanted was to amputate above the knee and then have to amputate again if the tumor recurred in the stump. Freddie met with LeMaistre. If the tumor was malignant, "What are the odds?" he asked. "Next to zero without the surgery," LeMaistre replied. Cure rates for bone cancers were dismal—20 percent for an osteosarcoma and only 5 percent for a Ewing's sarcoma. Freddie hoped for a reprieve, at least to postpone surgery until January 2, 1970, one day after the Cotton Bowl. Martin urged him not to play. Playing might fracture the leg and spew cancer cells everywhere. Like the afflicted Dyomka in Alexander Solzhenitsyn's *Cancer Ward*—who mentally sways back and forth wondering whether "to amputate it. . . . To give it up, not to give it up. . . . To give it up, not to give it up, To give it up, not to give it up"—Freddie weighed his options.

On Thursday, December 12, he went under the scalpel. Martin took one look at the tumor and knew that the leg had to go. Few surgeons had seen more bone cancer than Martin, and Freddie's tumor displayed every sign of malignancy. The pathologists quickly identified the tumor as Ewing's sarcoma. The densely packed, small and uniform cells, with their round-to-oval nuclei and coarse chromatin granules, left little doubt. Once Martin received the news, he amputated.

Three weeks later, on New Year's Day 1970, Martin in his old Chevrolet arrived at the Anderson Mayfair accompanied by John Healey, an expert in rehabilitation medicine. They picked up Freddie, drove to Hobby Airport, and flew to Dallas in UT's new Cessna King Air. A limo chauffeured them to the Cotton Bowl, dodging antiwar demonstrators along the way. Freddie was determined to watch from the sideline. Wearing a new suit, he hobbled on crutches, the right leg of his pants carefully pinned up. Heavy rains had softened the turf, and he had to be careful; the tip of his crutches punctured the sod. In the press box, broadcasters noted Freddie's presence, and as fans with portable radios listened in, an ovation pulsed through the Cotton Bowl. Freddie leaned on his crutches throughout the game. At halftime, Governors Preston Smith of Texas and Edgar Whitcomb of Indiana visited Freddie in the locker room. Smith introduced Whitcomb to Freddie and announced, "Mr. Whitcomb brought an Indiana hog down as part of his bet on this game and when we win it, I'm going to have you over for pork ribs." UT won 21 to 17. After the game, former president Lyndon B. Johnson showed up in the Longhorn dressing room to congratulate the team; he invited Freddie to visit the LBJ ranch.[5]

President Richard Nixon had taken an unusual interest in Steinmark. Soon after the amputation, he called Freddie at the hospital and followed up with a personal letter. Already worried about the 1972 election, Nixon was searching for an issue in the tradition of God, motherhood, and apple pie, a cause the media would be unable to resist and Democrats unable to contest. Cancer was perfect. No American family was untouched. Nixon decided to declare "war on cancer," a crusade comparable to the Manhattan Project to develop an atomic bomb or President John Kennedy's space race to the moon. On April 13, Nixon hosted Freddie at a White House luncheon to launch the American Cancer Society's annual fund drive. Steinmark gave the president a football autographed by every member of the team and thanked him for the phone call. "Well," Nixon replied, "you needed a lift then. You're up and going and you don't need any help now. It's good to see you trying to help other people."[6]

But Freddie needed a great deal of help. In February 1970, he returned for his first checkup. The clinics at M. D. Anderson sagged under the weight of collective anxiety. Day after day, patients waited to learn whether their diseases had recurred, whether they were soon to endure more debilitating treatments, or worse. Like every patient, Freddie sat for a while in a general waiting area before being escorted to an examining room. A nurse soon entered, took his vitals, and left. A muffled, rustling noise sounded as she placed Freddie's chart in a slot on the other side of the examining room door. Tension mounted. Minutes can seem like eternities in the M. D. Anderson clinics. More rustling sounds. Freddie's gut twisted. Someone had lifted the chart from the slot. In walked a resident, the long white coat signifying a physician, but not *the* doctor, not *the* man with *the* news. The fellow looked at the surgical wound and probed for lumps in the groin. Another knock. Freddie's pulse quickened. *The* man entered. Martin wasted no time telling Freddie that the chest x-rays were fine, *the* news that all sarcoma patients yearn to hear. Freddie exhaled and his vitals slowed. Martin moved to the sink and washed his hands. Steinmark reclined on the examining table. Martin examined the surgical wound, and then retraced the route the resident had taken. His touch seemed more informed. He stood up, told Freddie that all was well, and returned to the sink, where he washed again. Martin made small talk and scheduled Freddie for an appointment in two months. Few moments in life are as sweet as leaving the clinic with a good report.

Freddie returned in April 1970. Like most cancer patients, he came to dread "the trips. For several days before the time for me to report . . . I can't sleep. The doctors told me long ago to stop worrying; these are just routine checkups. But that's easy to say . . . when he's the one taking the x-rays, not the one on my side of the machine." The visit went well and thoughts about the disease retreated. Late in June, however, as another checkup loomed, the worries returned. Early in July, Martin revealed that tumors flourished in both lungs. Had only one lung been involved, he might have been able to remove it. With tumors pockmarking both lungs, however, the disease now transcended surgery. Martin and Robert Moreton met with Freddie, who asked, "It looks bad, doesn't it?" Moreton delivered crushing news. "It's up to the man upstairs now."[7]

Freddie's survival odds, slim though they were, now depended on chemotherapy. Martin referred him to pediatrician Wataru W. Sutow. At M. D. Anderson, Sutow had explored chemotherapy protocols for childhood cancers, identifying several regimens for treating Ewing's sarcoma, including the drugs vincristine and cyclophosphamide. Oncologists measure progress in weeks of life beyond existing statistical averages; over the years, as one patient after another gains a few weeks, long-term survival improves. Freddie lined up in oncology's deadly queue.[8]

In a few decades, new chemotherapy protocols would remove Ewing's sarcoma from cancer's roster of automatic death sentences. Norman Jaffe, a pediatric oncologist at Harvard destined for tenure at M. D. Anderson, would add actinomycin D to cyclophosphamide and vincristine. But all this was twenty years beyond the reach of Freddie Steinmark. He steadily lost weight, hair, and time, the chemotherapy cocktail unable to kill the cancer cells. In the eighteenth century, Europeans valued *artes moriendi*, the art of dying well. Freddie mastered the art. "I know he must have had his fears," remembered teammate Ray Dowdy, "but he didn't reveal them. He enjoyed us, and we him, until the end. He was gallant." On April 20, 1971, Freddie journeyed to Houston for the last time. The sarcoma overwhelmed him. He remained an inpatient for forty-eight days, spending the rest of his life with family, friends, and M. D. Anderson staff. "Show me a hero and I'll write you a tragedy," once penned the Nobel Laureate William Faulkner. Freddie Steinmark died on June 21, 1971. Texas mourned.[9]

—⟪⟫—

Ewing's sarcoma, a "peripheral primitive neuroectodermal tumor of bone," killed Steinmark. One tumor cell became 2, 2 became 4, 4 became 16, and 16 became 256, and so on and so on in a morbid progression until billions of the little round cells choked out his life.

Give or take a few hundred million years ago, the earth's first case of cancer killed an animal. Perhaps a virus penetrated the cell's membrane and parasitically employed its host to proliferate. Maybe a supernova somewhere in the galaxy a billion years ago bathed the earth in neutrinos, radiating all living tissues and wreaking havoc with genes and chromosomes. Fossil remains demonstrate that cancer afflicted dinosaurs. Or perhaps, in a convoluted way, the very process of natural selection and evolution found in cancer a survival advantage and then programmed it into the genetic hard drive of complex organisms, giving malignant cells, like a colony of fire ants, some teleological sense of themselves. That first neoplastic cell left no remains behind, but cancer certainly existed in other species hundreds of millions of years ago. In 1932, the anthropologist Louis Leakey found in the bones of an early hominid signs of Burkitt's lymphoma, a disease endemic to Africa. Whether the earliest humans ever suffered from Ewing's sarcoma will remain a mystery, but when the first *Homo sapiens* trekked out of East Africa, other varieties of the disease surely accompanied them, a legacy that in 2007 killed nearly 8 million people around the world, including 34,170 Texans.

—◦◦◦◦∩◦◦◦◦—

In the fall of 1929, a young medical student jostled for a good seat in the surgical amphitheater at the Medical College of Virginia. R. Lee Clark Jr. was crashing the party. Just two years into medical school, he had not been invited to the operation, which several upperclassmen mentioned as soon as they saw him, but they pulled no rank. They liked Clark and his easy camaraderie. Though raised in rural Texas, he displayed a cosmopolitan self-confidence. He was hard not to like. "People could not stay mad at him for very long," a contemporary would remember. "Those blue eyes and that smile always softened hearts," another recollected. Lee Clark did have a good smile, full of gleaming, cavity-proof teeth, their roots soaked for his first two years in the fluoride-rich underground waters of Hereford, Deaf Smith County, Texas, where he was born on July 2, 1906.[10]

From German ancestors on his mother's side, Lee acquired stubborn determination and prodigious work habits. He saw things through, persisting long after lesser hearts surrendered. As a boy in Wichita Falls, Clark and his close friend Russell W. Cumley often slept outside on sticky summer nights. While the hot night air chased away slumber, they talked and talked. For several weeks in a row one summer, lying side by side, they divided the sky into grids, trying to count the stars in the Milky Way. To compensate for the earth's rotation, the boys counted from one grid to the next at the same time each night, scribbling data into an old Indian Chief notebook. They kept counting, night after night, until Cumley came across an astronomy book and learned that the stars numbered in the millions. The two switched to dinosaurs.[11]

On his father's side, Clark traced his ancestry to England, Scotland, and northern Ireland, to men and women imbued with an ethos of individualism and hard work. They were highly mobile and uprooted themselves at every opportunity, moving west frequently, each person one drop in a tidal wave of settlers. While most pioneers were farmers, the Clarks took more to prayer than the plow and more to the classroom than the barn. They spawned generations of teachers and preachers, not just those who pounded pulpits and put chalk to blackboard but builders of churches, schools, and colleges. In 1869, Lee Clark's grandfather Randolph Clark and Lee's great-uncle Joseph Addison Clark launched Add-Ran College in Fort Worth and spent the next two decades keeping it alive. The two brothers also had a hand in launching the Jarvis Institute in Thorpe Spring, Texas, in 1896; in 1901, Randolph College in Lancaster, Texas; and Hereford College in Hereford, Texas, in 1907.

Clark's father—Randolph Lee Clark—followed in their steps. An ordained minister of the Church of Christ, he founded Wichita Falls Municipal Junior College in Wichita Falls, Texas. "Sad is the day," he often preached, "for any man when he becomes absolutely satisfied with the life that he is living, the thoughts that he is thinking and the deeds that he is doing; when there ceases to be forever beating at the doors of his soul a desire to do something larger which he feels and knows he was meant to do." In many ways, he symbolized the Progressive Era of the early 1900s, when Americans celebrated education, community service, and good government. He brought the Boy Scouts to Texas and made sure that his son became one of the state's first Eagle Scouts. Reflecting years later on the culture that had pervaded four generations, Lee remarked, "It seems as if the Clarks have been known [for] making schools for several years, just like bees make honey." If it is true that children dance to the unlived lives of their parents, R. Lee Clark Jr. had his life choreographed from birth.[12]

Clark received moral support from home but little money. In the summer, he sometimes hitchhiked to the Texas Panhandle and followed the threshing crews, working in the dusty clouds behind the big combines to bag and load freshly cut wheat. Back in medical school, he hustled jobs. Blessed with elegant, almost feminine handwriting, Clark earned extra money lettering diplomas. "You could do it on your own time. . . . You didn't have to shovel coal so many hours or drive a streetcar."[13]

A capacity to divine the feelings of others played to Clark's advantage, as did his congenital optimism and athleticism. As an undergraduate at Tarleton State College in Texas and at the University of South Carolina, Clark had boxed, wrestled, and played baseball, and in medical school, he continued to wrestle, as a sophomore winning the National Amateur Athletic Federation wrestling championship for the 155-pound division. Like most competitive wrestlers, Clark obsessed about his weight. An extra egg and biscuit for breakfast every day could in a couple of weeks push him into the 165-pound division, where he would lose the winning edge. Lee Clark counted calories as assiduously as he used to number stars in the summer sky. And he liked the way he looked. When walking by a mirror or catching his own reflection in a store window, Clark often paused to admire his visage. He enjoyed striking a pose and flexing his muscles. Throughout his life, he would tell others that his father always "took care of himself and looked good" and that his mother, Leni Leoti Sypert, was about five two, weighed about 110, and "had a very narrow waist." Supremely comfortable in his own skin, Clark felt no need to conceal his vanity.[14]

He also bubbled with big dreams, an attitude that came naturally. In 1893, his father visited the Columbian Exposition in Chicago, a world's fair celebrating America, the Industrial Revolution, and 1492. Part cowboy, part preacher, and full-time student at Add-Ran College, the senior Clark wanted to see the 686-acre "White City" of new, gleaming buildings, classical facades, and gardens erected in Prospect Park. As he visited the exhibits, Clark developed a great respect for Daniel H. Burnham, the Chicago architect responsible for the White City, whose motto eventually became the Clark family anthem. "Make no little plans," Burnham wrote. "They have no magic to stir men's blood and probably themselves will not be realized. Make big plans; aim high and work, remembering that a noble, logical diagram once recorded will never die, but long after we are gone will be a living thing, asserting itself with ever-growing insistence. Remember that our sons and grandsons are going to do things that would stagger us. Let your watchword be order and your beacon beauty. Think big."[15]

Perhaps the motto reverberated in Clark's head that day in the amphitheater. He had never intended to become a family doctor delivering babies, mending broken bones, and tending to cuts and scrapes. Within weeks of registering for classes, Clark noticed that everybody treated surgeons deferentially. Clark already had a taste for deference, and his destiny—to become the first permanent director and surgeon in chief of M. D. Anderson Hospital—would provide it. He stared down from the gallery. A woman was about to be cured of breast cancer.

More than forty years earlier, William Stewart Halsted, a surgeon at Johns Hopkins, had dramatically improved survival rates for breast cancer. The operation now bore his name—the Halsted radical mastectomy. In 1894, Halsted had confessed, "Most of us . . . rarely meet a physician or surgeon who can testify to a single instance of a positive cure of breast cancer." Four years later, at a meeting of the American Surgical Association, Halsted reported on 133 patients who had undergone the procedure. More than half remained disease free after five years. Had anyone but Halsted proffered such a claim, the assembled surgeons would have laughed. At the end of Halsted's remarks, one said, "In

Dr. Halsted's series are included cases once regarded as absolutely unfit for operation, and even in [these] cases lives have been prolonged by surgical interference and rendered more comfortable. Best of all, in some very serious cases the disease has not returned after a lapse of years. [Halsted] deserves and has our greatest acknowledgments for the brilliant light which he has thrown upon these dark places of surgery." The Halsted radical mastectomy, with its possibility of a cure and its certainty of less misery for end-stage patients, was in 1929 the best-known surgical procedure of all time.[16]

Clark leaned forward. Until the era of anesthesia, surgeons required gags to muffle screams and straps to keep patients from writhing out of control. In 1846, all that had changed. After seeing the effects of ether on a dental patient, physician John Warren of Boston wrote, "A new era has opened on the operating surgeon. His visitations on the most delicate parts are performed, not only without the agonizing screams . . . but sometimes in a state of perfect insensibility." Before the appearance in the 1860s of aseptic and antiseptic medicine, those who survived surgery often succumbed to massive infections. But on this occasion, the surgeons and nurses in the amphitheater had washed their hands, instruments, and patient. The three great A's—asepsis, antisepsis, and anesthesia—gave rise to modern surgery.[17]

Earlier in the semester, a retired professor of surgery had piqued Clark's interest in cancer. Working in a laboratory, the elderly man nurtured colonies of cancer cells and produced time-lapse movies of them dividing. Cancer was the great mystery "still to be solved," he told the students, and "that solution would reveal much of the secret of life." Cancer itself, however, occupied little space in the curriculum. "Most of the study of cancer would just be part of another subject," Clark discovered. "You would study all these diseases of the various organs, then at the end of the sentence you'd find a mention of cancer. You would also find that this unexplained disease had no control or treatment." With one exception—surgery in the treatment of breast cancer.[18]

The surgeon first removed a piece of the tumor and had it whisked it off to pathology. Before proceeding, he needed to know whether he was dealing with a malignant or benign tumor, and only a pathologist could tell him. Clark breezed through pathology classes, but the specialty bored him. After majoring in chemical engineering at the University of South Carolina, he had grown weary of laboratories, with their black-topped work benches, four-legged stools, and acrid aromas. Not that Clark criticized the discipline. After all, pathology is to cancer what an umpire is to baseball.

In the 1830s and 1840s, Rudolf Virchow had first exposed cancer at the cellular level. At the University of Berlin, he peered into microscopes and eventually concluded that all disease was just that—"pathology" originating in biochemical malfunctions. Virchow gradually learned to differentiate between benign and malignant lesions, taking note of disheveled nuclei in cancer cells and their penchant for proliferation. They eventually manifested themselves as a tumor; death came when the tumor disrupted critical physiological systems. He later identified carcinomas as diseases of epithelial cells and sarcomas as maladies of connective tissues. Virchow claimed that all tissues were composed of cells, that all cells grew from similar cells, and that all living tissues shared a single-cell ancestor.[19]

In a matter of minutes, the pathologists reported that the tumor was malignant. The surgeon performed an en bloc resection, removing the breast, the pectoralis major and pectoralis minor muscles, and the lymph nodes under the arm, cutting out tissues not piecemeal, one after another, but all in a bundle, making sure not to cut through residual tumor or scatter malignant cells with a cancer-drenched scalpel. The surgeon seemed graceful and powerful, a man working with hands and head. He closed the wounds gracefully, his fingers flying back and forth. Finished, the surgeon stepped back and announced, "This woman has a good chance to live for many years. She came to us with a relatively small tumor. I think we got it all." Surgeons fixed broken bodies. Some even cured cancer. R. Lee Clark would become a surgeon.[20]

—⚬⚬⚬—

Curing cancer had long bedeviled physicians. More than thirty-five hundred years earlier, Egyptian physicians grasped the distinction between benign and malignant lesions and recommended hundreds of treatments, including surgery, a host of religious incantations, and a dizzying array of poultices, herbs, and animal parts. One Egyptian physician told his colleagues, "If thou examinst a [woman] having bulging tumors on [her] breast, and if . . . thou findst them very cool . . . they have no granulation, they form no fluid, they do not generate secretions of fluid, and they are bulging to thy hand. Thou shouldst say concerning [her]: . . . 'An ailment with which I will not contend.'"[21]

Physiological explanations did not surface for three millennia. In the fifth century BCE, when the Greek physician Hippocrates looked at malignant tumors, he noted a hardened mass with tentacle-like extensions and blood vessels penetrating surrounding tissues. Over time, as the mass grew, the tentacles extended farther and farther. Tumors reminded Hippocrates of crabs—the hard-shelled exteriors, the legs reaching out, and the claws, with their tenacious, preda-tory grip. He also understood that tumors could reappear in distant sites, like the crustaceans, creeping along from place to place. He labeled cancer *karkinos* (carcinoma), the Greek word for "crab."

More than the Egyptians, the Greeks wondered about how the world worked and why. In sick people, Greek physicians observed, bodily fluids tipped out of balance. Hippocrates hypothesized that the human body possessed four fun-damental fluids, or "humors"—blood, phlegm, yellow bile, and black bile—and that all diseases originated in one of them. Cancer was a disorder of black bile. Because cancer cells proliferate faster than normal cells, tumors generate dead, necrotic tissue. They eventually ulcerate, the necrotic tissues secreting a dark, fetid liquid. For the Greek sage, cancer was obviously systemic, a disrup-tion of the entire body, not just the local site of the tumor.[22]

Treating cancer, therefore, had to be systemic. Greek medicine put some faith in surgery but assumed that it had little effect on the volume of black bile, a fact confirmed, they believed, by the frequency of tumor recurrence. Curing the disease, therefore, required physicians to address the "humors." In the sec-ond century CE, Greek physician Clarissimus Galen advised physicians to drain off excess bodily fluids in cancer patients. With laxatives, enemas, and the lancet to induce vomiting, diarrhea, and bleeding, cancer could be treated. Galen also prescribed a treasure of herbal tonics. Since Hippocrates had thought tumors resembled crabs, Galen suggested burning alive an odd number of the crus-taceans, mixing the remains with Cyprian oil, and with a bird feather swabbing the concoction over a tumor, all the while chanting for divine intervention.

Galen's assumption that *karkinos* could be treated with crabs seems ludicrous today, but between 130 and 200 CE, he wrote four hundred treatises on medi-cine and surgery, most based on clinical observation, and he was the first to distinguish cancer from warts, insect bites, cysts, boils, and pimples. From Galen, physicians learned that not all lumps were cancer. He issued special warnings about growing, knotty lumps painless to the touch. Soon after his death, Galen's texts were disseminated throughout the Roman Empire. Monks ren-dered the Greek *karkinos* into Latin words—including *canalis, canonous, can-dela*, and *cancer*. By 300 CE, Galen's ideas constituted medical canon, and the dread disease had a name.[23]

Galen's intellectual grip held tight until the late eighteenth century, when a few rebels began to ask serious questions. For cancer at least, his treatments ran out of steam. No paste or poultice erased it, nor did ingested concoctions. In the 1790s, George Washington's mother developed breast cancer. Her physician applied a locally recommended cure—"Hugh Martin's Special Powder"—but the tumor expanded. He consulted with Benjamin Rush, a signer of the Declaration of Independence and the nation's foremost physician. "Arsenic is . . . the basis of Dr. Martin's powder," Rush wrote bluntly. "I am afraid no good can be expected from the use of it. . . . I am disposed to believe that there does not exist in the vegetable kingdom an antidote to cancers."

Faith in Galen waned. More and more doctors saw in the "humors" evidence of disease but not causes. New explanations appeared. In 1776, Percivall Pott, a physician at St. Bartholomew's Hospital in London, rejected Galen and accused coal dust of precipitating scrotal cancer in chimney sweeps. "Golden lads and girls all must, as chimney-sweepers, come to dust," wrote William Shakespeare. Black dust, not black bile, Pott concluded, killed them. Because chimney sweeps bathed infrequently, coal dust accumulated in the groin and irritated the skin carcinogenically. In the nineteenth century, major medical schools jettisoned Galen, relegating him to the past.[24]

Surgeons migrated to the forefront. Armed with anesthesia, asepsis, and antisepsis, they waxed confident, and as their confidence soared, so did the extent of their procedures. Anesthetized patients operated on in clean surgical environments could tolerate procedures unthinkable a generation earlier. And in the 1870s, Rudolf Virchow elaborated on metastasis, noting with awe that the sarcoma cells in the femur of a young boy resembled the sarcoma cells inhabiting his lung. Virchow broadcast a powerful message to surgeons: when treating cancer, cut a wide swath around the tumor. Leave behind no cancer cells to form new tumors. Cancer metastasizes to distant locations via the lymphatic system, Virchow argued, and neighboring lymph nodes act as barriers, filtering out malignant cells and localizing them temporarily. Since in his view cancer was at first only a local disease, Virchow encouraged surgeons, when removing a tumor, simultaneously to excise adjacent lymph nodes, which would prevent the cells from disseminating any farther. The window of opportunity for a surgical cure would close, he argued, as soon as one malignant cell exited the lymph node. The sooner a surgeon cut out the tumor, surrounding tissue, and the lymph nodes, the better still.

The age of radical cancer surgery dawned. Between 1872 and 1891, Theodor

Billroth of the University of Vienna fashioned a cavalcade of firsts: resection for cancer of the esophagus, the laryngectomy for cancer of the larynx, the pylorectomy for rectal cancer, the gastrojejunostomy (reconnecting the stomach and small intestine after excising a tumor), and the hemipelvectomy (amputation at the hip for pelvic cancers). In 1890, Halsted performed his first radical mastectomy.[25]

Word spread about cancer surgery. In 1880, the American Surgical Association was organized. Four years later, millions of Americans read about former president Ulysses S. Grant. A habitual cigar smoker, Grant had a tonsil cancer that spread to his tongue, soft palate, throat, and lymph nodes. His physicians pondered radical surgery—cutting out the tongue, mandible, soft palate, and as many lymph nodes as possible—but then retreated. Grant was sixty-three and weakened by the disease. Journalists tracked the deterioration until his death on July 23, 1885. Eight years later, however, radical surgery cured President Grover Cleveland of a soft-tissue sarcoma of the hard palate. Surgeons removed the tumor, along with part of the soft palate and the entire left jaw, and installed a rubber prosthesis. Cleveland died fifteen years later. News of the operation leaked out slowly but demonstrated that radical surgery could sometimes cure head and neck cancer. Flush with success, surgeons launched a public health campaign that cancer was not necessarily incurable. In 1904, Dr. E. S. Judd of the Mayo Clinic wrote, "I wish to emphasize . . . that the surgeon can provide a definite cure in the majority of cases where the patients present themselves for treatment on the appearance of the first symptoms."[26]

Most Americans, however, viewed hospitals with fear and trembling, as places to die, not to recover. In 1876, hospital patients recovering from surgery were less likely to die of infection if they went home to recuperate rather than remaining in a hospital. Paupers without families might show up at a hospital, but only reluctantly, as would the mentally ill, wounded soldiers, accident victims, or patients with infectious diseases who needed to be quarantined.[27]

Not until the early 1900s did public opinion tilt. As preanesthesia horror stories faded, so did fear of surgery. Americans began to meet people cured of cancer. In 1913, a series of articles in the *Ladies Home Journal* addressed cancer. "No cancer is hopeless," wrote journalist Samuel Hopkins Adams, "when discovered early. . . . The only cure is the knife." This was a message Clark first heard in his teens. Cancer treatment and surgery were becoming synonymous. For farmers and ranchers in Texas, decades of mending fences, plowing fields, and chasing cattle under cloudless skies and a blistering sun wrinkled their skin until it resembled the cracks and fissures in the parched earth. Such exposure to ultraviolet radiation elevated the risk of skin cancers—basal cell and squamous cell carcinomas and melanomas. That many farmers and ranchers used chewing tobacco and snuff only raised the odds. Waiting with his father in a barber shop one day, the younger Clark noticed an old man take a seat and remove his hat, exposing a brown, leathery face capped with a shiny, white-as-cream forehead and crown. On one side of his face, the chin and jaw were badly deformed, double the size of the other. Lee quietly inquired of his father, and the senior Clark whispered, "Cancer."[28]

Coordinated efforts against the disease accelerated during the Progressive Era. Cancer deaths rose from forty-eight thousand in 1900 to sixty-five thousand in 1910, far outpacing growth rates in the general population. In 1907, a group of scientists and physicians established the American Association for Cancer Research (AACR). Four years later, Peyton Rous, a founding member of AACR and a faculty member at the Rockefeller Institute, discovered the virus responsible for chicken leukemia. He predicted, "I believe that a cure for the disease will be discovered before the cause of the disease is known, but whether this will be in our lifetime I am unable to say." AACR scientists dedicated themselves more to discovering the cause of cancer than to curing it, but most doctors, helplessly facing sick and dying patients, felt more urgency to find a cure. In 1913, physicians established the American Society for the Control of Cancer (ASCC) and inspired a national movement to educate the public about the merits of early detection and surgery.

To help spread its message, the ASCC turned to the General Federation of Women's Clubs (GFWC). Club women demanded improvements in water quality, schools, and municipal services; lobbied against child labor and for laws protecting working women; and pushed for pure food and drug laws and conservation. The GFWC had a membership of 1.1 million and considerable clout at the community level. In cancer, they found a cause worthy of their energies and saw that ASCC pamphlets received wide distribution.

In seeking out educated women, ASCC leaders also identified a target audience. Surgery was the only antidote, and early detection depended on knowledge of cancer's warning signs, awareness of one's body, and an elevated sense of hygiene, instincts that women displayed, so the sexist logic went, more conscientiously than men. Even vanity could be exploited. "I have never seen a beautiful woman in whom cancer of the face developed," claimed Johns Hopkins surgeon Joseph Bloodgood. "[Women are] jealous of their lovely skins, and therefore the moment a persistent blush of the skin becomes noticeable, they seek a physician." Since women bathed more frequently than men, they were also more likely to discover troubling lumps, keep track of unusual sores and lesions, and brood about unexplained bleeding.[29]

The anticancer campaign gathered momentum slowly. In 1927, the ASCC adopted "Fight Cancer with Knowledge" as an official slogan, with its publications and posters bearing the image of Saint George wielding a sword to slay the dragon "Cancer." Posters of Saint George appeared in subways, bus stations, and train depots nationwide. "While passengers are waiting in the tubes," blazed one poster, "the silver sword flames forth with its message of hope [that] . . . in early discovery lies the hope of permanent recovery." The ASCC and GFWC pushed hospitals to open cancer detection clinics, and the number climbed from 15 in 1920 to more than 180 in 1925. Efforts to enlist support from the federal government began in 1927 when Senator Matthew Neely of West Virginia sponsored legislation to award a $5 million prize for the first cure. It died in committee. In Texas, the state legislature in 1929 approved the establishment of a cancer, pellagra, and mental illness hospital, with Dallas as its home. The stock market crash doomed the proposal. Permission resurfaced in 1931 but not funding. Hunger and homelessness had crowded cancer off the public policy agenda.

As Americans took aim at cancer, medical education also become a target of Progressive Era reform. Back in 1910, the Carnegie Foundation for the Advancement of Teaching published Abraham Flexner's *Medical Education in the United States and Canada*, a blistering indictment of medical training. He condemned the existing proprietary medical schools, with their rank commercialism, virtual autonomy, and dismal curricula emphasizing rote memorization. Medical practice lagged far behind the state of medical knowledge. To reform the system, Flexner looked to the model of Johns Hopkins University and called for dismantling proprietary medical schools and expanding university-based medical schools, where faculty would conduct original research to generate new knowledge, and students would prepare to become physicians through lectures, laboratory work, and clinical training. Such reforms, he was convinced, would help medical practice catch up with medical knowledge and improve the health of millions of Americans. When Flexner became secretary of the General Education Board, John D. Rockefeller's huge philanthropic foundation, he had a pulpit and millions of dollars for medical schools built on the Flexner model, thus launching the "Flexner revolution."[30]

In the fall of 1932, the United States seemed destined for revolution. The money markets were in shambles, the stock market in the basement, and the doors of thousands of banks shuttered. For Lee Clark, though, it was the best of years. He graduated from medical school, and in a country where few Americans had any choices, Clark had several. He had to decide whether to get married and whether to seek a gold medal in wrestling at the Olympic Games in Los Angeles. Lee had fallen in love with Bertha Margaret Davis, the only woman in his medical school class. They shared a work bench in the lab. A native of Asheville, North Carolina, "Bert" sported a waist as slim as Lee's mother's and a keen intellect. Not easily intimidated and with a short fuse, she cursed like a stevedore.

In the fall of 1932, Clark won the Southeastern Regional Championship of the Amateur Athletic Union and a berth on the Olympic team. The opening ceremony, however, was scheduled for July 30, 1932, the day when hospital internships began. The uncertainty lasted until May, when Garfield Memorial Hospital in Washington, D.C., offered internships to the couple. Lee dumped his Olympic dream. They married on June 11, 1932, and settled in the nation's capital. One year later, Lee secured a surgical residency at the American Hospital of Paris, and Bert negotiated an obstetrical appointment at the Tarnier Hospital in the City of Lights' old quarter. They left behind crowded soup kitchens and idle men with empty eyes, to rent a furnished flat on the Left Bank. The apartment had a maid, big windows, and a million-dollar view of paradise.[31]

American Hospital residents enjoyed access to other clinics, and in his second year, Clark worked half-time among some of them, including the Pasteur Institute and the Radium Institute. In 1885, Louis Pasteur had developed a vaccine for rabies, and two years later the Pasteur Institute was founded to promote further research. No medical institute in the history of the world, before or since, enjoyed such success. During the next twenty years, Pasteur scientists revealed the causes of diphtheria, rabies, bubonic plague, malaria, and polio. Clark had signed on at the epicenter for the study of infectious diseases.[32]

At the Radium Institute, he learned that surgery might *not* be the only way to attack cancer. In 1909, the University of Paris and the Pasteur Institute had erected a laboratory for Marie Curie, winner of the 1903 Nobel Prize in physics. At the Radium Institute, she soon earned a second—the 1911 Nobel Prize in chemistry—for isolating radium crystals. Curie devoted her life to exploring the medical applications of radium, especially for cancer treatment. As far back as 1901, she had filled glass or metal tubes with radium and placed them in or near tumors, what a later generation of radiotherapists would call interstitial therapy, or brachytherapy. Some tumors regressed in the presence of radium. Radium was unstable at the atomic level, emitting alpha, gamma, and beta rays as it disintegrated. Those rays bombarded the atomic structures of nearby cells, benign as well as malignant, and dislodged electrons, leaving atoms with deadly electrical charges and cells with disheveled genetic codes. Cancer cells proliferate without the controls that stop normal cells. The challenge for the next four generations of radiation oncologists would be to walk a tightrope, delivering sufficient radiation to kill tumors without slaughtering normal tissues. Curie died of leukemia in 1934, perhaps from too many years of handling radium. The Clarks read the eulogies and felt the adulation.[33]

By the time Clark's residency brought him to the Radium Institute, radiation therapists had introduced external beam technologies. In 1895, the German physicist William Röntgen discovered x-rays and gave birth to diagnostic radiology. Some physicians soon learned that x-rays had the capacity to temporarily shrink tumors, spawning what would evolve into the medical discipline of radiotherapy. Clinical experience soon revealed, however, that voltage levels limited the effectiveness of x-rays. Low-voltage x-rays penetrated deeply seated tumors but also inflicted unacceptable damage to surface tissues. Between 1913 and 1925, physicists improved the effectiveness of radiotherapy. The Crooke's tube and then the Coolidge hot cathode tube armed physicians with hundreds of thousands of volts of electricity, encouraging treatment of deep tumors with less damage to surface tissues. When Clark arrived in Paris, the Radium Institute employed brachytherapy and external x-ray beams. Equally important is that Radium Institute scientists were laying the foundation of "fractionated" therapy—delivering radiation in small, daily doses over the course of several weeks instead of in large single doses—which helped preserve surrounding normal tissues while still hitting tumors with deadly loads.[34]

On the other side of the English Channel, Ralston Paterson of the Christie Hospital and Holt Radium Institute campaigned to substitute radiotherapy for radical surgery, especially in cancers where extensive procedures left patients badly disfigured. During the 1930s, Paterson won converts among such British surgeons as Robert McWhirter of Edinburgh and Geoffrey Keynes of St. Bartholomew's. They combined less radical surgical procedures with radiotherapy to determine whether existing long-term survival rates could be pre-

served. For women with small breast tumors, they recommended surgical removal of the lump followed up by external beam radiation to the surgical site
as well as to neighboring lymph nodes, not the Halsted mastectomy.[35]

Surgeons on both sides of the Atlantic heaped abuse on those with the gall
to question Halsted. They introduced Clark to a debate destined to roil oncology for much of the century. Another ten years would pass before survival data
in breast cancer demonstrated the advantages of surgery and radiotherapy, and
another forty before most American surgeons abandoned Halsted. For all the
excitement at the Radium Institute, radiotherapy in 1934 remained on the periphery of medicine, and Lee Clark had no intention of spending his life on the
periphery of anything. Surgeons dominated American medicine, and he considered himself a player. Late in 1934, he accepted a surgical fellowship at the
Mayo Clinic, leaving the Louvre and the Left Bank for the bucolic environs of
Rochester, Minnesota.

—⟪⟫—

When he accepted the fellowship, Clark entered the future of American medicine. The days of the small town doctor working seven days a week at the beck
and call of an entire community would not survive the twentieth century. In 1889,
William Worrall Mayo and his sons William J. and Charles—all three physicians—
built the first general hospital in southeastern Minnesota. Patients soon
crowded the corridors. The Mayos needed more staff to meet the demand and
to afford themselves a life outside the hospital. Instead of bringing aboard more
generalists, however, they hired specialists, which allowed them to exploit the
explosive growth in medical knowledge and to develop a culture of multidisciplinary team work. "It [has become] necessary to develop medicine as a co-
operative science; the clinician, the specialist, and the laboratory workers unit-
ing for the good of the patient," argued William J. Mayo. "Individualism in
medicine can no longer exist." In 1915, they founded the Mayo School of Graduate Education to conduct research and train medical specialists. Four years
later, they created the Mayo Foundation as a private, nonprofit charitable trust,
endowed it with the bulk of their personal assets, and put all physicians on salary,
scrapping the fee-for-service system held sacrosanct by the AMA. Lee Clark absorbed the Mayo system.[36]

At first, Bert and Lee thought the surgical fellowship would keep them at
Mayo for three years, but three stretched into five when Mayo persuaded him
to invest the first two years in what amounted to a residency in internal med-

icine. Mayo awarded Bert a staff position in pediatrics. During the next five years, she moved from pediatrics to gynecology, finally settled on anesthesiology, and gave birth to two children. In 1930, Hugh Cabot, the most illustrious urological surgeon in the nation, came to Mayo and brought with him a conviction that surgeons needed a basic understanding of biological processes. Lee Clark ended up a pioneer, a surgical fellow trained in medicine before ever lifting a scalpel. Not until 1937 did he assume more surgical responsibilities, performing approximately two thousand operations in his last two years at Mayo. At "least two-thirds of them were cancer operations," he remembered. Every day, he opened up patients only to discover disease so extensive that he closed right away and sent them home to die. Every day he operated on patients whose tumors had recurred after previous surgery. Every day, he learned of cancer's truculence. But he also followed up on long-term survivors. R. Lee Clark was his father's son. Hope sprang eternal. Cancer was not hopeless.[37]

Clark could draw on a growing public faith in medical science and medical education. By the end of World War I, the Flexner revolution had dramatically improved the training of physicians, and in the 1920s and 1930s, a variety of new drugs put at bay tetanus, meningitis, diphtheria, and diabetes. In 1930, Congress changed the name of the federal government's Hygienic Laboratory to the National Institutes of Health (NIH). Cancer's time had arrived as well. In 1936, the GFWC and the ASCC formed the Women's Field Army (WFA) to lead a "war against cancer." First Lady Eleanor Roosevelt served as honorary sponsor, and within months, the WFA opened field offices in twenty-five states and enlisted one hundred thousand "soldiers," an army of middle-class women to spread the word about warning signs and early detection.[38]

The New Deal seemed to have stalled the Depression; why not deal cancer a setback as well? In 1937, a rebel Texas congressman took up the challenge. In 1935, freshman Democrat Maury Maverick of San Antonio had taken Congress by storm, organizing thirty-five liberal Democrats who met weekly to push Roosevelt and the New Deal to the left. Reelected to a second term in 1936, Maverick took up some family business. His cousin Dudley Jackson, a San Antonio surgeon, was curious about cancer biology. Decades ahead of his time, he touted clinical research and felt that the federal government should commit resources to battling cancer. With the Women's Field Army enjoined, the time was ripe, and Jackson now had a cousin in Congress. He lobbied Maverick, and Maverick hustled his colleagues to add a National Cancer Institute (NCI) to the recently established NIH. It was an easy sell—a modest proposal of only $750,000 to construct research and treatment facilities on the NIH campus and to pur-chase one gram of radium; it would play well in the upcoming elections.[39]

The AMA in 1938 blasted proposals for national health insurance. A special White House conference called on the federal government to spend $850 mil-lion a year for ten years to provide health care for all Americans. Irvin Abell, president-elect of the AMA, likened the proposal to the socialism that had ru-ined European medicine. With the stealth of a big cat tracking prey, AMA lob-byists stalked Congress, ready to pounce on any attempt to place proposed bills on the docket. Abell filled waiting rooms in physicians' offices everywhere with pamphlets denouncing national health insurance.

At Mayo, Clark operated in the center of the controversy. A few months be-fore the White House conference, he was named first assistant emergency sur-geon, a position reserved for Mayo's most trusted fellows, and came under the tutelage of Hugh Cabot. Cabot believed that indigent patients deserved medical care as good as that extended to the wealthiest and that access to health care was a civil right. Soon after his arrival in Rochester, Cabot experienced an epiphany over Mayo's salaried, group culture and saw an alternative to fee-for-service. "As at present organized," Cabot argued, "the mainstream of standards of health care . . . are grossly unsatisfactory." He warned that selfish behavior on the part of physicians could bring down the wrath of the federal government.

Cabot's warnings rang true. In October 1937, more than one hundred Amer-icans, mostly children, died after taking the drug sulfanimide for venereal dis-ease and strep infections. Produced by a manufacturer of veterinary products, the drug caused liver and kidney damage. Responding to public outrage, Con-gress passed the Food, Drug, and Cosmetic Act, which expanded the authority of the FDA. Under the original Pure Food and Drug Act of 1906, drug manu-facturers faced truth-in-advertising constraints, with the FDA insisting that com-panies not exaggerate product claims. The new legislation also required the drug companies to convince the FDA of a drug's safety before it could be marketed. Congress approved the bill, and President Roosevelt signed it in 1938, giving the FDA a vicelike grip on the development and use of new pharmaceutical drugs.

The Food, Drug, and Cosmetic Act of 1939 spawned the modern pharma-ceutical industry. Federal law permitted companies to patent new drugs, and the need for an FDA stamp of approval created drug monopolies. And just then, the advent of sulfa drugs to treat several infectious diseases stimulated enor-mous demand. According to one historian, the pharmaceutical industry was about to go "from a handful of chemical companies with no interest in research and no medical staffs to a huge machine that discovered, developed, and marketed drugs of real use in treating disease."[40]

At the time, however, few people looked so far into the future. While one of the great revolutions in medical economics was just beginning, the Clarks left Minnesota for Jackson, Mississippi, where the Flexner revolution had finally penetrated the Deep South. The governor had recruited five physicians from Mayo, including Clark, to build a medical school in Jackson. Clark's gilt-edged résumé boasted of the best medical training in the world—the Medical College of Virginia, the Pasteur Institute, the Radium Institute, and the Mayo Clinic. In the spirit of his father and grandfather, Clark prepared to build a school of his own. Such large medical institutions as the Pasteur Institute and the Radium Institute, he believed, possessed the power to eliminate the scourges of the past, especially if their internal administrative structures and institutional cultures, like those of the Mayo Clinic, minimized economic competition among physicians, encouraged multidisciplinary cooperation, and viewed government as an ally, not an adversary. Lee Clark made no little plans.

2

Present
at the Creation

*We have every opportunity in the world to find here in Houston the answer
to the question of the cause and cure of cancer. . . . Some unknown boy
working in the research laboratory of the Anderson hospital may
find the answer.*
—DUDLEY WOODWARD, 1944

The warm, muggy weather in Jackson suited the Clarks better than the
cold winters of Minnesota, but 1939 was an unlikely year for building a
medical school and Mississippi an inhospitable place. At the state level,
where funding for education and social services was about as popular as civil
rights, Clark soon realized that he would not be building a first-class medical
school. In private surgery, on the other hand, his technical skills and bedside
manner won more patients than he could handle. As in Paris and Minnesota,
he operated mostly on cancer patients, and consistent with the prevailing consensus that cancer was primarily a local disease until it broke from its moor-
ings, Clark believed in prompt, radical surgery.

In Jackson, he displayed unusual agility in the operating room, his eye-hand
coordination exquisite and his speed unrivaled. Unlike most surgeons, Clark
had the training of an internist with an intellectual curiosity that transcended
tumor removal and tissue repair. He had performed a wide range of procedures, removing lungs, larynxes, stomachs, colons, gallbladders, arms, legs,

breasts, and rectums. A few operations stood out, like that of a 258-pound woman with a benign, 98-pound tumor. Even years later, he regaled others with descriptions of the unwieldy mass, trying to lift it only for it to slide, like a chunk of Jell-O, onto his feet and then watching an orderly try to carry away the slippery blob.

By 1941, Clark had grown restless. He had never wanted to be a country doctor, and in Jackson he learned that he did not want to be just a general surgeon either, even one with a stately brick home and a Cadillac in the driveway. He yearned to change the world. On December 7, 1941, Pearl Harbor gave pause to his ambition. At thirty-five, with a wife and two small children, Clark could have sat out the war, but Pearl Harbor was the defining event of his generation. He stewed throughout 1942, tugged between the demands of family and work and the call of duty. Late in the year, Clark enlisted in the army air corps medical department.[1]

Monroe Dunaway Anderson also hailed from Jackson—Jackson, Tennessee. In a way, Anderson and his friend William Clayton were twentieth-century apparitions of nineteenth-century heroes—Tennesseans like Sam Houston and Davy Crockett who moved to Texas in search of fame and fortune. Houston and Crockett found theirs at San Jacinto and in the Alamo, while Clayton and Anderson discovered fortune in the cotton fields. Clayton grew up in Jackson, and in 1895 he joined the American Cotton Company, rising to assistant general manager of the New York office. As a boy, Clayton came to know James and Ellen Anderson and their seven children. James Anderson was president of the First National Bank and wired into the regional economy. His sixth child—Monroe "Mon" Dunaway Anderson—was born on June 23, 1873. When Mon was six, his father died suddenly. After high school, Mon completed two years of college in Memphis and then returned to Jackson to work for People's Bank. Diligence and integrity earned him the nickname "Careful Cashier."

Monroe's brother Frank had already entered the cotton business. Cotton prices fluctuated wildly according to weather and political conditions, and cotton farmers needed financing in the spring to plant and in the fall to harvest. Monroe and Frank Anderson pooled their talents, with Frank supplying sales expertise and Mon the capital. Because cotton was produced and consumed all over the world, they needed access to international markets, a talent William

Clayton had perfected at American Cotton. On August 1, 1904, with three thousand dollars each, Frank and Monroe Anderson and William Clayton established Anderson, Clayton, and Company.[2]

At first, they located the company in Oklahoma City, but it soon became obvious that cotton grown in Oklahoma, southwestern Arkansas, and north and east Texas went to Houston and from there to freighters in Galveston. In 1907, Monroe Anderson moved to Houston and opened an office. The city was crawling with wildcatters, oil land men, and swindlers. The discovery of oil at Spindletop in 1901 had transformed the economy. Floating on clouds of cotton and a sea of hydrocarbons, Houston boomed.

Monroe served as the firm's "watchdog over finances," according to a Houston journalist, "providing the cautious reserve necessary in a business that relied on huge amounts of bank credit. . . . He was incessantly on the lookout for a way to save a penny." Monroe built huge reserves. Frank, in buying and selling on an international level, often had to act precipitously, either to take advantage of an emerging opportunity or to stave off disaster. Monroe's financial planning always left Frank money to do his job.[3]

For the global cotton trade, Houston was the nexus of opportunity. The hurricane of 1900 had killed six thousand people and derailed Galveston's economic future. Houston had the chance to eclipse its smaller rival. A delegation of businessmen led by Jesse H. Jones hatched a plan to dredge a ship channel all the way inland and transform Houston into the Gulf Coast's busiest port. In 1916, Anderson Clayton relocated its corporate offices to Houston and three years later loaded its first bale on an oceangoing vessel that steamed into the Gulf.[4]

Business boomed, fed by the boll weevil infestation and World War I. Between 1904 and 1919, cotton jumped from nine to thirty-five cents a pound, and Anderson Clayton was becoming the largest cotton broker in the world. Clayton and the Andersons were multimillionaires.

Despite his fortune, Monroe was still "careful" in his personal finances; he rejected hard-luck stories and easily discerned between deception and destitution. He defied the logic of the social theorist Thorstein Veblen—that American businessmen would engage in conspicuous consumption to give meaning to otherwise vacuous lives. If anything, he practiced inconspicuous consumption, often walking to work and carrying his lunch. "He never acted like the rich man he was," remembered a nephew. "A bellhop at the Bender Hotel, who my uncle talked to daily, said for many years he thought my uncle worked for the shoe store downtown." Monroe Anderson did have one vice—a taste for

Cadillacs. As his millions accumulated, he lived frugally in boarding houses and then in the Texas State Hotel and the Bender Hotel.[5]

⸺◦⸺

Now and then, when he needed medical advice, Anderson called Ernst W. Bertner, a prominent Houston physician born to German immigrant parents in Colorado City, Texas. As a child, Bertner found friendship and a window on the world at a local drugstore. The pharmacist dispensed pills to adults, peppermints to children, and advice to everybody. His seemed an ideal life. In 1906, Bertner graduated from the New Mexico Military Institute and opened his own store, soon learning that small-town success came only when the owner also had a pharmacy diploma. In 1907, Bertner locked up the store and enrolled in the University of Texas School of Pharmacy in Galveston. Most of his friends, however, were medical students at the University of Texas Medical Branch, and they persuaded him to transfer to the medical school.

In 1911, Bertner graduated and moved to New York City for a succession of internships and residencies in obstetrics and gynecology at Willard Parker Hospital, St. Vincent Hospital, and Manhattan Maternity Hospital. In 1913, a chance encounter with Jesse Jones altered Bertner's plans. Jones happened to be in New York on business when he ended up at St. Vincent needing emergency surgery. Bertner administered the anesthetic. Jones took an immediate liking to Bertner and hired him as house physician at the Rice Hotel, a new, seventeen-story Jones property in Houston. He offered Bertner a suite in the hotel, suggested that he open a private practice, and provided him office space downtown at the Second National Bank Building. Heaven had smiled on the twenty-four-year-old German American.[6]

Once Jones handed over the keys to the city, Bertner opened all the doors. He looked every inch the Prussian gentleman—hazel eyes, a thick, tightly cropped mustache, and ramrod straight posture that Kaiser Wilhelm would have admired, but when Bertner spoke, the veneer dissolved, revealing a handsome, congenial bachelor. His hair thinned in his twenties, leaving behind small wispy curls at the top of his forehead, a circle of shiny skin at the crown, and a full head of hair everywhere else. Bertner enjoyed flirting with women and they with him, although a longtime friend remembered, "There was not an ounce of mischief in the man." Never crass or uncouth, he epitomized the iconic physician endowed with concern, confidence, and competence. He became "Dr. Billy" to patients and "Billy" to friends.[7]

World War I complicated his life. German Americans, from isolated enclaves in Texas to core communities in the upper Midwest, felt a shear in the winds of tolerance. Even Bertner, for all his success, could feel the turbulence. His own English still bore the slightest trace of German, with an occasional throaty consonant or shortened vowel. As U-boats sank Allied ships in the North Atlantic, anti-German sentiments deepened. In April 1917, Congress declared war, and within weeks, some states outlawed the teaching of German in public schools, some libraries burned the works of Thomas Mann and Johann Goethe, and some orchestras discarded Beethoven, Mozart, and Brahms.

In May 1917, Bertner joined the U.S. Army Medical Corps and was assigned to the British army. Gustave and Anna Bertner waited anxiously for news that he had survived the U-boat-infested North Atlantic and arrived safely in England. When good news finally reached them, Gustave told a reporter, "My son gave up a fine practice at Houston to join the reserve corps. We feel greatly relieved when the cablegram arrived telling us he got safely through the submarine zone." The reporter then noted, "Although the elder Bertner is a native of Germany, he says he is glad his son is doing 'his duty.' The older Bertner is a naturalized U.S. citizen." Ernst "Billy" Bertner ended up in France on the medical staff of General John "Black Jack" Pershing. After the war, Bertner mustered out with the rank of major and returned to Houston.[8]

He did not stay for long. Battlefield medicine had taught Bertner that he needed more training, and in 1921 he accepted a residency in surgery, urology, and gynecology at Johns Hopkins. There he fell under the tutelage of Thomas S. Cullen, a pioneer of gynecological pathology. Cullen also promoted state cancer registries, the American Association for Cancer Research, and the American Society for the Control of Cancer (ASCC). Schooled by Cullen, Bertner treated ovarian, uterine, cervical, and breast cancer, and he adopted Cullen's conviction that concerted effort could make a difference. In 1922, Bertner left Baltimore, married Julia Williams, and returned to the Rice Hotel.

Like many German Americans, Bertner gradually assimilated, eschewing the Lutheranism of his father for the First Presbyterian Church, the Houston Country Club, and the Masonic Lodge. He eventually rose to become a thirty-second degree Mason, a Knight Templar, and a knight commander. Taking a cue from Jesse Jones, he amassed a fortune purchasing land between downtown Houston and what is now Hobby Airport, more than enough to finance Julia's expensive tastes. She always appeared in the latest Parisian designs, scented with the most expensive perfumes. She also monitored her husband's wardrobe, preferring wide-lapelled woolen suits, usually in dark blue or charcoal gray. Theirs was a storybook marriage without the distraction of children. Bertner was active in the ASCC, and Julia threw herself into the cause.[9]

⸺⸺⸺

Monroe D. Anderson had an interest in cancer too. Although warmhearted and witty, he had been as a young man self-conscious around women. As he matured, Mon dated women his own age, considering it unseemly for an older man to be seen with someone too much his junior. In middle-age, he dated widows. Anderson fell in love only once. He mustered enough courage to propose, but she turned him down abruptly and married another. Anderson never seriously dated again. She remained the one love of his life, and they kept in close contact. He became a sort of favorite uncle to her children. When she died of cancer, Anderson mourned as if he had lost his own wife.[10]

By 1936, Anderson's personal assets exceeded $19 million. Ill with kidney disease and congestive heart failure, and without children, he realized that his fortune was destined for the coffers of the U.S. Treasury unless he transferred it to a charitable trust. After exploring the matter with attorneys John Freeman and William Bates, he decided to establish the M. D. Anderson Foundation. Anderson asked Bates and Freeman to serve with him as trustees. "Uncle Mon had no idea a hospital would ever be named after him," remembered his nephew Thomas D. Anderson. As John Freeman recollected, "Mr. Anderson had some rather definite ideas as to [the trust's] purposes, but nothing regarding specific activities. . . . He came by the office almost every morning to talk things over. So Colonel Bates and I knew a great deal about what he had in mind." Hunting, fishing, and time with nephews had always been Anderson's preference over museums, operas, and the symphony. After the death of his brother Frank in 1924, Anderson ate dinner almost every night with his sister-in-law and nephews. Bates later recalled that "Anderson's foremost concern was that his foundation would serve people in the very best sense of what service to mankind truly meant."[11]

On October 7, 1938, while eating lunch at the Majestic Grill in downtown Houston, Anderson complained of numbness in his right arm. His nephew Thomas happened to be eating at the Majestic and urged Anderson to go immediately to a hospital. Mon refused at first but finally consented, and they took a cab to Baptist Memorial Hospital, where physicians, as his right side sagged, diagnosed a stroke in progress and gave him nitroglycerin. He spent weeks at the hospital recovering from the stroke before being moved to convalesce at his home at 1902 Sunset Boulevard. Anderson suffered a severe heart attack there and died on August 6, 1939.

The firm of Fulbright, Crooker, Freeman, and Bates probated the estate and established the M. D. Anderson Foundation. To fill Anderson's vacancy on the board, they enlisted Horace Wilkins, a trusted Houston banker. They gave the M. D. Anderson Foundation a somewhat generic mission—the "establishment, support and maintenance of hospitals, homes and institutions for the care of the sick" as well as the promotion of "health, science, education, and advancement and diffusion of understanding among the people." In 1941, the IRS gave its approval to the nonprofit organization.[12]

The timing was fortuitous. Freeman had for years resided at the Rice Hotel and often shared with Bertner details of the M. D. Anderson Foundation. For several years, Bertner had dreamed of creating a medical center and bringing together many of Houston's physicians in a central location where patients could be shuttled more conveniently between general practitioners and specialists, and doctors would have more opportunities to interact. Physicians with medical privileges at more than one hospital did not want to drive back and forth across town.

With tight logic and infectious enthusiasm, Bertner spread the gospel of a medical center, and Freeman was converted. During the late 1930s, the two men discussed the matter repeatedly, over lunches downtown and at the Rice Hotel. As they talked, the contours of the Texas Medical Center (TMC) congealed. Freeman also saw in the TMC an opportunity to diversify Houston's economy beyond cotton and petroleum. Bertner had endowed the M. D. Anderson Foundation with a cause worthy of its founder.

Bertner's work with the American Society for the Control of Cancer and the Women's Field Army persuaded him of the need for a state cancer hospital. Memories of a patient he had treated at Johns Hopkins still haunted him. A vivacious twenty-two-year-old woman had reported to the clinic complaining of sharp abdominal pains. When Bertner pressed his fingers above the right ovary, he felt a dense, hard mass. As he probed down toward the intestine, he detected more nodules. Bertner called in Cullen, who suspected ovarian cancer. Exploratory surgery confirmed it. "She survived, if 'survival' is the right word," said Bertner. "She looked like a skeleton when she died." Since then, Bertner had seen hundreds of equally hopeless cases. Only research could unlock cancer's secrets. A state cancer hospital should anchor the TMC. By 1940, as World War II erupted in Europe and the national economy rebounded, what had once been merely a noble idea was about to become plausible.[13]

⟨⟩

While Bertner dreamed, Arthur Cato ruminated. His town of Weatherford, Texas, could have been the little city of Zenith in Sinclair Lewis's 1922 novel *Babbitt*. A small-town druggist, Cato walked the streets of Weatherford engaged in the middle-class pursuits of civic improvement, better business, and Boy Scouting. Cato yearned to improve Weatherford and to turn a dollar at the same time. Located thirty miles west of Fort Worth, Weatherford produced peaches and cotton. As a scoutmaster, Cato wanted the largest and best-equipped troop in the district, with the most merit badges, the most jamboree prizes, and the most Eagle Scouts. But unlike Babbitt, whose world never reached beyond Zenith, Cato hoped to do "something big," in the words of his son. Fulfilling big dreams from the confines of a drugstore in Weatherford seemed unlikely, so he turned to politics. In 1939 he won a seat in the state legislature. Almost immediately after the swearing in, Cato began talking about cancer. He cornered other members in the hallways and cloakrooms of the state capitol. His preoccupation startled at least one fellow legislator. "I can't remember 'why' but Arthur had one obsession and that was CANCER. . . . From the first day I met Arthur he never let up until everyone knew that something could be done about cancer." Arthur Cato suspected that both his parents had died of cancer, and cancer ran in the family of his wife, Bettie Lillian. Early in February 1941, Cato introduced House Bill 268 to establish a state cancer hospital.[14]

The measure caught the medical community off guard. Dr. James Martin of Dallas had spent forty years campaigning for better cancer medicine, and in 1929, as head of the committee on cancer of the Texas State Medical Association, he had secured the first bill to create the hospital. Although the project was stillborn, Martin persisted, and in 1940, he got the endorsement of the Women's Field Army. Cancer killed four thousand Texans that year. With the assistance of the Texas Department of Health, Martin's committee drafted a bill. Martin was about to have the bill introduced when he learned of House Bill 268. He joined Cato and crafted a compromise. John Spies, dean of the University of Texas Medical Branch, urged them to affiliate the state cancer hospital with the University of Texas. For Spies, if the Flexner reforms taught anything, it was the preeminence of the university-based medical school.[15]

The legislation faced pitfalls. Representative Sam Hanna of Dallas bitterly opposed it. "You know you can't cure a cancer for $500,000," he preached. Dorsey Hardeman of San Antonio accused Cato of political smoke and mirrors, predicting that he would "use this money to lobby a big appropriation through in 1943." Hardeman argued in his own defense that he sympathized "fully with cancer victims" because he had "suffered from a dread disease—tuberculosis." Cato reminded Hardeman that his TB had been cured in a state hospital with a treatment discovered by a relatively obscure physician. Many private physicians believed that such government initiatives threatened their livelihoods. To ease their distress, the measure prevented the new hospital from accepting patients without a referral from a private physician. It also required that paying patients be billed at rates consistent with fees of other hospitals in the region. The bill specified no location for the hospital, leaving that matter up to the UT Board of Regents. No more than $250,000 of the $500,000 appropriation could be used for building and equipment. On June 20, 1941, the Senate approved the measure unanimously, and four days later the House of Representatives, by a vote of 80 to 35, created the Texas State Cancer Hospital and Division of Cancer Research, giving it the dual mission of patient care and research.

Throughout the year, political maneuvering over location intensified. Political and medical interests in Galveston fully expected the new cancer hospital to become associated with the medical school there. On June 25, 1941, the *Houston Post* claimed that "a new state cancer hospital will be established at Galveston if a bill sent to Governor O'Daniel Tuesday meets with executive approval. . . . It is generally assumed that the board [of regents] will recommend John Sealy hospital at Galveston for the purpose." Politicians in Dallas claimed hereditary right, since the original bill in 1929 had designated their city. Maury Maverick, now the mayor of San Antonio, lobbied for his city. San Antonio was centrally located.[16]

With a population of 528,961 in 1940, Harris was the largest county in the state. The Gulf Coast was the most rapidly growing region. Cancer incidence rates in Harris County were also the highest in the state, giving Houston a macabre advantage. In the end, however, the battle played out not in the state legislature but in the offices of the UT Board of Regents, and in that smaller venue, Monroe D. Anderson's fortune sealed the deal. Late in June, William Bates and John Freeman lobbied the board to put the hospital in Houston. On the back porch of Bates's home at 2128 Brentwood Drive, with a ceiling fan stirring up the summer air, the M. D. Anderson Foundation made the board a tantalizing offer. If the regents located the new hospital in Houston and named it after Monroe Anderson, the foundation would match the five hundred thousand dollar appropriation with a gift of five hundred thousand dollars, provide temporary housing for the facility, and donate land for a permanent home. In August 1941, the board approved the proposal.

When word of the decision leaked out, protest resounded. Politicians in Dallas and Fort Worth promised to revisit the issue during the next legislative session, but their threats were as empty as their pocketbooks.[17]

The M. D. Anderson Foundation moved quickly again. On May 16, 1942, the foundation purchased from the Rice Institute for sixty thousand dollars the old Baker estate, a six-acre compound at 2310 Baldwin near West Gray south of downtown. The property was known as "the Oaks." Concerned that interests in Dallas and San Antonio might still try to scuttle the location, the foundation began to remodel the two-story Baker estate, with its carefully manicured shrubs, broad stretches of mowed and edged grass, and large cypress trees tangled in wisteria. The main residence, its brown brick exterior draped in thick ivy, had a basement and two other floors. Behind the main residence, the two-story stable and carriage house could be refitted for research laboratories.

The board of regents needed an interim director. With political connections to the most powerful men in the state and moral capital accumulated through a lifetime of service, Ernst W. Bertner was really the only choice. That he had treated with compassion and skill the wealthiest women in Houston cemented his credentials. Bertner agreed to serve as a part-time, interim director without salary, and on September 25, 1942, the board renamed the hospital the M. D. Anderson Hospital for Cancer Research of the University of Texas. Two weeks later, Bertner moved into his office at the Oaks.

At the time, Lee Clark was still in Mississippi pondering the dismal prospects for a medical school there. In September 1942, while reading the latest copy of the *Journal of the American Medical Association*, he paid special attention to an announcement that the M. D. Anderson Foundation in Houston had donated five hundred thousand dollars to UT for a state cancer hospital, with Bertner as interim director. Lee confessed to Bert that he was envious: "Now that's something I really would like to do. It's too bad they've already appointed a director."[18]

To Bertner fell the chore of creating a hospital. He borrowed from the University of Texas Medical Branch at Galveston two biochemists, a biologist, and John Musgrove, a business manager destined to spend the rest of his career at M. D. Anderson. Well aware that academics ferociously protect their turf, Bertner repeatedly assured the board that his scientists "would in no way duplicate or compete with the biochemical work in progress . . . in Austin."

To attract patients, Bertner constructed a one-story cancer detection clinic. For those needing in-patient care, Bertner leased twenty-two beds from Hermann Hospital. In return, the M. D. Anderson Foundation built a new home for Hermann's residents and interns. Because Jim Crow still ruled, black and white patients had to be segregated, so Bertner leased six beds from the Houston Negro Hospital.[19]

Creating a physical plant vexed Bertner. In the early 1940s, defeating the Axis powers superseded everything else, even curing cancer. With the economy on a war footing, building materials were in short supply. Prices rose so quickly that contractors insisted on cost-plus rather than fixed-bid arrangements, and as the remodeling of the Baker estate proceeded, Bertner and Musgrove scrambled to cover the cost overruns. Musgrove almost worked himself to death supervising the remodeling and scrounging for extra two-by-fours, cement, paint, and shingles. State auditors harassed him at every turn. One even asserted, "I'll not allow another damn cent to be spent on this project." He relented when Musgrove convinced him of the project's importance. The clinic got built, the main residence remodeled, and the carriage house transformed, but purchasing x-ray equipment and hiring a radiologist were as difficult as squeezing a nickel out of the legislature.[20]

The remodeled Baker estate left much to be desired. Large gray wood rats infested the property. For several years, the night watchman killed so much time and so many rats with a .38 Smith and Wesson that Houston police learned not to answer calls about gunshots at Baldwin and Gray. Target practice was best early in the morning, when sunlight lit up the estate just as the nocturnal creatures paraded back to their nests. One groundskeeper regularly showed up for work with a .22 rifle. Banshee-like shrieks periodically erupted when the rats scurried across floors, tipped over garbage cans, or showed up in the kitchen.

The laboratories teemed with equally obnoxious critters. After decades as a stable and carriage house, they were home to hundreds of pigeons and millions of fleas. Some days were worse than others, the fleas resembling an Old Testament plague. Microbiologist C. P. Coogle exploited the menagerie. He focused his research on avian diseases, particularly those with fleas as vectors. Coogle was often seen scaling the carriage house, hoping to catch pigeons to sample and to study their blood. Convinced that the etiology of several pigeon diseases lay in the parasites inhabiting the feces of fleas, Coogle ingeniously devised a technology for administering enemas to fleas. M. D. Anderson was destined, by the end of the century, to become a highly successful research institution, and it all began with Coogle aiming the tiniest of tubes at the backsides of fleas. He soon switched to cancer, measuring the impact of endotoxins on malignant cells.[21]

While Coogle trapped fleas and everyone else swatted them, Bertner fashioned a culture for M. D Anderson. As a southern gentleman, he always displayed a polite graciousness and expected it of others. After decades as a gynecologist, he had learned to listen to women, a kind of discourse known to few surgeons. His patience and gentility became an early hallmark of the institution. So did a sense of mission. Bertner viewed the hospital with the love of a first-time parent for a new baby, and he shared that affection and hope with the staff. "My father [John Musgrove] worshiped the ground Dr. Bertner walked on," Musgrove's son remembers. "Dr. Bertner was equally kind to everybody. Dad became so committed to the hospital. If the gardener needed some help on weekends, Dad would be there pulling weeds. If the night watchman was sick, Dad would take a shift. Bertner made Dad feel that even though [he was] a bookkeeper, he played a key role." Another early associate of Bertner's recalled, "From the very beginning, Dr. Bertner created a sense of family among the staff and a devotion to patients." Bertner recruited twelve local physicians to work part-time, without salary. "Dad really wanted to help cure cancer," Musgrove remembered. "He really felt that."[22]

On February 17, 1944, the dedication ceremonies for M. D. Anderson attracted medical and political luminaries. Governor Coke Stevenson glad-handed the crowd, as did Arthur Cato. Homer Rainey, president of the University of Texas, attended, along with trustees of the M. D. Anderson Foundation and several members of Monroe Anderson's family. Also present were representatives of the American College of Surgeons, the American Association for the Control of Cancer, the Texas Medical Association, and Memorial Hospital for Cancer and Allied Diseases in New York. "The world," declared Frank Adair, chief of surgery at Memorial, "will be watching with closer scrutiny than you may realize, this experiment in combining private philanthropy and state funds, under the aegis of the University of Texas." Less than two weeks later, on March 1, 1944, M. D. Anderson accepted its first patient, a fifty-seven-year-old man with lymphoma. Full of hope and naïveté, Judge Dudley Woodward, chairman of the UT Board of Regents and a prime mover in the state Democratic Party, predicted a bright future: "We have every opportunity . . . to find here in Houston the answer to the question of the cause and cure of cancer. . . .

Some unknown boy working in the research laboratory of the Anderson hospital may find the answer."

The Flexner reforms emphasized education as a mission equally as important as patient care and research, but M. D. Anderson, even though part of the University of Texas, could not offer academic degrees. To give M. D. Anderson some visibility in education, Bertner launched the Symposium on Fundamental Cancer Research in 1946, which became an annual affair and added education to research and patient care as the institutional mission.[23]

Two years earlier, in 1944, Lee and Bert Clark had been in their second military posting. After he enlisted, they ended up at the thousand-bed army air corps hospital at Seymour Johnson Field in North Carolina, where Lee supervised thirty surgeons and conducted research into testicular varicocele—varicose veins in the blood vessels surrounding the testicle—a problem that left many recruits unfit for flight school. With a colleague, Clark developed a surgical fix, allowing the army air corps to qualify more men for flight school and winning Clark a publication in the February 1944 issue of *Surgery, Gynecology, and Obstetrics.*[24]

For three months in 1943, Clark was enrolled in the flight surgeon course in the School of Aviation Medicine at Randolph Field in San Antonio and formed close friendships with two other physicians. Edgar White, a native of Louisville, Kentucky, was slim and athletic, and the two men respected each other's physical prowess. In the surgical suite, they considered themselves equals, although Clark's interest in medicine had a more intellectual bent. Both men loved ranching, Clark preferring cattle and White, horses. White never managed to get the blue grass out of his blood. He rode horses for pleasure and exercise. He also smoked heavily. As soon as Clark felt comfortable enough to do so, he warned White about the dangers of tobacco. Bert Clark befriended White's wife, Mary Frances.

Clark and White forged a close friendship with one of their instructors, a specialist in tropical medicine. In 1939, Canadian Clifton Howe graduated from the University of Vermont Medical School and enlisted in the British army. Howe served with the Royal Army Medical Corps in Nigeria. Two years later, he was studying at the London School of Tropical Medicine. Once the United States entered the war, the Royal Army loaned him to the School of Aviation Medicine in San Antonio. Like Clark and White, Howe too had dirt under his fingernails. While growing up in Vermont, Howe had worked dawn to dusk on his grandfather's farm, milking cows, slopping hogs, cutting hay, and cooking for the hired hands. He took great pride in his sourdough bread, slathered in butter and distributed freely to friends.

At Randolph Field, the three men played on the same baseball team. Howe smoked as heavily as White, and Clark nagged him as well as White about it. The three also discussed life after the war and the possibilities of going into practice together. After his experience at Mayo, Clark had no intention of going it alone as a surgeon. "We used to talk about the future," Clark later recalled. He asked them, "If I do something after the war, if I can build a clinic somewhere, would you be interested in joining me?" Clark had no trouble selling them on the idea of a group practice, where specialists worked together in a multidisciplinary setting.

The army next transferred Clark to Wright Patterson Field in Ohio to direct programs in surgical education and edit the *Air Surgeons' Bulletin*. There Clark met Mavis Kelsey, an internist destined to become one of Houston's most prominent physicians. The constant activity of Clark's mind impressed Kelsey. "Clark was just brilliant. He was always thinking, always coming up with a new idea for this or that. . . . I had enormous respect for him."[25]

Others concurred. Early in 1943, the board of regents began searching for a permanent director of M. D. Anderson. They were not in a hurry. American troops were just beginning to island-hop across the Pacific, and George Patton's armored divisions were butting heads with Rommel in North Africa. Like a voracious predator, World War II consumed all the nation's resources. Ambitious programs of physical expansion were out of the question. Nor could staff be increased. Too many physicians and nurses wore uniforms. Bertner went into a holding pattern. In June 1944, however, when Allied troops landed at Normandy and began to sweep across northern Europe, victory no longer seemed so distant. Dudley Woodward of the board of regents took the initiative and traveled to prominent medical schools recruiting a new director. In Rochester, Mayo officials recommended R. Lee Clark. "One of your own Texans down there is very interested in cancer and was on our staff for five years," they told him. "He's right down there in San Antonio at the School of Aviation Medicine. He would be our first recommendation." Woodward shared the recommendation with Theophilus Painter, a geneticist and new president of the University of Texas. Painter invited Clark to his home in Austin. Over cornbread, turnip greens, and buttermilk, they discussed the job and the disease, both agreeing that the secret to cancer rested in the genes. Painter found Clark witty and well informed.

Painter scheduled a meeting with the board. Shortly into the session, two board members recognized Clark. Attorney Orville Bullington had served as a regent at Midwestern Junior College when Lee's father ran the place. "Mr. Bullington seemed to know me. That was when I was a senior in high school. . . . Maybe we had met and he did remember. . . . At any rate, from then on I was his boy." Cabe Terrell, a Fort Worth physician, recalled the Clark family role in founding TCU. Clark mentioned that former governor James Allred had been his scoutmaster in Bowie and Wichita Falls. When they mentioned the reason for the meeting, Clark was direct. No doubt thinking of his father and grandfather, and the dance he wanted to dance, Clark realized that "this was what I wanted to do—to build a cancer center. So I talked to them and told them that this was my objective, and if I didn't do it here, I was going to do it somewhere else. I can do it in Mississippi . . . but I'd prefer to do it in my own state." The board tendered the job.[26]

3

Designing a Dream,
1946–1950

I intend to discover the cause of cancer and build the greatest cancer
hospital in the world.
—R. LEE CLARK, 1946

On July 10, 1950, several M. D. Anderson staffers packed their bags for a journey to Paris for the Fifth International Cancer Congress. R. Lee Clark stayed home. He would have enjoyed once again seeing the Sorbonne, but events, including the ongoing plight of Ernst Bertner, kept him in Houston. When Clark first arrived in Houston, the two had anticipated a long working relationship, but in 1948 cancer ambushed Bertner—a deadly rhabdomyosarcoma in the muscles of one leg and a deadlier osteosarcoma in the other. In a series of operations, doctors removed the rhabdomyosarcoma and surrounding muscle bundles from one leg and amputated the other.

Within a month, Bertner was at the office every day, hobbling around on crutches and making plans for the medical center, but both the rhabdomyosarcoma and the osteosarcoma were on the loose, and in 1949 surgeons removed a lung. After Christmas, the other femur fractured and confined Bertner to bed for several weeks. "It surely was good to see you looking so well," Clark wrote Bertner on October 12, 1949. "After a study of the pathology, along with other findings, we have every confidence that you will be all right. It is our earnest prayer each day that this will be so." Both men knew otherwise. Within months, x-rays revealed tumors in the other lung. Bertner's apartment might as well have been a teaching hospital for all the physicians at his beck and call.[1]

In May 1950, his lung filled with tumor and his breathing grew shallow. Cornelius "Dusty" Rhoads of Memorial Hospital wrote Bertner. "The structure now at hand is important, but these things, outstanding, vital though they are, remain still subject to the ups and downs of life. The soul you have given this unit will go on forever, withstand any change and alter substantially the course of this country's medical and social progress." Bertner deeply appreciated the letter. "From the beginning of my work with the M. D. Anderson Hospital for Cancer Research and the Texas Medical Center," he replied, "I have felt in my heart that if I could accomplish some of the dreams . . . there could be something worthwhile chalked up on the credit side of my ledger. To have a man like you, unquestionably the number one man in this field, look at our work, and find it good, is a wonderful milestone." Bertner's condition deteriorated early in July, but not his ardor. A visitor saw "a bunch of architects' drawings of the Texas Medical Center spread out on his bed. He was planning to the very end so that his lifelong fight against cancer might continue." Clark was one of Bertner's attentive physicians. Paris beckoned, but so did Bertner; Clark remained at his side.[2]

While presiding over the start-up of M. D. Anderson, Bertner also brought life to the Texas Medical Center. He envisioned the cancer hospital as one of its cornerstones but also understood that a medical center without a medical school was like a telescope without a lens. Mike Hogg, the real estate developer responsible for the posh River Oaks subdivision, owned property south of downtown and hoped to entice the University of Texas Medical Branch to abandon Galveston for Houston. The advantages seemed obvious. Houston was vibrant and growing, while Galveston languished.[3]

For Galveston's medical community, Hogg might as well have suggested that Sam Houston was a Yankee. They reacted swiftly, predicting Armageddon—the wholesale exodus of physicians as well as hospital and pharmacy closings. Most faculty members at the Medical Branch worked part-time, using the affiliation to recruit patients to their private practices. Dean John Spies wanted to take the Medical Branch in the direction to which Abraham Flexner had pointed, a full-time research and clinical faculty. In Lee Clark's opinion, "John [Spies], dean of the Medical Branch, wanted to create an academic medical school with full-time people, and the part-time people didn't want him to do it. Survival is what a lot of people make their decisions on. They may use other high-sounding reasons." Only Spies saw merit in Hogg's proposal. "Move it to Houston," he urged."[4]

Spies had logic on his side but not the board of regents, who were embarrassed that the AMA had placed the Medical Branch on probation. At the time, Homer P. Rainey, president of the University of Texas, was battling conservatives on the board. Appointed president in June 1939, within a few months he had antagonized several board members by refusing to renominate the chair of the athletic council. Rainey backed Spies. In August 1942, the board fired Spies, and replaced him with pharmacologist Chauncey Leake. When the board terminated three economics professors with New Deal sympathies, Rainey condemned the action. He went public after the board banned from English classes John dos Passos's novel *The Big Money*. Rainey complained loudly when the board broached the subject of eliminating tenure. On November 1, 1943, they fired him. The Medical Branch stayed put.[5]

The decision to keep the Medical Branch in Galveston offered Bertner a fresh opportunity. With his vision of the medical school dashed, Hogg sold back to the City of Houston, at cost, the 134 acres that he owned south of Hermann Hospital. When John Freeman and William Bates learned of the sale, they moved. Working closely with the city council, they secured a special election to decide whether the city could sell the land to the M. D. Anderson Foundation, which would then donate it as the site of the Texas Medical Center and contribute money for the construction of city parks. On December 13, 1943, voters approved the measure.[6]

Hogg, Bertner, and the M. D. Anderson Foundation still wanted a medical school. The trustees of Baylor University College of Medicine, in the midst of a severe financial crisis, listened eagerly when the M. D. Anderson Foundation offered free land, $1 million, and $100,000 a year for ten years if they would relocate from Dallas to Houston. The trustees agreed, and the Baylor University College of Medicine became the second institution of the TMC. The revolution Flexner had unleashed had also engulfed proprietary dental schools. In 1943, Frederick C. Elliott, dean of the private Texas Dental College in Houston, persuaded the legislature to purchase and rename it the University of Texas School of Dentistry. M. D. Anderson and the dental school would share a site in the medical center. In 1944, the M.D. Anderson Foundation assumed title to the property. Ten months later, Bates and Freeman incorporated the Texas Medical Center, with Bertner its first president.[7]

—◆

Clark wasted no time erecting the institutional scaffolding necessary for M. D. Anderson's success. The distribution of power worried him, perhaps because he had been raised in a household where school boards and boards of regents were the stuff of dinner table conversation. Clark needed to know how M. D. Anderson fit into the University of Texas. Leake wanted the cancer hospital subservient to the medical school. The medical school handled every disease, not just one; a special cancer hospital independent of the medical school made little sense. Clark wanted exactly that independence, knowing that if "you were subservient to the medical school, every department would have its say about what you were doing. . . . If the school wanted to use the laboratory you were doing cancer research in for some other kind of research . . . you'd lose your research space. . . . The only way was to have autonomy so you could build the institution as a peer to the medical school."

To draw the lines clearly, Clark met repeatedly with UT's president Theophilus Painter, Leake, and various members of the board. "Would I go through the dean of the medical school at Galveston," Clark wondered, "or could I report directly to the board of regents?" Painter wanted to help Clark avoid the political quagmire in Galveston, and the board of regents awarded M. D. Anderson a status independent of and equal to that of the Medical Branch. The decision, made in August 1946, "really made M. D. Anderson different," remembered the physician William Seybold. "It always seemed that Clark could do anything he wanted."[8]

Clark then targeted part-time physicians. He intended to avoid a staff of part-timers preoccupied with their private practices, as had been common at medical schools. Patients might view physicians as healing demigods, but Clark knew them as men and women with families to feed, mortgages to pay, retirements to fund, and tuition to pay. Financial incentives sometimes shaped their medicine. The model at Mayo—full-time physicians working in group practice—remained Clark's ideal. He made his case, and Painter and the regents acquiesced.

Clark's commitment to full-time faculty was mirrored at medical schools throughout the nation, where part-timers were being swept away. By 1954, sixty-five of the nation's eighty medical schools had completed the transition to full-time faculty, and within a decade, the practice would be universal.

The virtual explosion in research funding after World War II explained much about the change. Research would soon rival and then swamp education and patient care in the mission of many medical schools. In 1948, Congress created the National Heart Institute and the National Institute for Dental Research, placed them with the National Cancer Institute (NCI) inside the National Institute of Health (NIH), and changed NIH's name to the National Institutes of Health. Each institute was charged with conducting its own research internally (intramural) and financing research at external (extramural) teaching hospitals, medical schools, research laboratories, and universities. Public faith in the medical profession had never been higher. Because of advances in controlling infectious diseases, life expectancy in the United States had risen from forty-seven in 1900 to more than sixty in 1944. The floodgates of federal funding opened. Some conservatives cautioned that federal money would come with strings attached, but the cornucopia was too much to resist. In 1947, Congress appropriated $87 million for medical research, an amount that would balloon to $2.05 billion within two decades, and the money did not include funds dedicated to construction and training. Adjusted for inflation, the government investment would be fifteen times greater.[9]

Clark then raised the issue of indigent patients. To subdue the suspicions of Houston doctors about a state-subsidized cancer hospital in their backyard, Bertner had only accepted patients unable to pay. Clark understood Bertner's logic, but he also envisioned risks. "We'd only have people in the lower income and charity brackets who naturally would wait as long as they could, and we'd get only late stages of cancer. . . . If we were going to build a great institution, it should be available equally to those who could pay for it and as well for those who couldn't." Clark needed three revenue streams—the state legislature, insurance companies, and paying patients.[10]

At the time, medical schools around the nation began to make the same case to their governing boards—to broaden their patient base from just indigents, the traditional constituency, to paying patients as well. Such proposals, of course, riled private physicians and exacerbated the town-gown rivalries.[11]

In his pitch to Painter and the board, Clark insisted on the "right of consultations for private physicians so that Anderson wouldn't be just an eleemosynary institution but could take all cancer patients. There would be no fiscal triage. We'd get the people who could pay and the charity patient as referred by the physician." Multiple revenue sources would allow for the delivery of advanced care to indigent patients, a premise that satisfied Clark's Jacksonian streak. Once patients walked through the front door, they deserved the best treatment medicine had to offer, regardless of financial circumstances. The board bought it, even though local physicians griped. Clark considered their arguments selfish, parochial, and annoying, especially when solemnly draped in the rhetoric of "concern for the best interest of patients."

Having pressed the regents this far, Clark tackled a fourth goal—dumping the fee-for-service model of medical compensation. All over the country, physicians were paid according to the number of patients treated, the number of tests ordered, and the number of procedures performed. Such a system, Clark believed, guaranteed competition for patients among physicians and the temptation to order more tests and perform more procedures than medically necessary. For Clark, the Mayo model of full-time, fixed-salary physicians without private practices was infinitely superior. Having no need to compete for patients, physicians readily consulted and referred patients to appropriate staff members. Reducing competition would cement the foundation for multidisciplinary care, to become a hallmark of M. D. Anderson. It would take another decade to become fully operational, but the Physicians Referral Service was born.[12]

Clark realized that no culture, regardless of its emphasis on excellence or the power of its mission, could in the long run trump self-interest. Financial incentives had to exist. By offering flush benefits packages, Clark hoped to recruit the finest physicians with the promise of comfortable retirement at an early age and a nice estate for their children.[13]

At most medical schools, individual departments controlled the money they earned. Cash-cow departments like laboratory medicine or radiology often gorged while such departments as medicine or pediatrics, intensive in labor and paltry in income, starved. Civil wars often raged. Clark wanted to put all income in a central pool and reallocate it according to larger institutional objectives. Centralizing the control of the budget would give the director more freedom of action. In recruiting superstar clinicians or basic scientists, for example, he did not have to siphon money from one department to give it to another. Clark intended to run M. D. Anderson with dictatorial power over budget. It would take him more than a decade to fully implement the idea, but he broached it first in 1946.[14]

In the giant chore that engrossed him, Clark's ambition melted into the cause, and his life went full throttle, assuming a pace that would never relent. M. D. Anderson was on its way to becoming his alter ego.

———

In 1946 and 1947, Clark undertook a special tour of major medical centers. Mayo renewed his passion for group practice, multidisciplinary care, and salaried physi-

cians. In Buffalo, New York, he visited the country's first hospital dedicate
clusively to cancer care—the Roswell Park Memorial Institute. In Bethesda, he
made a courtesy call at NCI.

Just a few years earlier, the age of cancer chemotherapy had dawned. On December 2, 1944, German fighters sank several Allied ships laden with nitrogen mustard gas in the port of Bari, Italy. The exploding ships released massive amounts of the deadly chemical into the air and the water, and when sailors abandoned ship, they fell into a toxic soup. The incident killed thousands. The army asked Col. Steward Alexander to investigate the disaster, and in autopsy reports he noted that blood cells in many victims had all but disappeared, as had a substantial portion of lymphatic tissues. "If nitrogen mustard could do this," Alexander thought, "what could it do for a person's leukemia or lymphosarcoma?"

Alexander sent the data to Yale, and Louis Goodman and Alfred Gilman began experimenting on lymphoma patients, theorizing that rapidly proliferating lymphoma cells might also be vulnerable. They were. Many patients experienced temporary remissions, the first documented time in human history that a medicinal agent had corralled a cancer. In 1940, a New York physician remarked, "Throughout the centuries . . . the fields and forests, the apothecary shop and temple have been ransacked for some successful means of relief from this intractable malady. Hardly any animal has escaped making its contribution in hide or hair, tooth or toenail, thymus or thyroid, liver or spleen in the vain search [for] a means of relief." Perhaps history was about to turn.[15]

Sidney Farber, a pathologist at Children's Hospital in Boston, drafted the creation story of modern oncology. Intrigued by the effect of nitrogen mustard on lymphomas, he speculated that folic acid might have a similar impact on leukemia, the most dreaded malady in cancer's lexicon. The bone marrow in leukemia patients produces excess white blood cells that clog the bloodstream, damage the immune system, and induce severe hemorrhaging. The drug company Lederle had just come out with aminopterin, which slowed the proliferation of leukemia cells. In November 1947, Farber treated the first of sixteen children, and several achieved temporary remissions.

Clark happened to be in Boston just as Farber noted the remissions. They met at Farber's top floor office in the Jimmy Fund Building on Binney Street. Not until Farber's landmark article in the June 3, 1948, issue of the *New England Journal of Medicine* was the evidence unequivocal. Although all of Farber's children eventually died, ten had responded, setting the stage for further research on doses and scheduling to extend survival time.

From Boston, Clark took the train to New York City and then a taxi from Penn Station up to East Sixty-seventh and Sixty-eighth streets, where new buildings financed by the Rockefeller Foundation housed Memorial Hospital for Cancer and Allied Diseases. Presided over by Cornelius "Dusty" Rhoads, Memorial was considered the best in the world. Convinced that chemotherapeutic agents could be applied to cancer cells, Rhoads had just taken Memorial into the new modality, directing a cohort of young clinicians that included Joseph Burchenal and David Karnofsky. Most cancer patients died from systemic disease, not locally confined tumors, and chemotherapy offered the only hope.[16]

For years, Lee Clark would measure M. D. Anderson against the Memorial standard. In 1948, after attending the dedication of the Sloan-Kettering Institute as a research arm of Memorial, he mailed a Memorial pamphlet to Theophilus Painter and copies to the board of regents. The pamphlet, he told Painter, "will give you an idea of the possibilities of the development of the M. D. Anderson Hospital. . . . Memorial Hospital is the only one of its kind in this country. While there are other cancer hospitals, their progress is limited in some of the fields. Memorial is the only one that has a complete program of Prevention, Research, Education, and Treatment."[17]

In New York, Clark also visited Mary Lasker, a woman about to become the press-dubbed "fairy godmother of medical research." The two could hardly have been more different, one a child of rural West Texas and the other, the Upper East Side of Manhattan; one an alumnus of Tarleton State College and the other a graduate of Radcliffe. She was married to Albert Lasker, an advertising tycoon whose "Reach for a Lucky Instead of a Sweet" made the cigarette brand name a household term and condemned millions to untimely deaths from lung cancer. In 1942, they founded the Albert and Mary Lasker Foundation to support medical research. When her maid died of cancer, the grieving Mary Lasker paid a visit to the American Association for the Control of Cancer (AACC).

Cancer took more than 150,000 Americans in 1940, but Lasker found the AACC a shoe-box operation dominated by conservative physicians. Energetic and dynamic, with a vision as deep as her purse, Lasker dreamed of eliminating cancer, which most of her physician colleagues considered impossible. That she had a close friend in First Lady Eleanor Roosevelt gave Lasker instant cachet. She transformed the AACC board, engineering the addition of Lewis Douglas, the president of Mutual Life Insurance Company and FDR's former secretary of the treasury. Lasker took over fund-raising. In 1941, the AACC budget totaled only $120,000. She raised $832,000 in 1944 and convinced the board to change the name to the American Cancer Society. In 1945, Lasker collected $4.29 million. The physicians formerly in charge felt upstaged, but money flowed and Lasker gripped the spigot tightly.[18]

Clark's tour included a week in St. Louis for the Fourth International Cancer Congress. He booked a room at the Jefferson Hotel and spent time with Theophilus Painter, who was there to learn about the latest in cancer genetics. Both men told everybody about M. D. Anderson and met the premier figures in European oncology—Alexander Haddow of Great Britain and Antoine Lacassagne of France—and Peyton Rous, Jacob Furth, Sidney Farber, Cornelius Rhoads, Albert and Mary Lasker, and Eleanor Roosevelt of the United States. Leonard A. Scheele, assistant chief of NCI, had piqued Clark's ambition with the news of $1.8 million in grant money and plans to partner with the American Cancer Society.[19]

After returning to Houston, Clark assembled a permanent staff. To lead a department of medicine and direct the clinic, he approached Clifton D. Howe, his old friend from Randolph Field. Howe knew more about tropical medicine than cancer, but Clark considered him a fine physician, up to the task of doing what internists at the time did for cancer patients—diagnose their disease, prep them for surgery, and redeem them, if possible, when surgery went awry. Over drinks at the Warwick Hotel early in 1948, they sealed the deal.

Clark also wanted to start a medical publishing company, and he admired *What's New*, a monthly scientific news magazine published in Chicago by Abbott Laboratories. Clark set up an appointment with the editor R. W. Cumley. Clark had no idea that Cumley was Russell W. Cumley, his childhood friend from Wichita Falls, now with an interdisciplinary Ph.D. in science from the University of Texas. Late in the afternoon, meeting at Clark's hotel for whiskey, dinner, and cigars, the reunited friends reminisced and talked of curing cancer. Both agreed that it would become the biggest news in history. The next morning, they met for breakfast. Clark asked Cumley to become director of publications at M. D. Anderson, with the possibility, on the side, of establishing a medical publishing company. Cumley hated Chicago, especially the icy winds screaming in from Lake Michigan. Texas winters beckoned. Cumley took the job.

For several years, the two ran M. D. Anderson, with Cumley wielding more power than outlined in the job description. An individual of continuous mirth with a bottomless reservoir of anecdotes, Cumley composed jingles, poetry, and short stories, read fiction and nonfiction, kept current with major developments in science, and nurtured a keen interest in the flora and fauna of the Southwest. And he could speak truth to power, sometimes reminding the boss that there were too many stars in the sky to count. The two men established Medical Arts Publishing Foundation, with Cumley as executive director. In 1948, the foundation published the *Cancer Bulletin*, M. D. Anderson's nonrefereed publication for practicing physicians. Within a few years, the *Cancer Bulletin* had more than a hundred thousand subscribers, elevating M. D. Anderson's profile and supplying a tidy sum to Medical Arts.

As surgeon in chief, Clark maintained an active presence in the operating room. He hired as assistant director Roy C. Heflebower, a physician and old friend. A retired brigadier general once in charge of the Medical Replacement Training Center and Hospital at Camp Barkeley near Abilene, Heflebower acted as if one star still glittered from an epaulet. Pompous and used to being obeyed, he marched around the Baker estate, chewing a cigar and, as he liked to say, "policing the grounds and inspecting the troops." The staff nicknamed him "the General," and he made few friends, treating most people like army recruits.

Heflebower had a near fetish for cleanliness, inside the buildings and out. As he walked around the Baker estate, the General noted lawns not mowed or edged, shrubs not clipped, and table tops not dusted. Custodians learned to check restrooms several times a day, groundskeepers to remove every speck of litter, and maids to make a tight bed. "He could smell urine from a mile away," according to one staff member. Heflebower was not above picking up litter himself and depositing it in trash cans or applying paper towels to smudged sinks. He instructed nurses that when they emptied bedpans into toilets or bodily fluids into basins, they must leave behind a fresh restroom, even if they had to do a little cleaning. He wanted M. D. Anderson as orderly and as clean as West Point.[20]

Clark selected Arthur F. Kleifgen, an old Mayo associate, to serve as an administrative assistant and establish a patient care service department. Best friends with Kleifgen's wife, Molly, once a nurse in the anesthesiology department at Mayo, Bert Clark wanted the couple in Houston, and Lee obliged. Kleifgen often tucked away tidbits of gossip and fed them to Molly, who conveyed the juiciest to Bert. Sometimes news circumnavigated the hospital in one day.

Kleifgen helped instill a patient-friendly attitude that seeped deeply into the M. D. Anderson culture. In memo after memo, he urged the staff to have concern for patients. "If a patient has a complaint [it] offers us an opportunity to make a new friend, and friends made this way are the kind who will gladly prove themselves when the occasion arises. . . . Little or much may be done to satisfy a complaint. The main thing to keep in mind is to make him feel that his complaint has been listened to with attentive consideration. . . . At times complaints may even be welcomed because of the opportunity it offers to strengthen the ties between patient and institution." In the ensuing years, Kleifgen made a mantra of this.[21]

With administrative staff in place, Clark turned to the physical plant. Patient volume already outstripped the capacity of Hermann Hospital and the Houston Negro Hospital. In 1947, the gift of a four-unit apartment complex on Webster Street relieved some pressure. Clark remodeled the Webster Street Annex to house thirty patients needing intermediate nursing care. The next year, he purchased twelve war-surplus buildings. Constructed of pine and painted white, the buildings gave the Baker estate the ambiance of a military camp. The buildings provided space for an operating room and three hospital wards with sixty-six beds. On June 27, 1949, the first patient checked into the Baker estate facilities.

From the beginning, Clark made sure that M. D. Anderson adhered to the demands of Jim Crow. He had grown up in a segregated world, and Texas was not yet ready to change. In 1950, the number of patients at the Negro Hospital exceeded the ten rented beds. A black woman had to be seen in the barracks, and Clark was not pleased. He reminded Kleifgen, "Will you see that the allotment is not exceeded in the future without specific permission?" When it happened less than a year later, Clark told Kleifgen, "We must utilize those facilities [Negro Hospital] for our colored patients. . . . This policy is to be followed except in the case of life and death." When another physician outlined more exceptions, Clark reacted with a stony stare. "It is necessary to make a limitation," he replied, "and to enforce it to avoid its abuse by making exceptions." He even segregated blood. M. D. Anderson phlebotomists made regular trips to Huntsville, where prison inmates eagerly donated blood in return for a thirty-day commutation of sentence per pint. Consistent with practices throughout the South, each pint was then labeled according to blood type, Rh factor, and race.[22]

———

Convinced that radiation would someday complement surgery in cancer treatment, Clark looked for a radiotherapist. From Hiroshima and Nagasaki, the world knew of the destructive power of radiation. Research intensified for positive uses of the force. Cancer seemed perfect. Physicians had learned that the radiotherapy devices generating the most power also possess the most promise; they treat deep-seated tumors without turning skin into mush. An obsession to build the most powerful radiation machine infected physicists. In 1930, Cal Tech boasted of its 750,000-volt x-ray tube, and in 1934 Mercy Hospital in Chicago heralded its 800,000 volts. Perhaps the answer to cancer, some thought, lay in multimillion-volt contraptions. Journalists hyped each new arc. Physicists at the University of California at Berkeley bragged of generating hundreds of millions of electron volts in a linear accelerator. A 1950 issue of *U.S. News & World Report* claimed that "millions of dollars and hundreds of scientists, and careful planning are being used in what authorities regard as medicine's counterpart of the war time atomic bomb project." M. D. Anderson needed a radiotherapist.[23]

In Gilbert H. Fletcher, Lee Clark found the man about to become prince of a new medical discipline. Born on March 11, 1911, in Paris, to a wealthy importer, Fletcher grew up under the tutelage of a domineering mother. Whenever Fletcher or brother Henri talked about getting a job, she urged them to stay in school. In 1929, Fletcher graduated from the Sorbonne with a bachelor's degree in Latin, Greek, and philosophy. He then earned a degree in civil engineering at the University of Louvain. Instead of finding a job, he transferred to the University of Brussels early in 1935 for a master's degree in physics and mathematics. Medical school then beckoned. In 1941, he received a doctor of medicine degree; one year he later started an internship in obstetrics at the French Hospital of New York City. Later in the year, eager to fuse his training in physics with medicine, Fletcher undertook a residency in radiology at New York Hospital. There he met Mary Walker Critz, a physician from Mississippi. They married in 1943. At the conclusion of the residency, Fletcher joined the army and deployed to the Veterans Administration Hospital in Pittsburgh.[24]

He soon gravitated to the radiation side of the discipline. At the time, radiologists occupied a low rung on the ladder of medical disciplines, as if all they did was identify broken bones and locate tumors. Some surgeons compared them to shoe salesmen sizing customers by placing feet into a fluoroscope. In radiation therapy, on the other hand, radiologists ascended the ladder by treating patients. Radiation treatment, increasingly being called "radiotherapy," also trafficked in the sophisticated discourse of mathematics and physics. Upon his discharge from the army, Fletcher entered the job market. Back in Mississippi in 1939, the Clarks had befriended Mary Critz. Eight years later, Mary learned that M. D. Anderson sought a radiotherapist; she contacted Bert and Lee and arranged an interview for her husband. Impressed with Fletcher's training, Clark asked him to head up a department of radiology.

Fletcher accepted, but only after squeezing from Clark enough money to visit leading radiation treatment centers in Europe. On October 1, 1947, Clark dubbed Fletcher a Traveling Anderson Fellow. Fletcher spent three months touring the Curie Foundation in Paris, the Radiumhemmet in Stockholm, the Christie Hospital and Holt Radium Institute in Manchester, and the Royal Marsden Cancer Hospital in London. Everywhere, he noticed that radiotherapy departments employed physicists, unlike the United States, where physicians wielded x-rays like cowboys hefting six-guns, shooting from the hip with little understanding of dosages or scheduling. If radiation doses are excessive, the damage can be disabling, even lethal. Many early patients ended up with gaping, necrotic wounds vulnerable to infection, the cancer gone, but at a steep price. Insufficient doses, on the other hand, left tumor cells thriving. Fletcher envisioned radiation therapy as a science, and he needed a physicist to realize that vision. "Success," he argued, "will depend upon our combined efforts" because without physicists, "clinical radiotherapy is left with inadequate empirical methods."[25]

In London, Fletcher met the physicist Leonard Grimmett, and over a cup of tea, they discussed the state of radiotherapy. In 1934, Grimmett had opened his own laboratory at Westminster, specializing in the application of physics to medicine. He believed that radiologists and physicists needed to link up. Only with good data could standardized protocols develop. The two connected immediately, Grimmett impressed with Fletcher's grasp of physics and mathematics, and Fletcher certain that Grimmett embodied the future of radiotherapy.

In January 1948, Fletcher moved into the basement of the Baker estate, which Clark equipped with a gram of radium for interstitial treatment, a low-voltage device for x-ray treatments, and an x-ray machine for diagnostic radiology. Mercurial and feisty, Fletcher had the visage of a falcon and a temperament to match. He could at once terrorize staff and residents and then shrink obsequiously in front of Clark, like a puppy in the presence of an alpha male. His French accent modulated according to mood. The more frustrated he became, the more French he sounded, code-switching back and forth from French to English or, at peak moments, producing an unintelligible patois. Fletcher sold Clark on the need for a physicist, assuring him that M. D. Anderson, not Memorial or Roswell Park, would be the first in the United

States to do so. Clark needed no convincing; the prospect of upstaging Memorial was reason enough. Early in February 1949, Grimmett arrived to establish the new department of physics.

As fastidious as he was brilliant, and given to bow ties and horn-rimmed glasses, Grimmett sported a natty mustache and a head of thick, dark hair combed straight back and lathered in a shiny pomade. He set up a physics shop just above Fletcher's laboratory but tired of the dust. Grimmett wanted his shoes to shine as brightly as his hair, but no amount of vacuuming could eliminate the dust filtering up through slats of the wooden floor. Unable to extract money to renovate, Grimmett raised his own. He played piano with concert-level prowess, staged several performances, and with the proceeds replaced the dusty pine floors with mahogany.[26]

The collaboration between Fletcher and Grimmett was a volatile one. An Anglophobe convinced that "all of England could be seen on one square mile of downtown London," Fletcher often needled Grimmett on the deficiencies of English cuisine. In return, Grimmett now and then butchered the pronunciation of Fletcher's given name "Gilbert" rendering it closer to "Gerbal." Their shouting matches often reverberated throughout the Baker estate. When Fletcher needed something from Grimmett, he expected it immediately. In a 1950 letter to Clark, he wrote, "I have had during the last eighteen months many heart to heart conversations with Dr. Grimmett. . . . I have had to reach the conclusion that it has been entirely in vain. I am willing to cooperate and do team work with Dr. Grimmett but it will be meaningless as long as Dr. Grimmett does not make a true and honest change of heart." For all their friction, however, the two men held each other in high intellectual regard and jointly undertook the task of making radiotherapy as important as surgery in the treatment of cancer.[27]

The multimillion electron volt behemoths at Chicago and Berkeley made for good newspaper stories, but they meant little in 1947 to the 150,000 Americans diagnosed with cancer. Radiologists needed therapy units that had millions, not thousands, of electron volts and were accessible to cancer patients. Radiotherapy was not like popping a pill or having surgery; it had to be delivered in daily "fractions" over the course of five to six weeks, with the effects of the treatment—positive and negative—accumulating over time. Patients could become quite debilitated, and great distances made them more likely to stop before completing treatment or not to start treatment at all. Cancer patients needed access to megavoltage devices near home.

And they needed safe, cost-effective machines. Generating powerful radiation beams required high-energy radioactive materials that posed health threats to patients and to medical professionals. Physicians, nurses, and technicians wanted to know that radioactive materials were safely housed. Although relatively cheap to produce, x-rays could not generate sufficient energy. Radium had been commonly used as a source of radiation, but its production required processing massive volumes of earth to extract a single gram. Expanding access to radiation therapy demanded a relatively inexpensive radiation source. And therapists needed a radiotherapy device simple in design to reduce the frequency of breakdowns.

Fletcher and Grimmett decided to design a new radiotherapy unit. For a safe, cost-effective, high-energy source, they selected cobalt-59. A brittle, nonradioactive metal with a grayish bluish hue, cobalt occurs naturally in a variety of minerals. Two 2×2 centimeter squares of cobalt-59 would be buried in a radioactive pile until they became cobalt-60 and absorbed 1,000 curies, enough to emit 1.3 million electron volts. Cobalt-60 readily sloughed beta particles, emitted powerful gamma radiation, and decayed into nonradioactive nickel. It provided a source of enormous energy at a relatively low cost. Grimmett and Fletcher promised to place the cobalt-60 inside a tungsten alloy head thick enough to reduce external surface radiation. A trunnion mounting would permit remote-control vertical motion of the head and angulation through 120 degrees in one vertical plane. When out of use, the cobalt could be self-screened by rotating a tungsten disc on which it was fixed. Four adjustable tungsten diaphragms would shape the radiation beam. Simple in design, it safely produced megavoltages of energy at relatively low cost. They correctly estimated a price of $150,000 for a new cobalt-60 machine, a technology affordable to many large hospitals.

The two men took the design to Clark, who sought permission from the Atomic Energy Commission (AEC) to build a prototype. The lead AEC physicist rejected the proposal, telling Clark, "This is as impractical, as wild as saying we'll go to the moon." Bypassing the AEC, Clark entreated the Oak Ridge Institute of Nuclear Studies, where Marshall Brucer, head of the medical division, bought into the idea. In December 1949, Clark, Grimmett, and Fletcher met with officials from AEC and the Oak Ridge Institute. The officials approved the design and agreed to ship a small volume of cobalt-60. The Oak Ridge Institute then invited research laboratories throughout the nation to submit designs for a cobalt-60 machine. Twelve major universities submitted proposals, and the Oak Ridge Institute selected for development the M. D. Anderson design. Six months later, M. D. Anderson signed a contract with General Electric X-Ray Corporation to build the prototype.[28]

—◁◁◁◁◁\◁▷▷▷▷▷—

The winning design brought M. D. Anderson some recognition in radiotherapy circles, but surgery was still the treatment of choice. To lead a department of surgery, Clark hearkened back to Randolph Field and his old friend Ed White. Surgeons needed pathologists, and Clark recruited William Russell, a graduate of Stanford University Medical School. Tall and lanky, with a squeaky, high-pitched voice incongruous with his stature, Russell had already published articles on a variety of tumors in a variety of journals when he first met Clark at the Fourth International Cancer Congress.[29]

After World War II, the explosion of medical knowledge and the increasing specialization in disciplines placed new demands on the education mission of medical schools. To meet the demand, the phenomenon of postgraduate training for physicians in the form of residencies and fellowships emerged. With the advent of specialty boards and board certification exams in the 1930s and 1940s, a premium had been placed on sound postgraduate medical education. Clark desperately wanted to develop M. D. Anderson as a "peer institution" to the Medical Branch and to the Baylor University College of Medicine. All over the country, in the wake of the Flexner revolution, medical schools had adopted "research, education, and patient care" as their mission, and so did Clark. Patient care would take care of itself, and he intended to build basic science and clinical departments worthy of NIH and NCI research funding. Education, however, was more problematic. At the time, most hospitals considered education a nuisance. The medical schools, because they trained physicians and awarded medical degrees, had their educational mission easily defined. Clark had to be more creative; he led a hospital that was not a teaching hospital, and a branch of the University of Texas that enrolled no students and awarded no degrees. He also worried that M. D. Anderson's lack of an academic venue would hamper recruitment of top-quality scientists and clinicians. Taking his cue from the Mayo Clinic's postgraduate program, Clark persuaded the state legislature to establish the University of Texas Postgraduate School of Medicine as part of M. D. Anderson in 1948. Theophilus Painter named him dean, over the protests of Chauncey Leake and the heads of several science departments in Austin. Through the Postgraduate School of Medicine would flow to the Texas Medical Center a legion of interns, residents, and fellows, and Clark was in charge of them all. Alando J. Ballantyne became M. D. Anderson's first resident.

━━◁◫◯◫▷━━

As a young medical student returning to Virginia after a summer in Texas, Clark was hitchhiking through Atlanta when one building at Emory University caught his attention. Clad in Georgia Etowah pink marble, it seemed to his artistic fancy a beacon of peace and hope. The drabness of the wooden barracks at the Baker estate offended his aesthetic taste, and he dreamed of building a new home for M. D. Anderson, one with the rosy hues of Georgia Etowah pink marble.

M. D. Anderson needed a new home. Among other things, Clark wanted to go into pediatrics, and the Baker estate was ill equipped for treating children. Cancer treatments required long hospital stays for children and their families and more elaborate facilities for supportive care. In 1947 M. D. Anderson treated just three children. Clark ached to employ chemotherapy for childhood leukemia, but he refused to treat patients in the absence of excellent supportive care. He began dropping hints to politicians. In January 1947, when the legislature convened, Clark took a room in Austin at the Stephen F. Austin Hotel. At 6:30 every morning, he met in the dining room with elected officials and shared eggs, bacon, grits, and vision. The guests included Alexander McCauley (A. M.) Aikin Jr., chairman of the Senate Finance Committee. In Aikin, Clark found a booster. After his father's miserable death from laryngeal carcinoma, Aikin had vowed to do something about the disease.[30]

In June 1949, with the approval of the board of regents, Clark hired the architectural firm of MacKie and Kamrath to draw up preliminary plans. He was a testy client, often hovering around the architects, checking on their progress, and suggesting design changes. A neophyte in cancer medicine, Clark believed that a cure might be on the horizon, and he worried lest the new hospital be architecturally unsuited for change. In 1943, Selman Waksman of Rutgers had synthesized the antibiotic streptomycin. First administered to a tuberculosis patient in 1944, the drug wiped out the bacteria. In 1949, as Clark promoted his own plans, the legislature closed the state tuberculosis hospital, leaving a vast, sprawling complex of buildings adaptable to little else. Clark wanted to make sure that M. D. Anderson could retool after he wiped out cancer.[31]

While the architects drew up plan after plan, Clark raised $3.8 million from public and private sources. To finance the portion of the building housing the radioactive materials and radiotherapy machines, he approached the state legislature for $1.35 million. Budget-conscious conservatives stalled the bill and in doing so badly miscalculated public sentiment. They had denied Texans the most recent technological innovations in cancer medicine. Governor Allan Shivers called on Frances Goff, the state budget director. Goff had worked for the House Appropriations Committee, the Senate Finance Committee, and the Texas Railroad Commission. Active in the women's club movement, she mobilized the Women's Field Army of the American Cancer Society, the General Federation of Women's Clubs, the American Legion Auxiliary, the Business and Professional Women's Clubs, the Veterans of Foreign Wars, and the Order of the Eastern Star. Goff had the bearing of a sergeant-major, stiff-postured and attired in suits and sturdy, short-heeled shoes. Opponents disappeared before her onslaught faster than TB bacteria in the presence of streptomycin. During a special session in 1950, the legislature passed the bill after Shivers threw in the weight of his office: "Cancer is a public problem," he said. "Scientists tell us that no one is ever really immune from the disease. There is no cancer germ, no human cancer virus. There is no vaccination, no immunization against it. . . . Cancer takes educator, scientist, businessman, and beggar without considering human values." On March 1, 1950, Shivers signed the bill.

Clark went out for bids in October 1950, anticipating a low one of $5.25 million, but the Korean War had spiked the price of construction materials. Clark's taste for marble did not help. The low bid came in at $7.45 million, leaving him $2.2 million short. "It was a blue day," Clark recalled. "It was hard to bear after we had waited from 1946 to 1950 and had planned it and were ready to go and had gotten that extra money from the state." Clark returned to the drawing board, literally, working with regents, architects, and general contractor to downsize the plans from 310 to 150 beds by eliminating one wing. Late in the month, the UT Board of Regents awarded a contract for $5.24 million to the firm of Farnsworth and Chambers. Two months later, shovel-wielding club women broke ground.

Clark extracted from the contractor a promise to build the deleted wing if Clark could raise the $2.2 million during construction. The contractor agreed to hold the line for eighteen months, a dicey promise given the soaring rate of inflation. But he wanted to cure cancer too. Returning to the M. D. Anderson Foundation, Clark secured $1.2 million. Several other Texas philanthropies donated money. Newspapers buzzed with M. D. Anderson's need, and donors responded by the thousands. Boy Scout and Girl Scout troops collected money door to door, and chambers of commerce, Rotary Clubs, and Kiwanis Clubs tapped members. Clark secured the money, and the contractor built a hospital with 310 beds.[32]

————⫘⫘⫘————

In the midst of the legislative maneuvering, Grimmett and Fletcher were invited to present a paper at the Fifth International Cancer Congress in Paris and at the Sixth International Congress of Radiology in London. Both would return home in glory—London for Grimmett and Paris for Fletcher. M. D. Anderson had designed a new technology for radiation treatment, and Clark's hope of building the greatest cancer center in the world was no longer merely wishful thinking.[33]

Fletcher and Grimmett delivered the paper in London on the afternoon of July 27, 1950. Within hours they received a telegram announcing Ernst Bertner's death. Almost to the end, Bertner had talked of little but M. D. Anderson and the Texas Medical Center. In earlier generations, a sarcoma patient with lung metastases had no options once surgery failed. Death was as inevitable as sunset. By 1950, however, oncology offered a sliver of hope, and Bertner was ready to experiment. A determination to "treat to cure" would eventually flower into an M. D. Anderson hallmark, and during the last weeks of his life, Bertner planted the seed. As an institution with a profound research mission, M. D. Anderson had to evolve into a place where patients went to be cured, and the only path to that destination was clinical experimentation. Bertner, of course, hoped to save his own life but also to make a point: Treat to cure. "During his illness," one reminiscence records, Bertner "allowed himself to become a 'guinea pig' for various cancer experiments. He eagerly tried out every new medical development that offered some hope of cure for the disease." On July 13, 1950, the medical director of the AEC flew to Houston with a vial of the radioisotope gallium and treated Bertner. Although an autopsy revealed that both the rhabdomyosarcoma cells and the osteogenic sarcoma cells had absorbed the radioisotope, Bertner did not live long enough to tell whether the treatment had slowed tumor growth. But on the day he died, his beloved M. D. Anderson was the toast of cancer medicine.[34]

4

The Pink Palace,
1950–1955

*Cancer will be brought under control probably within your lifetime
and mine.*
—CHARLES P. HUGGINS, 1953

Early in the 1950s, chemotherapy became the new frontier. Unlike surgery and radiation therapy, which addressed only local disease, chemotherapy offered hope for treating widespread illness, especially for such systemic cancers as the lymphomas and leukemias. In Boston, Sidney Farber had demonstrated some effectiveness of a folic acid antagonist against childhood leukemia, as had Cornelius Rhoads and Joseph Burchenal at Memorial Sloan-Kettering. For the parents of leukemia patients, Boston and New York offered glimmers of hope, and in the spring of 1953, George and Barbara Bush picked Memorial Sloan-Kettering. They had just settled in Midland, Texas, for George to assume the reins of Zapata Petroleum. Their three-year-old daughter Robin awoke one morning and told her brother, "I don't know what to do this morning. I may go out and lie in the grass and watch the cars go by, or I might just stay in bed." With curly blonde hair and sparkling eyes, Robin usually kept her mother busy all day long, but not that morning. As Robin languished in bed, Barbara intuitively sensed danger. Within hours, she took Robin to a pediatrician, who drew blood and, as Barbara later remembers, "suggested that I might want to come back without Robin, but with George. That sounded rather ominous to me, but I wasn't too worried. Certainly Robin had no energy, but nothing seemed seriously wrong."

Nothing could have been worse. Late in the afternoon, the Bushes learned of Robin's leukemia. "Neither of us had ever heard of it," Barbara recalls, "and George asked her how [do] we cure her?" In spring of 1953, all children with leukemia declined rapidly, their last days filled with bruises, bleeding, and blood transfusions, death arriving before parents could even begin to fathom the tragedy. The doctor, Barbara writes, "talked to us a little about red and white blood cells and told us as gently as possible that there was no cure. Her advice was to tell no one, go home, forget that Robin was sick, and make her as comfortable as we could, love her—let her gently slip away. She said this would happen very quickly, in several weeks."

The Bushes rejected the counsel. Pediatric oncology was still in its infancy, but it did exist, and they knew where. M. D. Anderson did not even cross their minds. George's uncle John Mercer Walker had just joined the surgical staff of Memorial Sloan-Kettering. As Robin declined, the Bushes moved into Walker's apartment at Sutton Place in New York City and checked Robin into Memorial, where she received what George at the time called "radically different treatments that saved her life."

In August, Robin rallied. "I am hoping that Barbara and Robin will come back to Texas in the next couple of days," George wrote a friend. "Robin apparently is making headway, or at least has not lost ground. . . . She is still full of fun and we hope that she will have many more months of active life." In 1953, however, leukemia was stingy with time, dishing it out in tiny portions, and chemotherapy put patients and their families on a roller coaster of remission and recurrence. Robin's condition soon deteriorated, and she died on October 11, 1953. "One minute she was there," as Barbara remembers, "and the next she was gone. I truly felt her soul go out of that beautiful body. For one last time, I combed her hair and we held our precious little girl. I never felt the presence of God so strongly as at that moment."

Robin's death left the Bushes acutely aware of cancer's greatest toll, the anguish of losing a child. Still able to count her blessings, Barbara remembered other families at Memorial who had found the stress and expense of cancer overwhelming, sending some to divorce court, bankruptcy lawyers, and unemployment offices. "In a strange kind of way, we learned how lucky we were," she later said. "We loved each other very much. . . . We had the most supportive family. . . . Financially, we were very lucky, as our insurance covered almost everything."[1]

People like Sidney Farber and Cornelius Rhoads observed that some leukemia patients responded positively to chemotherapy, but those patients also died, surviving only a few weeks or months longer than untreated patients. In Robin Bush's case, as the disease rebounded and her health imploded, the physicians tried the drug again, hoping to repeat their earlier success. The leukemia cells, however, had evolved and now defied the drug. They proliferated wildly, clogging Robin's circulatory system and stealing her life. Robin Bush had been in the vanguard of cancer patients for whom chemotherapy offered hope. Physicians treating childhood leukemia began experimenting with schedules and dosages. Slowly, they eked out a few more weeks of survival, enough to give the doctors a sense of progress, but of little consolation to the leukemia patients and their parents.

—

Fletcher and Grimmett had little time to enjoy the limelight of winning the contract to build the cobalt-60 radiotherapy unit. On the way home from Europe in 1951, they speculated about supervoltages, electron therapy, and the day when radiation would take its rightful place alongside surgery. Once in Houston, they went to work on the cobalt-60's progression from design to production. The two decided that the cobalt-60 needed company, a new betatron to harness electrons and generate megavoltage high-energy beams. With the excitement of little boys planning an adventure, they looked to the day when M. D. Anderson would enter the pantheon of major radiation treatment centers with Radiumhemmet, the Curie Foundation, the Royal Marsden Cancer Hospital, and the Christie Hospital and Holt Radium Institute.

Personal relations between Fletcher and Grimmett remained tempestuous. On May 25, 1951, the physics laboratory erupted, Fletcher shouting in frustration about Grimmett dragging his feet, Grimmett shaking his head in dismay. It was a Friday afternoon, and Fletcher stormed out of the building. He never saw Grimmett alive again. At 1:10 on Sunday morning, Grimmett succumbed to a massive heart attack. The Fletchers had invited friends to their farm in Hempstead—including the Grimmetts—for a Sunday brunch prepared by a talented French chef. Instead of a party, they staged a wake.[2]

—

Lee Clark seized every opportunity to promote M. D. Anderson, but touting its merits at professional meetings had limits. Too many eastern physicians could imagine nothing of value in Texas except rib eye steaks. Clark often bristled.

"There are still people who think that if you're not in New England or Pennsylvania or in the East, or maybe Berkeley, that you're not really in the intellectual center of American medicine. . . . Boston can't be dispelled until another generation come[s] along because people have thought that way for two hundred years." An impatient man, R. Lee Clark refused to wait.

With the new hospital under construction and with the blessing of Governor Allan Shivers, Clark brought Frances Goff to M. D. Anderson as his special assistant. She was drill sergeant tough. Clark often sent her to deal with contractors and construction workers. She spoke their language, and because she was a woman, they held their tongues, unable to speak as bluntly to her as she did to them.[3]

Clark brooded about more than bricks and mortar. Although M. D. Anderson had research and educational missions, too, many people considered it just a hospital. The AMA had delayed accrediting its residencies and fellowships, and UT officials hesitated to acknowledge M. D. Anderson as an educational institution. "We have continually had difficulty in explaining that we are not simply a cancer treatment hospital," Clark wrote UT chancellor James Hart in January 1953, "but, in addition, have the equally important functions of cancer research and cancer education." Clark looked to Memorial, with its associated Sloan-Kettering Institute for Cancer Research, and he asked the board of regents to establish a Tumor Institute in Houston as part of M. D. Anderson. An institute "would be particularly advantageous in attracting men of the desired academic stature and interests." Within a month, the board agreed. The M. D. Anderson Hospital for Cancer Research of the University of Texas became the University of Texas M. D. Anderson Hospital and Tumor Institute.[4]

On March 19, 1953, to great fanfare, the board of regents laid the cornerstone for the new hospital. Clark invited cancer luminaries from around the nation. Bertner's widow agreed to endow a prize and a lecture in her husband's name, to be staged annually in conjunction with M. D. Anderson's Symposium on Fundamental Cancer Research. Back in 1950, from his deathbed Bertner had concurred when Clark first proposed the lecture. In 1953, the Bertner Award went to Charles Huggins of the University of Chicago for his work on the relationship between hormones and prostate cancer and for developing antiandrogenic therapies. Some prostate cancers feast on hormones. Huggins hypothesized that a hormonal approach might arrest prostate cancer as well. He prescribed diethylstilbestrol to suppress testosterone levels and, in some cases, tumor growth and metastasis.

About the time Huggins won the Bertner Award, Clark learned that the M. D. Anderson coffers did not contain enough money to furnish and equip the new building. The hospital would have a beautiful, marble exterior and empty rooms. Again Clark approached friendly donors. Jesse H. Jones donated $25,000, and with the help of James Anderson, a nephew of Monroe D. Anderson, Clark added $725,000 from donations in memory of deceased friends and relatives. Benefit football games, barbecues, stock car races, rodeos, and quarter horse races also helped raise funds.

To open wallets, Clark and others preached of a cure. Huggins told journalists that "cancer was many diseases, not just one, and little by little [we are] liquidating it disease by disease. . . . The successful treatment of cancer by pills is now possible. . . . Cancer will be brought under control probably within your lifetime and mine." From New York, Cornelius Rhoads echoed the giddy optimism, assuring reporters, "Inevitably, as I see it, we can look forward to something like penicillin for cancer, and I hope within the next decade." A. M. Aikin bought into the promise and arranged periodic tours of the construction site for state legislators. Several wondered out loud whether marble exteriors were not extravagant, but a journalist dubbed it the Pink Palace. Aikin successfully promoted the appropriations bill, proclaiming the "hospital [as] the greatest thing in Texas." Since its inception, M. D. Anderson had treated more than thirteen thousand patients. The exterior walls soon glowed in pink and the cobalt-60 and 20 betatron found a home in the basement. On March 19, 1954, ambulances transferred forty-six patients to the new hospital.[5]

―――

The Pink Palace finished and furnished, Clark filled the research labs, hospital, and clinics with talented people. In 1954 he hired A. Clark Griffin to breathe life into a comatose biochemistry program. At the time, biochemists were glorified technicians supervising daily routines in several medical center laboratories. The arrangement earned income, but it did not allow for serious research. "I firmly believe," Griffin argued, "that we must discontinue the contractual arrangement . . . [which] is degrading to the medical school and the biochemistry department. . . . We could and should serve at a much higher level." Griffin wanted the department to identify environmental carcinogens, isolate chemical markers of cancer in human blood, and test various compounds for anticancer properties. None of this could happen without new biochemists.[6]

In physics, Warren Sinclair inherited the heavy mantle of Leonard Grimmett. A British-trained physicist with clinical experience, Sinclair joined M. D. Anderson in 1954 hoping to have a nuclear reactor at his disposal. Clark wanted Anderson to become a center for nuclear medicine, which Sinclair felt required a reactor. He told Clark that the use of radioisotopes in cancer diagnosis and treatment and the increasing volume of patients receiving radiation treatments made M. D. Anderson an ideal candidate for one. He envisioned research in radiology, radiation biology, biophysics, health physics, and radiation treatment. Sinclair claimed that distance from an AEC laboratory had stalled research into the medical uses of radioactive materials. It was impossible to deliver isotopes with short half-lives. The reactor would allow Anderson to manufacture isotopes for sale in the medical center. Finally, Sinclair argued, "M. D. Anderson . . . has available to it a large volume of cancer patients at all stages of the disease. Many . . . are terminal cases who would be available for research study with isotopes and radiation beams."[7]

The more carefully M. D. Anderson physicists analyzed the details, the less enamored they were to acquire a nuclear reactor. One obstacle was American pop culture. Most Americans in the 1950s still associated nuclear reactors with Hiroshima, atomic bombs, and mass death. On March 9, 1954, the United States tested its first hydrogen bomb at Bikini Atoll in the Pacific. A thousand times more powerful than the device that leveled Hiroshima, the hydrogen bomb produced a mushroom cloud one hundred miles wide. The prospect of a nuclear reactor in Houston unsettled many people. While Sinclair wrote letters, Americans thronged to films portraying the horrors unleashed by scientists run amok: *Them*, *Tarantula*, *Godzilla*, and *The Fly*. Millions of Americans were concerned about radiation. Edward Munger had sold his laundry business in 1954 to come to M. D. Anderson and was soon waving Geiger counters over pillow cases, blankets, sheets, and uniforms. Safe storage and disposal of radioactive waste materials had to be monitored. Even the morgue had problems handling radioactive corpses.

Economic problems loomed large. Sinclair estimated a price tag of $1.5 million for the reactor and its associated laboratories, with annual operating costs of $500,000, a sum that made even the usually effervescent Lee Clark blanch. Both men, however, took comfort in the possibility that M. D. Anderson could offset the cost with external grant funding and pioneering, patenting, and marketing medical applications.

Concerns about the scientific merits outweighed political and economic worries. In 1958, a representative of the Rockefeller Foundation told Clark that the reactors already in place at MIT, the University of Michigan, and the Brookhaven National Laboratory might be sufficient for preliminary investigations of the medical use of radioactive products. The Argonne National Laboratory declined to join M. D. Anderson, as did the AEC. Dissension erupted among physicists at M. D. Anderson. In 1954, after pestering Clark about the need for a radiobiologist, Sinclair had hired Robert J. Shalek, a recent Rice Ph.D. Shalek voiced concern about the reactor idea. In one letter, he predicted that "a medical reactor is a venture of little promise . . . that will yield . . . little fundamental knowledge in radiation chemistry and biology. . . . The possibilities for radiation therapy [are] a very long gamble." Shalek also worried about the expense and the likelihood that a medical reactor would "demand a major fraction of the energy and concentration of this department, to the exclusion of work that is more promising and more fundamental."[8]

Shalek prevailed. Brookhaven National Laboratory officials argued that the estimate of $500,000 in annual operating costs was far too low. They confessed that their own medical reactor had "required a tremendous amount of developmental work with so far little return." Eventually, Sinclair scuttled the project, concluding that for the time being, "developing a reactor research program may be fine for a national laboratory concerned with the development of nuclear energy, but . . . [might not] be the best basic approach to the cancer problem . . . and pursuing it will . . . color and perhaps engulf many other aspects of our basic work to the detriment of the overall research program."[9]

At the suggestion of Theophilus Painter, Clark hired University of Texas geneticist Felix Haas to head up the biology department. Haas's interest in cancer genetics, especially the genetic mutations induced by radiation in bacteria, seemed to Clark a perfect fit for M. D. Anderson. Haas concurred. He was from Alvin, Texas, and eager to return to the Houston area.[10]

━━━━━━━━━━

In 1955, Clark scored his biggest coup since the cobalt-60. At the UT Medical Branch, T. C. Hsu labored in the biology laboratories. Born in 1917 in Zhejiang Province, China, Hsu had cultivated childhood interests in collectibles, first stamps and then insects. In 1936, he matriculated at National Zhejiang University to major in entomology. Hsu's undergraduate career stalled in 1937 when Japan invaded China. The Zhejiang faculty relocated to the interior. Hsu soon tired of entomology. If he stayed in the field, Hsu might finish life with a beetle or two named after him but not understand biology at its fundamental level. Two courses in genetics and cytology taught by C. C. Tan (Jia-Zhen Tan) changed his life. Tan liked Hsu and introduced him to experimental biology, more specifically to the genetics of *Drosophila*, the fruit fly. *Drosophila* possesses only two pairs of chromosomes, produces large numbers of offspring with readily discernible mutations, and completes its life cycle in only twelve days. In five years, a geneticist could examine more than 150 generations and observe the processes of mutation and natural selection. When Tan offered Hsu a job, he accepted. Tan did not have much of a lab, however. Without electricity and incubators, he could not keep mutant stocks alive. But microscopes worked, and plant and insect species were abundant, so Tan and his students switched to cytogenetics.

Hsu married in 1946 and in 1948 departed for the United States, expecting to spend several years in Austin earning a Ph.D. and following it with one or two postdocs before returning to China. The triumph of Mao Zedong in 1949 changed everything. The new Chinese regime adopted as its party line in science the ideas of the Soviet biologist Trofim Lysenko, whose theory that evolution proceeds by the inheritance of acquired characteristics denied the existence of genes. In China, geneticists were not welcome, so Hsu stayed in Texas. For his dissertation, he explored the cytogenetic relationship between various *Drosophila* species and produced a well-received chromosome map of their salivary glands.[11]

In 1951, Hsu received a Ph.D. and accepted a postdoctoral fellowship at the UT Medical Branch. He soon switched from *Drosophila* to human tissue cultures. One year into the fellowship, Hsu revolutionized cytogenetics. At the time, biologists believed that the nucleus of every human cell contained forty-eight chromosomes. Distinguishing among chromosomes was difficult because they clump together like sticky worms. In preparing chromosomes for microscopic examination, an assistant in Hsu's laboratory employed a rinsing solution designed to make the chromosomes swell. He mistakenly added distilled water to the salty rinse, which rendered the solution hypotonic instead of isotonic. The chromosomes swelled and then separated, revealing themselves completely. For the first time, geneticists could analyze chromosomal markers, exchanges, and breaks linked to specific genetic diseases and learned that the human chromosome complement was forty-six, not forty-eight. The chromosomes, in Hsu's description, "were suddenly beautiful." Word spread and labs around the world turned to the hypotonic solution for analyzing chromosomes.

At the Medical Branch where he taught histology. Hsu felt isolated. In 1953, the arrival in Texas of his wife and the six-year-old daughter he had never met tilted the axis of his life. Clark pursued Hsu, promising a better salary, a better laboratory, and better colleagues. Generous of spirit, Hsu yearned for a collegial setting. In 1955, he left Galveston. Clark had lured to M. D. Anderson the man soon to be recognized as the founder of mammalian cytogenetics. Just two years earlier, James Watson and Francis Crick had described the DNA molecule and the mechanism of genetic inheritance, and molecular biology had begun its conquest of the discipline. With Hsu in a lab on the fifth floor of the Pink Palace, M. D. Anderson took its first tentative steps in that direction.[12]

At the time the biology faculty were deep into the "Cancer Eye" project studying squamous cell carcinoma in Hereford cattle, which afflicted more than two hundred thousand animals a year. Squamous cell carcinoma affected cattle and people, and at the next legislative session, Clark cornered legislators from South Texas and the Panhandle, touting the hospital's pioneering effort. When cattlemen discussed "cancer eye," they could speak about the disease intelligently, and Clark listened intently.

During the 1950s, M. D. Anderson scientists worked with more than twenty thousand head of cattle. Veterinary epidemiologists demonstrated that grassfed cattle developed less cancer of the eye than those raised on grain-rich diets. The biologists identified ultraviolet light as a source of cancer eye. And after David Anderson joined M. D. Anderson in 1957, the research veered into genetics, noting that Hereford cattle seemed more vulnerable than other breeds and that older cattle were more susceptible than young ones. The research implicated genetic, environmental, and viral factors in the etiology of the disease.

The project taught Clark how politics and science are often yoked together. M. D. Anderson enjoyed grants from the American Hereford Association (AHA). AHA members wanted M. D. Anderson to censor publication of all data reflecting poorly on the breed. In advance of the 1960 World Hereford Conference, Paul Swaffer of the AHA declared to pathologist William Russell, "I feel a little 'ouchy' in dwelling a great deal on this cancer eye subject in a meeting of Hereford representatives from all over the world. . . . Some of our competitors have used this cancer eye propaganda to advance their own interest." Swaffer chastised Russell for spreading the word that cancer eye was common in Hereford cattle: "I don't mean to leave the impression that we want to retreat in our efforts to solve the difficulty, but we can and should be awfully careful not to give ammunition to our competitors."[13]

—⟨⟨⟨⟨⟨⟩⟩⟩⟩⟩—

The glimmer of hope chemotherapy offered leukemia patients sent researchers scrambling for new drugs. In 1953, NCI began clinical trials, and in 1955 Congress established the National Cancer Chemotherapy Service (NCCS). C. Gordon Zubrod, formerly a specialist in antimalarial agents for the U.S. Army, assumed direction of NCCS and started a vigorous search for anticancer drugs.

NCI then released spectacular news. Roy Hurst and M. C. Lin, two physician-scientists at NCI, had cured patients of choriocarcinoma, a fast-growing cancer of placental tissues, using the drug methotrexate. The drug worked at the site of the tumor and wiped out metastatic lesions, successes neither surgery nor radiotherapy could match. Before methotrexate, hysterectomy had been the only treatment, leaving women infertile without addressing metastases. Methotrexate offered hope for cure, even for women with widely disseminated disease. Survivors might even hope to bear children again. Lin had at first noted regression of the tumors upon administration of the drug, only to see them recur. He also observed high levels of gonadotropin in the blood and urine of patients, even among those whose gross tumors had disappeared. He concluded that occult malignant cells still survived. At the suggestion of his young colleague Emil J Freireich, Lin kept patients on the drug long after apparent remission had been achieved.[14]

For all the excitement, chemotherapy treatments for childhood leukemia had produced only temporary remissions. Some malignant cells, by virtue of their genetic composition, enjoyed immunity to the chemotherapeutic agent. The drug killed most of the cancer cells, but only one needed to survive. At the most microcosmic level, chemotherapy demonstrated why Darwin's theory of natural selection was the foundation principle of modern biology. The lingering cancer cell, equipped with a mysterious capacity to survive while its companions died, then underwent mitosis, dividing into descendent cells inheriting the survival advantage of their ancestor. Methotrexate killed every single cancer cell.[15]

At NCI, Memorial, Roswell Park, and a few other research institutions, physicians tested methotrexate on dozens of other cancers, but none responded like choriocarcinoma. Its hormone-fed cells proliferate very rapidly, leaving them more vulnerable because this type of chemotherapy devastates cancer cells during mitosis. Investigators experimented with other drugs too. Scientists did not yet realize cancer's extraordinary diversity, each distinctive according to tissue origins, molecular properties, and the immunological status of hosts. A clinical studies panel was formed to focus NCI efforts on leukemia, where the effects of folic acid antagonists had already been demonstrated. Rather than concentrate the research at Bethesda, the panel thought it wise to establish regional groups where clinical scientists could share results and prevent duplication of effort. NCI created the Southwest Cancer Chemotherapy Group, with headquarters at M. D. Anderson.

M. D. Anderson now had firm credentials in radiation treatment and a foothold in chemotherapy, but surgery still remained the treatment of choice. Surgeons swaggered home from the field hospitals of World War II convinced that their country had rid the world of fascism while their own skills had saved hundreds of thousands of lives. New technologies permitted new operations. In the past, radical surgery had often killed patients because of massive blood loss. Blood transfusions rendered surgery safer. During the Spanish Civil War in the 1930s, the technique of adding citrate and glucose to donated blood allowed for cold storage, and surgeons had abundant supplies. And after 1941, the other mortal side effect of radical surgery—massive infections—retreated in the face of the development of injectable penicillin. During the next fifteen years, pharmacologists synthesized other antibiotics. Young, ambitious surgeons itched to extend the discipline. "Those were heady days," recalled surgeon Richard Martin, who came to M. D. Anderson in 1951. "We were ready to conquer the world. Everything was possible."[16]

For surgeons who were convinced that cancer was a local rather than a systemic disease—as embedded in the then-prevailing medical paradigm—recurrence and metastasis signaled lost opportunities to contain tumor cells. William Stewart Halsted had considered metastasis a mechanical process, with malignant cells exiting the tumor relatively late in the disease process. The key to cure, therefore, was early detection and radical surgery. Logic held that tumors be approached aggressively, and the best surgeons searched for new ways to remove more and more tissue—to achieve wider and wider surgical margins in order to cut out every aberrant cell.

In the 1940s and 1950s, three surgeons at Memorial Sloan-Kettering opened the door to extended radical surgery. Jerome Urban fashioned what later became known as the "extended radical mastectomy," an expansion of the Halsted procedure that removed the breast, the lymph nodes under the sternum and clavicle as well as in the armpit, the pectoralis minor along with the pectoralis major muscle, and sometimes even portions of the sternum and rib cage. He justified so radical a measure on the grounds that "we should increase our salvage of early operable cases" and, astonishingly, felt that "it [is] possible to perform such a procedure without adding to the operative mortality, morbidity, or the patient's postoperative discomfort and disability." George Pack, who cut off hips and legs and shoulders and arms with the ease "lesser" surgeons performed tonsillectomies, enjoyed his moniker "Pack the Knife." Alexander Brunschwig rounded out Memorial's radical triumvirate. For women with gynecological tumors, Brunschwig pioneered pelvic exenteration—removal of the vagina, cervix, uterus, ovaries, bladder, and some intestinal tissues. He also became known for the most radical surgery of all—the hemicorporectomy—amputation of the lower half of the body. For some surgeons, performing increasingly radical procedures became a rite of passage. At meetings of the American College of Surgeons, they bragged about just how much tissue they had removed while leaving behind patients who, though mutilated, survived the operation.[17]

So when Lee Clark arrived at M. D. Anderson and designated himself surgeon in chief, he chose an apt title considering the state of American oncology. For all intents and purposes, oncology was surgery. From the most sophisticated teaching hospitals in urban centers to the smallest community hospitals, cancer patients were treated by general surgeons who removed tumors and hoped for the best.

As radical surgery evolved into the gold standard, the need for anesthesiologists capable of keeping patients unconscious for many hours increased. Clark found one in William Derrick, head of anesthesiology at Peter Bent Brigham Hospital in Boston. Derrick had already invented the first commercially available respiratory assister, which helped patients breathe during general anesthesia. An astute businessman and stock market afficionado, Derrick also served as an informal securities analyst.

Finally liberated from the confines of the Baker estate, Clark took M. D. Anderson into pediatrics, at first with four beds on the third floor of the east wing and later moving to the sixth floor of the west wing. To lead pediatrics, he hired H. Grant Taylor. In 1949, Taylor had become medical director of the Atomic Bomb Casualty Commission (ABCC), a federal agency charged with assessing the medical impact of the atomic bombs on Japan. He asked Japanese pediatrician Wataru Sutow to join the ABCC, and they formed a tight bond. Sutow had a special gift with children. In 1954, they read a report in *Time* about the opening of the new hospital in Houston. A few months later, in Washington, D.C., Taylor met Clark, who invited him to initiate a section of pediatrics. Taylor persuaded Clark to include Sutow in the deal. Accompanying them was Margaret Sullivan, a Duke-trained pediatrician with the ABCC.[18]

The venture into chemotherapy for children required the services of a hematologist, and in 1954 Charles C. Shullenberger headed up the first section of hematology in Clifton Howe's Department of Medicine. Shullenberger had arrived at M. D. Anderson in 1949 after completing a fellowship in internal medicine at Mayo. At the time, chemotherapy involved little more than administering Sidney Farber's protocols to children with leukemia. Within a few years, as chemotherapy was extended to adults and to other forms of cancer, and as the protocols became more potent, Shullenberger helped shape an institutional culture committed to multidisciplinary treatment, high quality patient care, and aggressive therapy as long as M. D. Anderson enjoyed in the Departments of Pediatrics and Medicine the supportive resources to handle any negative effects of surgery, radiotherapy, and chemotherapy.[19]

⟶⟳⟵

Nobody knew better than Lee Clark the cruelties of cancer, how often it produced intractable pain, disability, and death. For patients and their families, cancer could also bring depression and feelings of impotence, as if trapped in a universe unmoved by human suffering. Cancer patients often ended up destitute, unable to pay for rent, utilities, and transportation, even when hospitalization and treatment came free. Clark needed someone to adapt social work to the specific demands of cancer. In 1949, during a visit to Temple, Texas, he met a visionary woman trained to run such a department. Edna Wagner, born to a well-to-do Louisiana family, had a master's degree in medical social work from Tulane. Her friendly demeanor, experience, and well-informed opinions impressed Clark. He hired her to lead social services.[20]

Wagner first explored why so many black and Hispanic patients left M. D. Anderson before completing treatment. Many black patients lived outside Harris County and could afford neither the expense of transportation to Houston nor the cost of food and housing in the city. The Houston Negro Hospital had recently terminated its nursing education program, leaving empty a dormitory on Holman Street. Wagner made her case to Clark, and late in 1950, the Negro Hospital board opened the facility to black patients from M. D. Anderson.

Most Hispanic patients were Mexican Americans from South Texas who found Houston impersonal and expensive. They often grew lonesome and yearned for family and home-cooked meals. Only a residential facility serving familiar food and accommodating family members, Wagner concluded, could ease their plight. She felt an obligation "to provide decent maintenance and care for Mexican cancer patients while they are receiving treatment at this hospital." Wagner contacted prominent Mexican Americans in Houston and Anglo businessmen with Hispanic clientele. They organized the Emergency Relief Committee for Cancer Patients' Aid and staged fiestas, benefit dances, and country club dinners. In 1951, the committee purchased a building at 2702 Helena Street. They remodeled it to accommodate sixteen patients and dubbed it La Posada, or "the shelter." La Posada opened on April 6, 1952.[21]

Because of their efforts to raise money, committee members presumptuously assumed authority, such as determining which patients were eligible to stay and which had to pay for their meals. Albino Torres, who chaired the committee, even asked Wagner to supply him with "a copy of the report M. D. Anderson social workers prepare on each of these people, as we understand they do for every patient who enters your hospital." Wagner did not want amateurs "going into La Posada deciding for themselves who should and should not be given assistance, humiliating and embarrassing the patients."

Wagner arbitrated disputes between the committee and other Hispanic groups. A variety of Mexican clubs in Houston donated food, clothing, and toiletries to La Posada, but many committee members, in Wagner's words, had "exaggerated social ambitions. . . . They like the publicity and recognition which their charitable work for cancer patients has brought them." At the same time, the committee wanted to keep some distance from club members. "The La Posada Committee," Wagner told Clark, "readily accepted the help of the clubs, but feel that socially, most of the club members are beneath them, so they will not give the clubs representation on the Committee." Gradually she steered the clubs into volunteer services. La Posada helped reduce dropout rates. The number of Spanish-speaking patients would soon outgrow Helena Street, and in May 1959, with ten thousand dollars from the M. D. Anderson Foundation, a new La Posada went up at 406 Webster in Houston.[22]

For millennia, cancer patients had shrouded the disease in secrecy, ashamed to discuss it in polite company. Cancer became synonymous with filth and decay. When Anne of Austria, the mother of Louis XIV of France, was dying of breast cancer in 1666, she felt dirty. The tumors had ruptured and spilled necrotic tissues and fluids onto her gowns and bedding. A male friend told Anne that cancer must be a "great inconvenience, especially for You who loves perfumes, because at the end these illnesses stink terribly." Tumors of the bladder, colon, rectum, prostate, penis, and vagina could disrupt menstruation, urination, and defecation, leaving patients incontinent and in need of diapers and frequent bathing. The beds of cancer patients, especially those with leukemia and gynecological neoplasms, were often stained with blood. The disease—so intractable, so malodorous, so interwoven with images of blood and excrement—became synonymous with evil and death.[23]

Many Americans in the 1950s spoke of cancer in whispers. It was not uncommon to keep patients in the dark. In 1949, when Babe Ruth was diagnosed with throat cancer, his physicians did just that, even though he was being treated at Memorial Sloan-Kettering. When Ruth entered Memorial, journalists kept a tight lid on the story, identifying his affliction as a cold or congestion or the flu. Not until he died did they employ the C word, as if it were pornographic. Clark hated the cancer culture. "It's not the bubonic plague or gonorrhea," he once complained. "It's not immoral. It's just a disease. It's not dirty . . . and [those] who get it aren't dirty either. They're just sick." Clark also hated to hear physicians identify patients by diseases instead of names. "Nothing could get you on the wrong side of Dr. Clark quicker than saying, 'I have to meet with a colon cancer today' or 'I had a melanoma' this morning," recalled a longtime employee. "Soon after we moved into the new hospital . . . he fired a surgeon who repeated the sin after being warned. 'Patients are people, human beings, not diseases,'" Clark pronounced. As long as Americans associated cancer with death and disapproval, and robbed cancer patients of their identities, public health campaigns urging early detection would fail.[24]

It took a gutsy woman to put a face on cancer. In 1954, the most famous cancer patient in Texas was being treated at John Sealy Hospital. Mildred "Babe" Didrikson Zaharias in 1932 had achieved worldwide fame with two gold medals and a silver medal in track and field at the Olympic Games in Los Angeles. She took up professional golf in 1935, and word spread of her 240-yard drives. Babe staged personal tours to demonstrate her skills in swimming, billiards, diving, and basketball, eventually earning more than a hundred thousand dollars a year as a professional athlete. Babe became a household name and an icon of victory.

She developed rectal and colon cancer in 1953 and underwent a colostomy. Babe tossed her disease into the public domain. She held a mini press conference and asked fans, in lieu of flowers, to make a contribution to the Damon Runyon Cancer Fund. "I'm tired of being on the sports page; put me on page one," she urged a sportswriter. Her own physician predicted, "I don't know yet if the surgery will cure her, but I will say that she never again will play golf of championship caliber." But she staged a stunning comeback, winning the Babe Didrikson Zaharias Golf Tournament, a charity event for the Damon Runyon Cancer Fund. With Babe gracing the cover of its August 10, 1953, issue, *Time* magazine announced, "The Babe Is Back." *Look, Life,* and the *Saturday Evening Post* also highlighted her return, often with illustrations of Babe wielding the American Cancer Society's Sword of Hope and slaying the foe. The American public expected no less of her.[25]

Journalists pioneered in her what would later become the pop culture phenomenon of turning cancer into a contest to be won. "If anybody can follow up a bogey with a hole-in-one, it's Babe," wrote one. Variants abounded: "Cancer is the hurdle she can leap," "the ball she can hit," "the basket she can make," "the course she can run"—ad nauseam in a country soon to be bloated with celebrity heroes and muddled ideas about the disease. Born in a country of inveterate optimists, a nation where people considered success an entitlement, Babe mouthed all the platitudes. She promised fans that she would "beat the disease" and "win the race," as if cancer were an easily intimidated opponent on the final hole of the U.S. Open. In a few decades Americans would fashion a seductive but unproven faith that patients, by marshaling all of their emotional resources against killer diseases, could bend biochemistry, physiology, and genetics to their will. Babe led the way into the modern culture of an unrealistic assumption: that optimism and willpower could cure cancer. She signed on as a poster girl for the American Cancer Society, appearing at major fundraising events, recording radio and television messages, and touting early detection and generous donations. During the spring of 1955, in town for golf clinics at Memorial Park as part of Houston Open activities, Babe visited M. D. Anderson and lunched with Clark. She was an American heroine.

She was also very sick. In September 1955, surgeons discovered widespread metastases to the spine and pelvis. "That's the cut of the green," she replied. She established the Babe Didrikson Zaharias Fund "to help . . . in the fight against cancer." Babe remained unswervingly optimistic to the point of denial, insisting to the press, "I expect to be shooting for championships for a good many years to come. My autobiography isn't finished yet." But it was. In March 1956, she checked back into John Sealy for the last time. Journalists described her decline in excruciating detail, with such insipid headlines as "Babe's Grit Praised by Physicians" or "Babe Loses Strength, Weight but Continues to Put Up Fight."

They wallowed in clichés, and her death became the most graphic, publicized case of cancer in history. Babe shrank from 170 to 62 pounds, and she endured pain so unrelenting that surgeons finally severed her spinal cord to provide relief. Zaharias died on September 27, 1956.

Her death, however, precipitated a catharsis in the cancer culture. No disease played more to the metaphors of filth and loathing than colorectal cancer, but when journalists inquired about her condition, Babe readily discussed the devastation of her rectum, anus, and colon. When she won a tournament and was asked about how the stoma affected her swing, she could be disarmingly honest. After winning the 1954 U. S. Women's Open, she said, "It will show a lot of people that they need not be afraid of an operation and can go on and live a normal life." In the most delicate language, her comments appeared in newspapers and did much to mitigate the stigma of colon cancer and render it just a disease, not a metaphor for filth and disgust.[26]

5

Changing Paradigms, 1956–1963

She was the first patient that I know of with bone sarcoma
who was cured with chemotherapy.

—R. LEE CLARK, 1976

In 1961, long before Michael Crichton conjured up *Jurassic Park*, T.C. Hsu resumed an old hobby, collecting specimens. Concerned about the extinction of species, he started a "frozen zoo" collection of DNA from threatened mammals. "In 1961," he later recalled, "we were searching for mammal cell lines suitable for certain studies and were unable to find any standard research animals whose cells had the right chromosome properties." With the help of the Houston Zoo, Hsu acquired tissues from the big cats, and as word spread, samples arrived from all over the world, until the lab had specimens from more than five hundred species. Cloning species "sounds like science fiction now," Hsu said in 1978, "but not too long ago science fiction was rockets landing on the moon. Such reconstitution from a cell may eventually be done, maybe two hundred years from now, maybe a thousand." He proved correct about cloning but was off on the timing. In 1997, not 2178 or 2978, Scottish biologists cloned the first mammal—a sheep they named "Dolly."[1]

Rapid growth prompted Clark to identify new revenue streams, and in two cases, he drew on his experience at Mayo. In the emerging world of academic medicine, philanthropy was already becoming the difference between excellence and mediocrity. Just as the Mayo brothers had established the Mayo Foundation, Clark in May 1957 founded the University Cancer Foundation, with a board of visitors. James Anderson, a nephew of Monroe D. Anderson and head of personnel at Anderson Clayton, chaired the board. For many years, he had supported the hospital, working to build a proud family legacy. Anderson enjoyed entrée to Houston's elites. He loved the hospital but had little time to tap his friends for funds. Leukemia killed him on July 10, 1958, at the M. D. Anderson Hospital and Tumor Institute. Also in 1957, Clark formally established the Physicians Referral Service (PRS), M. D. Anderson's physicians practice plan to compensate its doctors. It had taken ten years to secure IRS approval. The PRS was based on the model Clark had absorbed at Mayo. All income from physician services at M. D. Anderson went into the coffers of the PRS, and from that centralized pool, all physicians received fixed salaries. Doctors' income did not depend on how many patients they saw. This became the economic foundation for multidisciplinary care.[2]

Clark needed more room. In 1956, the hospital registered its twenty-thousandth patient, and the next year another 3,810 were registered. Total clinic visits jumped from 89,480 in 1957 to 120,000 in 1960, imposing on Clark a constant pressure. To maintain good ambulatory care, M. D. Anderson needed more room for laboratories, beds, clinics, and offices. Between 1958 and 1961, Clark added nearly ninety thousand square feet of space. In seemingly no time, an administrative assistant saw him carrying around blueprints, a sure sign that he was brooding about the future. "When Dr. Clark pondered expansion in the physical plant," remembered an associate, "he thought in terms of acres, not square feet."[3]

Top scientists required sufficient laboratory space, but recruitment also depended upon academic credentials. The UT Postgraduate School of Medicine had provided an early academic veneer, but many agreed that "academic appointments for M. D. Anderson did not have sufficient substance in the Postgraduate School." It never enjoyed the full cooperation of other medical center leaders. Many Baylor faculty believed that the "Postgraduate School is under the domination of the M. D. Anderson Hospital and might be used to the advantage and disadvantage of some other doctors." They had resented Clark's appointment as dean, and Grant Taylor's becoming dean in 1954 did nothing to dampen animosities. Convinced that no need existed for a third party to contract residencies and fellowships, other institutions pulled out of the Postgraduate School, which began to hemorrhage.

Clark tried in vain to apply a tourniquet, but the deans of the medical schools called for its dissolution. The boom in research characterizing American medicine had increased demand for clinical scientists, men and women who possessed both an M.D. and a Ph.D. and worked in the laboratories and in the clinics. Clark took advantage of the trend in 1955 by proposing that M. D. Anderson scientists be allowed to offer courses leading to UT MS and Ph.D. degrees. The proposal raised hackles in Austin, but distance was an ally. The main campus played no role in the training of scientists in Houston. During the next eight years, Clark's proposal for a UT Graduate School of Biomedical Sciences gathered momentum. In 1963, he warned Harry Ransom, chancellor of the UT System, that without a degree-granting graduate school of biomedical sciences in Houston, M. D. Anderson stood to lose its best minds. Later in the year, the board of regents created the University of Texas Graduate School of Biomedical Sciences (GSBS) and transformed the postgraduate school into a continuing education venue.

The new GSBS faculty consisted of M. D. Anderson and UT Dental Branch staff, and the new school could offer graduate degrees in physics, biochemistry, and biology. Typical academic ranks were soon developed. Clark finally had his precious assistant professors, associate professors, and professors working under a UT degree-granting umbrella. Ransom directed him to develop a tenure system. Clark wanted tenure about as much as he wanted hoof-and-mouth disease at his ranch, but without tenure, recruiting good scientists would be problematic. Eventually he crafted "term tenure," in which a faculty member's right to continuing employment had to be renewed periodically.[4]

———

During the late 1950s and early 1960s, M. D. Anderson's major contributions to clinical oncology emerged from Gilbert Fletcher's department. In the spring of 1956, during a treatment planning session for a woman with large, dense breasts, Fletcher stewed about designing a target for the radiation beam. A fellow from the Curie Foundation was holding breast x-ray films where calcifications appeared as white flecks. He mused out loud whether better technology might improve diagnosis. Robert Egan, a resident in radiology, listened intently. Fletcher challenged him: "You're a diagnostic radiologist now. Why don't you figure out how to get x-rays of breasts like these so we can know what we're dealing with?" Egan did not reply, but Fletcher had planted a seed.

Almost casually, Egan set out to find the answer. Egan was perfect for the task, a man outspoken and frustrated over the nation's schizophrenia about breasts. Gerald Dodd, later the head of radiology at M. D. Anderson, explained the ramifications: "That was part of our culture back then. You didn't talk about breasts, and women weren't comfortable bringing problems to their doctors' attention because they were afraid it would lead to an examination, which most didn't want."

Although physicians had been x-raying breasts for decades, the images tended to be grainy and resistant to replication. Few radiologists wanted to grapple with the technical and cultural challenges. Before going into medicine, Egan had worked as a metallurgical engineer at a Pittsburgh steel mill calibrating, through trial and error methods, the furnace temperatures to the type of steel being manufactured. He adopted the same approach to breast x-rays. He correlated each image with the equipment, radiation levels, electrical voltages, and types of film used. Fletcher encouraged Egan while some of the surgeons lampooned him as the "titty man." Congenitally stubborn, Egan worked diligently with William Russell and the pathologists to determine which anomalies on x-rays had actually been malignant. He experimented with the best way to position a breast for the x-rays. He learned to distinguish some benign from malignant lesions on film and undertook an informal education campaign. As his wife rode shotgun in a blue Ford sedan, the backseat stuffed with clothes, food, and x-rays, Egan toured Texas repeatedly, teaching local physicians and radiologists how to examine breasts and read films. He often endured jokes and snobbish skepticism. In Gerald Dodd's words, "Mammography was a neglected procedure until Egan lit a fire under everyone. Eventually other physicians started catching the fever. They got their inspiration from his work because it backed up what he said and made mammography a reality."[5]

On the radiotherapy side, Fletcher garnered international prominence for aggressive treatment. More than anybody else in the 1950s and early 1960s, he created the "treat to cure" culture. M. D. Anderson could justify its existence, Fletcher insisted, only by pushing beyond palliation toward cure. For most cancer patients, according to Fletcher, boldness entailed fewer risks than caution. Many cancers, especially breast, head and neck, and gynecological tumors, had been prematurely declared resistant to radiation, not because they actually were biologically immune but because they had in fact absorbed insufficient doses. Fletcher administered radiation in doses others considered excessive. As a secondary objective, he hoped to preserve function. In laryngeal cancer, for example, he aimed to cure the disease *and* preserve speech, something the head and neck surgeons often could not achieve. For cervical cancer, radiotherapy could sometimes effect a cure without radical surgery, leaving vaginal tissues and sexual function intact. Men with penile carcinoma might avoid radical surgery as well. And Fletcher considered cosmetic issues, especially for breast, and head and neck cancers.

He had undertaken a complex task, establishing the parameters of radiotherapy with multiple variables, including tumor pathology and stage, radiation dosages and scheduling, and device used—cobalt-60, betatron, or the recently installed cesium-137. As results accumulated, Fletcher revealed the vulnerability of squamous cell carcinomas, as well as other varieties of gynecological and head and neck tumors. Some women with cervical cancer he saved from radical surgery that would have laid waste to the entire pelvic region; instead he cured them with lesser surgery and radiation. Radiotherapy allowed for treating many head and neck tumors without radical neck dissections or disfiguring surgery. Some of the results seemed beyond belief until replicated by others. "For instance," Fletcher beamed, "the MDAH [M. D. Anderson Hospital] results in floor of the mouth are 50 percent survival at five years, the best in the world." The betatron was especially effective with squamous cell carcinomas of the cervix, infiltrating bladder tumors, adenocarcinomas of the uterine fundus, carcinomas of the female urethra and vagina, and ovarian and rectal tumors. In 1961, for his work developing the cobalt-60 and its aggressive application, Fletcher became president-elect of the American Radium Society.[6]

Ever since Marie Curie, women physicians in Europe had found in radiotherapy a hospitable specialty. Mary Fletcher was a physician, and her husband, though a curmudgeon, actively recruited promising women. He was as effusive in praise as he was unsparing in criticism. An early protegée was Lillian Fuller of Canada. Fletcher hired her in 1956, and she staked out lymphomas as a special domain. Three years later, Eleanor Montague came to M. D. Anderson on a training fellowship. Her talent impressed Fletcher, and he groomed her for a permanent appointment. Over time, Montague studied the uses of radiation treatment combined with conservative surgery in the treatment of breast cancer. Norah Tapley constituted the third figure of Fletcher's female cohort. He brought her aboard in 1963 to focus on the clinical applications of electron beams.[7]

Radiotherapy at M. D. Anderson glittered, and Clark wanted nursing to sparkle too. At the Ellis Fischel State Cancer Hospital in Missouri, he found Renilda Hilkemeyer, a tough, enlightened woman who would help professionalize the career and midwife the birth of oncology nursing. Her appointment to the National Nursing Advisory Board of the American Cancer Society in 1955 brought Hilkemeyer to Clark's attention. After hearing his invitation to come to M. D. Anderson, she agreed, with one condition. As director of nursing, she would hire and fire all nurses. Physicians had to be removed from the loop. Clark agreed, and "Hilke" left Fischel for Anderson.

Blunt and direct, Hilkemeyer won over the nurses effortlessly. She responded to their needs and protected them from abusive physicians. Several days after starting work, Hilkemeyer learned that Clifton Howe, chair of the Department of Medicine, had fired a nurse. Hilkemeyer stormed into Clark's office with Howe in tow. To Howe's astonishment, Clark backed her. She also ferociously protected nurses. In 1959, for example, Grant Taylor complained that too many nurses hesitated to assist physicians in administering experimental drugs. Hilkemeyer reminded Clark that nurses had medical liability and that "they should not administer drugs that are experimental in nature since they do not know the background about the drugs." Hilkemeyer also paid close attention to the emotional challenges of caring for cancer patients—the ravages of the disease and its treatment, the trauma of severe pain, and the dismal survival rates.[8]

After World War II, surgeons grappled with rapid change in their discipline. Over the years, William Derrick, head of anesthesiology, had witnessed miserable deaths among pain-wracked cancer patients. In the worst cases, they exited life screaming, unable to focus on anything but the sweetness of death. Derrick wanted to relieve pain. In 1956, he developed what became known as subarachnoid pain blocks—injecting alcohol into localized points in the subarachnoid area, a fluid-filled space between tissues lining the central nervous system. The alcohol disrupted the sympathetic, sensory, and motor components of the nerve, offering, in some cases, pain relief without destroying function. When Derrick published his first paper on the technique in 1959, anesthesiologists throughout the country took note.[9]

In pathology, William Russell invented a new cryostat to produce frozen sections. The open-top device kept a constant low temperature, with a microtome to cut very thin tissue slices. Compared to existing models, Russell's offered better accessibility, greater speed, and clearer slides. Surgeons spent less time waiting for pathology reports, patients remained unconscious for shorter periods, and pathologists examined a broader range of tissues, including many previously ill suited to frozen section. And at half the price of existing technologies, the cryostat was eminently affordable and helped expand the availability of cancer surgery from major medical centers to community hospitals.[10]

John Stehlin forged one of M. D. Anderson's most important surgical innovations. He approached chemotherapy with surgical insight. Success with chemotherapy depended on dosage—the higher the dose the more effective the drug. Danger, however, lurked in high doses because of damage to other organs. Chemotherapy pumped volatile drugs into the bloodstream, and some damaged liver, kidney, heart, lung, or endocrine tissues. Stehlin searched for a way to deliver larger doses to the tumor site while sparing vital organs. For patients suffering from melanoma of an extremity, for example, he experimented with "regional perfusion"—blocking, or occluding, the major arteries and vessels feeding a tumor site and then drenching the area with drugs, delivering huge doses to the tumor and surrounding regions but not to vital organs. Perfusion improved melanoma survival rates and reduced the frequency of amputations.[11]

General surgery was falling victim to increased specialization and innovations in technique so complex that no single surgeon could master them all. Neurosurgery had long since stepped out on its own, but gynecological, urological, and head and neck surgeons prepared to leave as well. In 1954, Hayes Martin of Memorial Sloan-Kettering became the first president of the new Society of Head and Neck Surgeons. One of his students, William S. MacComb, had become chief of head and neck surgery at Roswell Park. Because Martin required his surgical residents to train in radiotherapy as well, MacComb was prepared for multidisciplinary care, and the opportunity to work with Gilbert Fletcher was enticing. In 1959, MacComb arrived in Houston to lead the section of head and neck surgery, and M. D. Anderson's Fourth Annual Clinical Conference that year was devoted to head and neck tumors, with MacComb and Fletcher in tag team fashion celebrating combined surgery and radiation. A New York gentleman always impeccably dressed and measured in tone, MacComb's collaboration with Gilbert Fletcher would produce major advances in the treatment of head and neck malignancies as well as *Cancer of the Head and Neck* (1967), the standard text for decades.[12]

Surgery also faced philosophical change. Ian McDonald, an iconoclastic Canadian-trained surgical pathologist at St. Vincent Hospital in Los Angeles, preached medical sacrilege in a controversial 1951 article by arguing that early detection of breast cancer was hardly a cure-all. The size of a tumor, he claimed, had less to do with time than with the tumor's cellular structure and its likelihood to spread. Canadian biometrician Neil McKinnon similarly argued that small lumps were small not because they had been discovered early but because they were biologically indolent in the first place and less likely to metastasize. Tumors "cured" by radical surgery were, most of the time, nonlethal anyway. Such notions undermined treatment paradigms based on early detection and on radical surgery. Ed White considered McDonald and McKinnon medical subversives, but Clark invited McDonald in 1956 to keynote the first annual clinical symposium—Management of Carcinoma of the Breast.[13]

In 1955, George Crile of the Cleveland Clinic heated up the debate. Known as "Barnie" to close friends, the Harvard-trained Crile was a gentle man with serious reservations about radical surgeries. Intent on curbing them, he wrote *Cancer and Common Sense* (1955) and rejected the conviction that solid tumors represented only local disease, claiming instead that some "cancers are incurable long before [they] can be recognized, no matter how often or how thoroughly the patient is examined . . . because they spread into the bloodstream before they are ever detectable. . . . The natural course of this type of cancer could not be affected by even the most perfect diagnosis and treatment." Logically then, the "rush to surgery" mentality needed revision. "There is no clear evidence," Crile continued, "that immediate treatment is any more effective than treatment given a little later . . . the factor of time has been so overemphasized."

Leaders of the medical establishment denounced Crile as an apostate. AMA president Elmer Hess accused him of tendering "a dangerous, fatalistic philosophy of cancer. We fear it may lead readers . . . to reject steps they can take for their own protection." NCI head John Heller labeled Crile's medical opinion as contrary to "the teaching of the country's 81 medical schools and to the experience of physicians and surgeons." Alfred Blalock, president of the American College of Surgeons, accused him of making "statements which can be misconstrued by the public and cause great harm." Crile irritated them even more by pointing out that American surgeons still worshiped Halsted because "partial mastectomies that removed the affected part of the breast and reconstruct the rest are not only more time consuming and difficult to perform, but medical insurance pays surgeons less for doing them than for removing the breast. In short, a surgeon is paid 2 to 3 times as much for performing a mutilating operation than for one that leaves a woman relatively intact."[14]

At M. D. Anderson physicians reacted predictably. Ed White, a proponent of radical mastectomies and adrenalectomies for women with breast cancer, rejected Crile at the time. Fletcher chortled with glee to see surgeons cannibalizing one another. Radical surgery was obsolete and the domain of men of dull intellect, he thought.

Richard Martin was less sanguine. A Korean War veteran who came to M. D. Anderson in 1951, he too backed radical surgery, but as time passed in the 1950s, he was increasingly puzzled by the mystery of why women with very small breast tumors sometimes died. Perhaps metastasis was not such a clear-cut process of malignant cells mechanically escaping the tumor site and then hugging tissue lines into axilla nodes, where they bided time before embarking on a deadly journey through the circulatory system. Perhaps the most biologically malignant cells were metastatic early on. As the data mounted, so did Martin's doubts. In January 1959, for example, Jerome Urban abandoned the extended radical operation. "Unfortunately, the extended radical procedures do not represent a major breakthrough in the treatment of breast cancer."[15]

Advances in radiotherapy and chemotherapy steadily undermined radical procedures. Between 1944 and 1965, M. D. Anderson surgeons performed 91 leg/ arm amputations, 62 hips, 41 intrascapular thoracics, 35 hemipelvectomies, and 14 forequarters, with declining frequency over time. In 1948, the gynecologist Felix N. Rutledge came to M. D. Anderson and began a long collaboration with Gilbert Fletcher, who considered pelvic exenteration as barbaric as the Halsted radical mastectomy and equally unnecessary, except in the cases of advanced locally recurrent disease. In temperament, the two men were polar opposites. A sweet soul lionized by surgical residents for his intellect and gentle demeanor, Rutledge quickly became a popular figure. When Lee Clark gave the section of gynecology departmental status in 1954, Rutledge was the natural choice for chair. Fletcher, in comparison, was mercurial and temperamental, but their collaboration, massaged repeatedly by Rutledge's Herculean patience, became a truly multidisciplinary assault on radical pelvic surgery. They worked out protocols to determine dosages, schedules, and whether to perform pre- or postoperative radiotherapy. As their data accumulated, the two men demonstrated the capacity of radiotherapy to reduce the need for radical surgery. In 1958, at the annual meeting of the American College of Obstetrics and Gynecology, Rutledge presented a paper on radiotherapy and conservative surgery in the management of gynecological cancers.[16]

Fletcher pushed his protégée Eleanor Montague in the same direction. European data seemed to indict the radical mastectomy and to endorse lesser surgical procedures when combined with radiotherapy. At M. D. Anderson, Richard Martin was the first surgeon to embrace Montague, and they eventually did to breast cancer surgery what Fletcher and Rutledge had done to the pelvic exenteration. Women came out of the operating room with more of their bodies intact and no diminution of their chances for long-term survival.[17]

In 1955, cancer flinched when NCI announced the methotrexate breakthrough. Other forms of cancer might be just as vulnerable. Perhaps there was a magic bullet, a drug that would do for cancer what antibiotics did to infectious diseases. At NCI, Memorial Sloan-Kettering, and Roswell Park, physicians tested methotrexate on dozens of other cancers, but none responded like choriocarcinoma. They experimented with other drugs as well. Grant Taylor began using vincristine on children and immediately developed concerns about cardiac side effects. Wataru Sutow worked on leukemia. Clifton Howe and Richard Martin tried phenylalanine mustard (PAM) on patients with osteosarcomas and melanomas. Meanwhile, Rutledge's results with methotrexate in curing metastatic choriocarcinoma exceeded 98 percent.[18]

Clark remembered Gloria Belsha, an effervescent twelve-year-old with an osteosarcoma. In 1962, she went to a local hospital, limping and complaining of pain in her swollen left leg. "It was in the summer and my [left] leg was swollen above the knee," she later recalled. "My parents thought I'd just bruised it." Pathologists diagnosed the malignant bone tumor and surgeons amputated at the hip. Within a few months, around Christmas 1962, Gloria awoke in the middle of the night short of breath. X-rays at M. D. Anderson revealed widespread lung metastases. "Thinking back," she said fifteen years later, "it seems like a dream. They told my parents I'd just have a short time to live." Several chemotherapy drugs administered in succession had no effect. As a last resort, M. D. Anderson pediatricians tried PAM. "Her chances for being alive for a year," Clark remembered, "were essentially nil. . . . Up to that time [PAM] had practically been of no use for generalized cancer but only for local melanoma." The tumors melted away and Gloria returned to her seventh-grade class. "She was," Clark said, "the first patient that I know of with bone sarcoma who was cured with chemotherapy."[19]

As a treat-to-cure culture imbued the doctors at M. D. Anderson, the chemotherapy regimens grew more aggressive. Increasing use of chemotherapy and radiotherapy demanded sophisticated standards for patient selection and informed consent. In 1963, the surgeon Robert C. Hickey arrived at M. D. Anderson from the University of Iowa Medical School, where he had served as director of research. Hickey understood the conflicts of interest inherent in aggressive research employing human subjects. Promotion and merit pay increases depend upon grants, and grants depend upon securing enough human subjects. Many incentives exist for clinical researchers to hedge their bets, to withhold full disclosure of risks and side effects, and to conceal the dark truth that most clinical trials fail.

Hickey's arrival at M. D. Anderson coincided with the triumphal moment of the civil rights movement. The Reverend Martin Luther King Jr. staged a massive demonstration on the mall in Washington, D.C., and with the Lincoln Memorial as a backdrop, he delivered his "I Have a Dream" speech, which embedded civil rights into the American consciousness. The transformation of individual rights into a national crusade would soon bleed into American medicine, especially in the area of clinical trials involving human subjects.

Auschwitz still cast a long shadow. In 1947, the Nuremberg tribunal outlawed medical experiments conducted on prisoners of war without their consent. Descriptions of barbaric Nazi medical experiments on Jewish prisoners forced scientists everywhere to reexamine their own practices with human subjects. In December 1957, Clark called for a more systematic approach to informed consent, a goal more urgent two years later when Senator Estes Kefauver of Tennessee opened hearings into the behavior of pharmaceutical companies, which paid physicians to dispense experimental drugs to unsuspecting patients. The doctors collected data on efficacy and side effects and turned the results over to the companies. Under Hickey's direction, the human experimentation committees tightened guidelines for securing consent and insisted that in all studies, "appropriate biometrical design [had to be] used."

The final requirement energized a proposal that had floated around M. D. Anderson for years—the need for a department of biomathematics. The thoracic surgeon Clifton F. Mountain chaired a committee that recommended a department of biomathematics to serve the needs of the proposed Graduate School of Biomedical Sciences and to provide integrated methodological support to institutional investigators, as well as training in research design, statistical techniques, mathematical models, and computer programming.[20]

⊸⊸⊸⊸

Clark had defined research as a fundamental mission at M. D. Anderson, but during the late 1950s, the enterprise only lumbered along in the departments of biochemistry, physics, and biology, and in the section of medical psychology in the Department of Medicine.

In biochemistry, A. Clark Griffin now had a high profile in chemical carcinogenesis, and in 1961 the American Cancer Society awarded him a coveted professorship of biochemistry, which guaranteed research funding for the rest of his life. Lee Clark hailed the award as evidence of quality research going on at M. D. Anderson.[21]

Within a year, carcinogenesis became a cause célèbre. Rachel Carson's book *Silent Spring*, which blamed pesticides for the rising incidence of cancer, rocketed up the best seller lists. "Can anyone believe it is possible to lay down such a barrage of poisons on the surface of the earth," she wrote, "without making it unfit for life?" President John F. Kennedy read *Silent Spring* and ordered a formal investigation of DDT. The chemical industry denounced the book as "science fiction, to be read in the same way that the TV program *The Twilight Zone* is to be watched." Unintentionally, Carson had crystallized the modern environmental movement. "Man alone, of all forms of life," she wrote, "can *create* cancer-producing substances." Her own battle with breast cancer bestowed a secular martyrdom on Carson. She died on April 14, 1964.

—⬛〰⬛—

M. D. Anderson physicists worked closely with Gilbert Fletcher. Physicists working in medicine often felt like stepchildren; they were viewed condescendingly by academic physicists and lacked a professional identity. At Anderson, Warren Sinclair built medical physics as a discipline. Early in 1958, he called on American physicists in biology and medicine to do the same, and in November, he helped establish the American Association of Physicists in Biology and Medicine, which later became the American Association of Physicists in Medicine.[22]

In biology, T. C. Hsu's investigations in cytogenetics continued, as did the cancer eye project, but Leon Dmochowski and virology were in the limelight. Virology was rapidly emerging as an exciting subdiscipline in biology, and scientists were scrambling to unravel its mysteries. Dmochowski's European demeanor made him hard to like, and a heavy cigarette habit stained his teeth and fingers a dirty yellow, but he made news in 1957 after identifying viral-like particles in the tissues of a patient with acute lymphocytic leukemia. The search for a viral etiology in cancer began to accelerate, and in 1963, M. D. Anderson's Annual Symposium on Fundamental Cancer Research highlighted viruses.[23]

In 1961, Dmochowski had read with interest an article in the journal *Cancer*. British surgeon Denis Burkitt described a lymphosarcoma with an unusually high incidence in Uganda, Kenya, and Tanganyika. Tumors appeared most often in the face, neck, and abdomen and expressed a virulent biology. Epidemiologists noted that the disease primarily affected children and seemed concentrated to a specific geographic region. That two other species of primates also contracted the disease prompted Dmochowski to wonder if a virus had jumped species. The United Nations, the Imperial Research Fund of Great Britain, the African Research Foundation, and the government of Kenya invited M. D. Anderson to participate in a research project. Clark was particularly interested because Joe C. Thompson, a Dallas industrialist and member of the UT Board of Regents, had just succumbed to a lymphosarcoma.[24]

Although Dmochowski did not become deeply involved in the project, the search for the cause of the lymphosarcoma accelerated. In 1961, Burkitt lectured at Middlesex Hospital in London about the disease, now increasingly referred to as Burkitt's lymphoma. The pathologist Joseph Anthony Epstein listened intently. He asked for tissue samples. For several years Epstein and his assistant Y. M. Barr tried unsuccessfully to isolate viral particles. In 1964, they identified what soon would be known as the Epstein-Barr virus—a cousin to herpes. They later speculated that the virus invaded B lymphocytes and triggered carcinogenesis in patients with immune systems weakened by malaria or yellow fever. Virologists would later establish an association, if not a cause, between Epstein-Barr and a number of other cancers, including T-cell lymphomas, Hodgkin's disease, lymphoepitheliomas, and nasopharyngeal carcinomas. Other laboratories around the world stepped up their efforts to find cancer-causing viruses.[25]

Medical psychology was a scientific train wreck. In February 1951, Beatrix Cobb arrived at M. D. Anderson, and her career soared like bottle rockets on the Fourth of July. Psychologists were infatuated with the psyches of cancer patients. In April 1953, Cobb addressed the American Cancer Society, discussing the economic, gender, and educational variables affecting patients. Wire services picked up the story. Within days, her name, and M. D. Anderson's, surfaced in magazines and newspapers throughout the country. Prominent psychologists trekked to meet her, and the ACS and NCI took notice.

Edna Wagner thought differently. She read human nature like Albert Einstein fathomed the cosmos. Cobb was "neurotic" and potentially "dangerous" when working "directly with sick, helpless people, because, sooner or later, she [will not be able] to resist the urge to use and exploit these people to satisfy some of [her] neurotic needs." Devotion to trendy personality theory did in Cobb. In the early 1950s, Austrian-born and Sigmund Freud–trained Wilhelm Reich became the rage in psychotherapy circles. He claimed to have identified a link between sexual inadequacy and a predilection for cancer. Cancer appeared, he claimed, in the lives of people experiencing "deep anxiety, deferred hope, and disappointment." They had a "bio-emotional disposition to cancer" because of "orgone depletions," possessed mild emotions, and lived in a state of perpetual, "painful acquiescence." At the core of their being, cancer patients were sexually repressed, and for Reich, aversion to sex was carcinogenic. Most physicians considered Reich deranged, father of what one sarcastically dubbed the "genital utopia."[26]

The notion of a "cancer personality" flourished in Cobb's sexually obsessed medical psychology section. Like a true believer, she had moved beyond assessing the emotional impact of cancer on patients to divining the psychological etiology of neoplasms. If the causes of cancer were emotional, Cobb concluded, then every M. D. Anderson patient needed psychotherapy, a prospect that terrified Wagner.[27]

Cobb pursued cancer personality theories relentlessly. In one memo, she insisted that cancer patients suffered a variety of emotional maladies, including a "loss of self-respect," "self-pity," "fear of death," and anxiety disorders. She circulated claims that "stress and separation anxiety," along with feelings of "hopelessness" and "helplessness," triggered leukemia and lymphomas, and that cancer patients possessed personalities that "thrive on dependency. . . . Some men hospitalized with cancer of the prostate," she intoned, "seem to be non-aggressive, compliant, co-operative, almost effeminate." "Melanoma patients . . . [are] hyperactive people either emotionally or mentally [and] precipitating factors which bring about a neoplastic transformation involve psychological processes as well as phenomena usually studied by experimental and clinical medicine." She even postulated that "repressed emotionality . . . could be one of the elements underlying the self-propagation of neoplasms."

In the eyes of M. D. Anderson's most influential physicians, especially the surgeons, the only thing worse than a psychiatrist was a psychologist. "It has been well stated," Ed White told Clark, "that ten minutes of the treating physician's time spent in answering the patient's questions and bringing assurance to him is worth more than hours of psychiatric study or psychological testing. M. D. Anderson's money, time, personnel, and space would better be devoted to other fields." He contemptuously dismissed psychology and psychiatry, citing the "wholly unsatisfactory position of this field as a science and as an art in medicine to-day." In July 1958, weary of the bickering, Clark handed down a death sentence, terminating hospital services in psychiatry and psychology. Decades would pass before the two recovered at M. D. Anderson.[28]

In 1960, Clark brought to M. D. Anderson one of the most influential physicians in cancer medicine, Murray M. Copeland, as director of education. Fully immersed in the whirlpool of medical politics along the axis of power in oncology—from NCI in Bethesda, Maryland, to Mary Lasker, Memorial Sloan-Kettering, and the American Cancer Society in New York City—Copeland knew everyone. One M. D. Anderson physician recalled dining with him at the Navy Club in Washington, D.C. They could hardly finish their meal for the constant interruptions from diners talking with Copeland. He greeted everyone by first name and inquired about their families. When Copeland arrived at M. D. Anderson, he was serving as chair of the executive committee on cancer for the American College of Surgeons, vice president of the Ewing Society, chair of the American Joint Committee on Cancer, and director at large of the American Cancer Society.[29]

At M. D. Anderson, Clark and Copeland became best friends, with Copeland eclipsing Russell Cumley as Clark's confidante. The two were very different. Clark was a fitness fanatic; Copeland considered exercise akin to death. Clark avoided elevators, preferring to climb seven flights of stairs to his office every day, and expected others to do the same. At Monday morning staff meetings, Copeland usually arrived early and took the elevator up first or lurked covertly nearby, waiting for Clark to enter the stairwell and then beating him upstairs on the lift. Clark was always trim and athletic, while Copeland struggled with his waistline. "He was obese," remembered an M. D. Anderson administrator, "but a big teddy bear of a guy." Copeland often seemed old-fashioned, his wife always following two steps behind and calling him "Dr. Copeland." In 1963, Copeland became president-elect of the American Cancer Society.

As director of education, Copeland insisted that M. D. Anderson's reputation in cancer medicine depended on the training of as many residents and fellows as possible. Only as M. D. Anderson–trained physicians scattered throughout the country and the world, singing the anthem of patient care, multidisciplinary care, treat-to-cure approaches, and excellent supportive expertise, would M. D. Anderson's reputation go national. Conferences and articles in professional journals would elevate M. D. Anderson's reputation, Copeland insisted, but not as effectively as filling the infrastructure of American medicine—from community hospitals to major university medical schools—with M. D. Anderson alumni preaching the "M. D. Anderson way."[30]

Clark wanted to hype the M. D. Anderson way on a global level. Early in the

1960s, his dream of bringing the Tenth International Cancer Congress to Houston in 1970 began to take shape. The Union Internationale Contre Cancer, a coalition of government agencies and private organizations, sponsored a congress every four years.[31] Few took Clark seriously. To people in New York City, Boston, London, and Paris, Texas seemed provincial, home to cowboys, Indians, and gunslingers. They underestimated Clark. Intent on raising M. D. Anderson's profile, Clark at the Seventh International Cancer Congress in London in 1958 wrangled for Houston to host the tenth annual congress.

Cold war politics complicated his plan. On October 7, 1957, the Soviet Union had placed into orbit an artificial satellite named Sputnik. Its launch caught America by surprise and sent the country reeling into a national identity crisis. If the Soviet Union could insert a satellite into orbit, could it not also lob a nuclear warhead at an American city? To keep pace, Congress in 1958 established NASA (National Aeronautics and Space Administration).

The identity crisis sharpened on April 12, 1961, when Soviet cosmonaut Yuri Gagarin orbited the earth. The Soviet edge in space appeared to be overwhelming and was a political liability for President John F. Kennedy. On May 25, 1961, Kennedy fired a salvo in the race to the moon: "If we are to win the battle . . . between freedom and tyranny . . . it is time for a great new American enterprise. . . . I believe that this nation should commit itself to achieving the goal, before this decade is out, of landing a man on the moon and returning him safely to the earth."

Within a few months, international tensions grew more strained. In August 1961, President Kennedy and Soviet Premier Nikita Khrushchev had stumbled through a summit meeting in Vienna, managing to escalate rather than ease tensions. Two months later, the Russians constructed the Berlin Wall, cutting off West Berlin from East Berlin and again lifting the lid on the possibility of nuclear holocaust.[32]

In July 1962, as the cold war heated up, the Eighth International Cancer Congress in Moscow attracted physicians and scientists from around the world. For the previous five years, Copeland had interacted with Soviet oncologists at international meetings, and the Russians considered him the czar of cancer medicine in America. Copeland went to Moscow for two reasons—to deliver a paper Clark had written and to lobby to bring the Tenth International Cancer Congress to Houston. The cold war percolated inside the international cancer community, but if Russian oncologists backed M. D. Anderson, the tenth would be in Houston.

Copeland had targeted three people: physiologist Nikolai N. Blokhin, a powerful figure in Soviet oncology; Alexander Haddow, the man behind the Chester Beatty Research Institute in London; and Tomizo Yoshida, an internationally known pathologist scheduled to preside in 1966 over the Ninth International Cancer Congress in Tokyo. Copeland assured them that M. D. Anderson would be an excellent host. All three agreed. Haddow noted that "we all sensed the reality of a communication between our nations which, as a contribution to world peace . . . made us all rejoice" and Blokhin admitted that "nothing other than benefit can flow from the meeting together in common cause of scientists of differing political persuasions." Copeland courted every Russian presenter at the congress, securing copies of abstracts in English, reading them carefully, and at every turn praising the authors effusively. The seeds that he planted would flower four years later in Tokyo.[33]

Copeland still feared the unexpected. Several months after the eighth congress, Khrushchev began constructing facilities to deploy nuclear-tipped missiles in Cuba. When Kennedy demanded their removal, the world edged toward global thermonuclear war, and Americans braced for Armageddon. M. D. Anderson disaster planners announced that in the event of a nuclear blast in Houston, staff members were to move as many people as possible into radiation-protected areas of the hospital, shut down the air conditioning, close the doors to rooms with broken windows, drape open doorways with sheets, and make sure that nobody smoked indoors. Kennedy and Khrushchev, however, reached a diplomatic solution. The world heaved a sigh of relief. So did Copeland.

One year later, the United States glowed like the city on a hill, able to solve any problem and conquer any foe. UT won the national football championship and was scheduled for a Cotton Bowl showdown with Navy. Even the cold war did not seem as frigid, with the Berlin Wall and the Cuban missile crisis retreating into the past and the 1963 Nuclear Test Ban Treaty gracing the headlines. The hospital registered its fifty-thousandth patient, and NCI awarded to M. D. Anderson a $3.5 million grant in clinical pathology, on top of the ongoing, annual grant of $575,500 for the Clinical Cancer Research Center. Construction contracts worth $6.5 million had been signed to add 209,000 square feet to the physical plant. Science writers noted that since 1930, the overall five-year survival rates for cancer had climbed from 20 percent to 33 percent, from one in five people to one in three. Richard Martin later recalled the early 1960s, when everything seemed possible. "We used to sit around in the lounge debating whether it would be five or ten years before the cure was discovered. We were very naive." Cancer killed more than 225,000 Americans in 1963.[34]

6

———๛ๆ๛———

M. D. Anderson and
the Rise of Medical Oncology,
1964–1969

*Cigarette smoking is a health hazard of sufficient importance
in the United States to warrant appropriate remedial action.*
—SMOKING AND HEALTH: REPORT OF THE ADVISORY
COMMITTEE TO THE SURGEON GENERAL, 1964

On January 11, 1964, the surgeon general of the United States put the topic of lung cancer on the front pages. During World War I, the tobacco companies had dispensed free cigarettes to soldiers, intentionally hooking them on tobacco to boost postwar sales. By the late 1930s, the incidence of lung cancer was escalating without prospects for a cure. Although a heavy smoker destined to die of lung cancer, surgeon Evarts A. Graham in 1950 had joined hands with pulmonologist Ernest L. Wynder for an article in the *Journal of the American Medical Association* that indicted smoking as the culprit in lung cancer. More than eighteen thousand Americans died of lung cancer in 1950, and most of the deaths could have been prevented.[1]

A mountain of data implicated smoking, and U.S. Surgeon General Leroy E. Burney in 1957 announced the correlation. Over the course of the next five years, the American Cancer Society (ACS), the American Heart Association (AHA), the National Tuberculosis Association (NTA), the American Sanitarium Association (ASA), and the American Public Health Association (APHA) lobbied the federal government for action. In June 1962, U.S. Surgeon General Luther Terry decided to establish the U.S. Surgeon General's Advisory Committee on Smoking and Health. Terry accepted nominations for committee members from an alphabet soup of groups—the ACS, ASA, APHA, NTA, FDA, and AMA, and the Tobacco Institute, a powerful trade group. The youngest of the appointees was Charles A. LeMaistre, a Texas physician who had specialized in diseases of the chest but who had not made enemies in the tobacco companies. Not on the Tobacco Institute's radar screen, he joined the committee uncontested.

Like the founding fathers in 1787 who sealed the windows of the Pennsylvania State House in Philadelphia to keep pesky journalists out of the deliberations of the Constitutional Convention, LeMaistre and the other committee members sequestered themselves in the subbasement of the National Library of Medicine. Two breaches of security, one by columnist Jack Anderson and another by a *Newsweek* photographer, increased the committee's vigilance. Terry wanted no leaks, since the Tobacco Institute was prepared, at a moment's notice, to muster in its defense the congressional delegations of North Carolina, South Carolina, Kentucky, Tennessee, Virginia, and Georgia. Nor did Terry want Wall Street speculating on tobacco futures. Armed guards protected the entrances to the library. All documents were locked up as tight as the gold bars in Fort Knox.

The committee reviewed more than seven thousand scientific articles and drafted the 387-page *Smoking and Health: Report of the Advisory Committee to the Surgeon General*, which attributed to tobacco the recent 70 percent higher mortality rates of smokers over nonsmokers and identified smoking as a cause of lung cancer. It was the only civilian document in the history of the U.S. Government Printing Office to be published under top secret conditions. Terry released the report on a Saturday morning—January 11, 1964. Armored trucks delivered it to the State Department, and copies were distributed to journalists behind locked doors. They were given one hour to read the report before Terry's press conference. Early in the morning, Terry delivered a copy to the White House. Unequivocal and unapologetic, the report proclaimed, "Cigarette smoking is a health hazard of sufficient importance in the United States to warrant appropriate remedial action." The media frenzy stunned Terry. "The report," he later recalled, "hit the country like a bombshell. It was front page news and a lead story on every radio and television station in the United States."[2]

Within a few months, lung cancer again grabbed the headlines. When surgeons cut into the chest of actor John Wayne, instead of finding healthy, pinkish tissues, they encountered a lung more resembling an old, dried-out sponge, speckled with black nicotine patches left behind by the more than 1 million cigarettes Wayne had smoked to satisfy his four-pack-a-day, forty-year habit. They removed the tumor and the lobe. John Wayne—the cowboy who conquered Indians, the soldier who defeated Japan, the lawman who killed desperadoes— had lung cancer, with its 5 percent five-year survival rate. Late in December 1964, Wayne told a syndicated columnist, "There's a lot of good image in John Wayne licking cancer—and that's what my doctors tell me. . . . I had the Big C, but I've beaten the son of a bitch. . . . I want people to know that cancer can be licked." He spoke openly, pleading with Americans to "stop acting as if cancer [were] leprosy." More than a hundred thousand letters poured in to Batjac, Wayne's production company. On January 4, 1965, Wayne headed to Durango, Mexico, to film *The Sons of Katie Elder*, humbled by his brush with death but unwilling to reenact Babe Zaharias's curious interaction with cancer. He turned down Mary Lasker's invitation to become part of the American Cancer Society's crusade. "The cancer societies want me on their campaigns. They're welcome to use my case, but I don't want to make a profession out of this. Before I know it I'll be 'The Man Who Had Cancer.' Thanks to the Man upstairs and my doctor I've got my life back and I want to go on living."[3]

————

The nutritionist wanted to serve attractive, palatable food, but cooking germ-free complicated the task. For one special patient, she autoclaved the tray, utensils, plates, and cups, sterilizing them as if they were surgical instruments. Before handling the meal, she donned surgical garb and scrubbed. She then placed the food in plastic bags and the bags in Mason jars and, like a farmer's wife bottling vegetables, sterilized them. After putting the food in dishes and the dishes on a tray, she delivered the meal to the patient's room, where the food was exposed to ultraviolet light. A nurse carried the tray to a bed shrouded in a plastic bubble. Through a slit in the bubble, she passed it to a thirteen-year-old boy suffering from acute lymphocytic leukemia (ALL). He had a weakened immune system and was vulnerable to infections. Because the disease suppressed blood platelets, he bruised and bled easily. Tiny red dots, the products of pinpoint hemorrhaging known as petechiae, spotted his pale skin. A spleen stuffed with leukemia cells distended his abdomen, and he complained of bone pain, where the malignant cells had clumped in the marrow. At a time when he should have been outside playing ball, he was whimpering in a germ-free bed.

Physicians treating ALL ran a life-and-death race against infections. At M. D. Anderson in the late 1960s, some leukemia patients lived in special rooms with germ-free beds shrouded in plastic. The sick boy's parents stayed hour after hour, reading and telling stories, playing games, and trying to distract him, but they could not kiss or hold him; the dangers of infection were too great. Then one day, just as the boy began to eat dinner, he started to hemorrhage from his nose, mouth, and eyes. He cried and writhed in pain, the blood splattering the pillow, sheets, and plastic liner. At first his parents and the staff stood paralyzed, anxious to help but afraid of exposing him to germs. When the father could stand it no longer, he climbed through the bubble into the bed, cradling and kissing his son and soaking in blood himself. "Blood was everywhere," remembered an M. D. Anderson employee. "The boy died a terrible death."[4]

That incident, along with clinical work by hematologist Gerald Bodey, led to building on one floor of the planned Lutheran Pavilion a laminar flow area to filter all outside air before it reached patients with compromised immune systems. Curing ALL required advances in infection control, and Bodey would play an important role, through the use of prophylactic antibiotics and the laminar flow rooms, in bringing it to pass.

But in spite of such incidents, M. D. Anderson had an unusually hospitable culture. Leaders at NCI repeatedly acknowledged the fact. Clark had built a public hospital with all the amenities of the finest private, for-profit institutions. In 1964, when representatives of the Public Health Service toured M. D. Anderson, they were visibly impressed. In a letter to Renilda Hilkemeyer, an NIH nurse wrote, "It was gratifying to experience such high morale among the patients and the personnel. Your outstanding staff indicates that they have been carefully considered and selected. . . . This institution [M. D. Anderson], from every angle, has the highest degree of patient-centered care that I have ever experienced."[5]

—◄░░░◖❙◗░░░►—

During the mid-1960s, a struggle of seismic proportions shook M. D. Anderson. Gilbert Fletcher had grown increasingly frustrated with some surgeons' unwillingness to cooperate with radiotherapists. The use of preoperative radiation could often reduce tumor size before removal, leaving surgeons with better margins and patients with more healthy tissue. Fletcher complained about the steady decline in the number of curable cases referred to radiotherapy, from 380 in 1964 to only 294 in 1968. Fletcher held surgeons responsible. He had high praise for Richard H. Jesse and William MacComb, whose cross training in surgery and radiation therapy had placed them in the forefront of surgical oncology, but disdain for their stubborn, uncooperative colleagues. John Bardwil attracted Fletcher's special wrath. Residents and fellows complained that he had turned the weekly planning clinics into "anti-irradiation session[s] to the point that the Radiotherapy Fellows have expressed that they are uncomfortable during the Conference." On one occasion, Bardwil removed a soft palate lesion without referring the patient to radiotherapy. During surgery, he discovered that the lesion extended deep into the pharyngeal wall and required an extensive procedure that would have been unnecessary with preoperative radiation. Fletcher accused him of ignoring the "known behavior of such lesions."[6]

Fletcher also singled out Alando Ballantyne. "This little note," he wrote to Ballantyne in 1967, "is not to pass judgement on your superior surgical management, but to bring [to] your attention that if you find little use for radiotherapy . . . you have found and still find it very useful to build up your bibliography as a significant percentage of your publications are connected with radiotherapy. I just want to let you know this is the end of your using Radiotherapy and myself to further expand your bibliography." Two years later, Fletcher complained of an operation that Ballantyne had performed on a patient with an adenocystic (adenoid cystic) carcinoma of the paranasal sinuses. Tracing the disease to the gasserian ganglion, Ballantyne had requested radiotherapy to the ganglion only. Herman Suit, a young radiologist Fletcher was grooming, thought the wider surgical area needed radiation. When Suit intervened, Ballantyne replied, "I have never had a recurrence with adenocystic carcinoma and, therefore, only the ganglion should be treated. . . . The radiotherapists here are just lackeys." In a blistering letter to Clark, the crotchety Fletcher groused, "Nothing will be achieved unless one is fully aware of the problem at hand which is essentially the personalities of Dr. Ballantyne and Dr. Bardwil. One cannot get to them; Dr. Ballantyne because he is walled off in his infallibility complex and Dr. Bardwil because of his mental limitations."

Anxious to diffuse the controversy, Clark alerted Robert Hickey, recently promoted to assistant director of the hospital. "Please review this situation as to cause and effect. Is this being done for the best interest of the patient? A change in material, or due to unilateral decisions on the part of the surgical service?" Hickey asked William MacComb to investigate, and MacComb concluded that the real explanation had less to do with the intransigence of surgeons than with the unmitigated success of Fletcher's cobalt-60 regimens. As the cobalt-60 machines and Fletcher's methods expanded to other hospitals, competition increased, and M. D. Anderson received fewer referrals of curable patients. The rivalry had blazed into open enmity.[7]

Fletcher reiterated his discontent. In 1965, he informed MacComb and Clark, of 750 patients admitted to the head and neck service, 211 had received primary radiation therapy, while in 1966 the number was 87 of 806 patients. Fletcher wanted to see more curable cases. "The majority of the patients undergoing radiotherapy for head and neck lesions," he insisted, "were post-operative either for questionable lack of surgical margins, definite cut-through, or residual or recurrent disease." Fletcher warned of the danger to the hard-won reputation of M. D. Anderson as a center for multidisciplinary treatment. "It is obvious," he bemoaned, "that in oncoming applications and yearly reports to the NIH, the numbers of patients with head and neck lesions treated by radiotherapy will be drastically less than in previous years and it is to be assumed that adjustments in the amount of money granted will have to be made." And a lack of patients could undermine M. D. Anderson training programs in radiotherapy and surgery. Observing that the "management of head and neck patients at the M. D. Anderson hospital was in the lead at some time and has justifiably received proper recognition for it," he feared that "in years to come no new knowledge will be acquired. It will be increasingly difficult to give talks or write papers except on limited subjects." Clark also handled queries about the rivalry from NCI and other major cancer centers. In 1969, for example, Fernando G. Bloedorn of Tufts University discussed with Clark the widening rift, and NCI complained about the "slow accession of patients" to the hyperbaric radiation treatment program. Clark reminded the surgeons that multidisciplinary care distinguished M. D. Anderson.[8]

By the end of the decade, Fletcher had established radiotherapy at M. D. Anderson as a leader of the discipline. At the heart of his success rested an uncompromising respect for clinical data, a commitment to aggressive therapy and multidisciplinary cooperation, fastidious attention to care of the patient, and detail in treatment. Many apparent failures in radiotherapy, he insisted, could be attributed to conservative therapy. "There [are] an increasing number of patients who never finish initial radiotherapy treatment on the basis that when seen in the fourth or fifth week of treatment the head and neck surgeon decides that the lesion is not responding. . . . Herman Suit's . . . study shatters . . . quick conclusions based on palpatory findings toward the end of therapy." Surgeons and radiotherapists jointly needed to analyze surgical failures. "The management of advanced disease [in the neck] is not going to be solved magically. Such an attitude is not conducive to further progress in the combinations of the two disciplines."

After a generation of aggressive research, Fletcher had established many dosages and schedules for the cobalt-60, the betatron, and combination treatments of photons, electrons, and gamma rays where appropriate, as well as the relative merits of interstitial and external beam treatments. His group enjoyed unusual success in curing some cancers alone or in combination protocols with surgeons, and in doing so with hundreds of patients in each disease category, enough to develop statistical norms and a common set of definitions and parameters.

In Gilbert Fletcher, radiotherapy found one of its most powerful voices. Much of his research was pivotal. He demonstrated that recurrences of squamous cell tumors of the tongue and floor of the mouth could be attributed to insufficient doses; that the rate of tumor regression is not necessarily an indication of the degree to which the cancer is under control; that electron therapy was superior to the cobalt-60 for some lesions; that survival rates for many lesions of the larynx and hypopharynx were considerably better with multidisciplinary treatment than with surgery or radiation alone; and that recurrences of anaplastic, or highly aggressive, lesions in the neck were lower when irradiation preceded radical neck dissection, especially once the tumor had spread to the lymph nodes.[9]

Fletcher's results startled many oncologists. Cervical cancer patients undergoing radiotherapy after conservative hysterectomy survived longer than women who postponed radiation treatment until after local recurrence. For endometrial, uterine, and vaginal cancer, radiotherapy was effective treatment. Preoperative radiation for breast cancer reduced the likelihood of local recurrences among women who received surgery alone. In Hodgkin's disease and other lymphomas, Fletcher demonstrated the value of maximum radiation doses to involved lymph nodes in the head, neck, and mediastinum and the prophylactic treatment of apparently uninvolved nodes, where occult cancer cells lurked. Among patients with Ewing's sarcoma, irradiation of the tumor site improved local control.[10]

In many ways, Gilbert Fletcher had also presided over the democratization of radiotherapy. Before he appeared on the medical scene, radiation oncology was reserved for a few cancers and for a few people, those with financial and geographic access to high energy devices. The cobalt-60 decentralized radiation oncology from major urban teaching hospitals to large community hospitals. In addition to improving accessibility, Fletcher had greatly expanded the variety of cancers being successfully treated with radiation, from superficial lesions of the skin to deep-seated tumors of the head and neck, breast, ovary, cervix, uterus, rectum, penis, vagina, bone, and other soft tissues. In many cases, he demonstrated radiation's curative capacities, and he played a leading role in the demise of radical surgery.

A prominent radiologist in 1970 put into words what had become consensus in oncology circles. "I am particularly impressed by the meticulous attention to detail, precision, and care in which treatment is planned and executed," wrote H. D. Kerman of the University of Louisville. "The philosophy of 'common sense' radiotherapy based on clinical experience, application of basic radiological physical principles and radiobiological concepts documented by close follow-up of patients and, most importantly, the attitude of interest and concern for the patient himself was most significant. . . . You can be most proud of the esprit of your staff and the development of the 'Anderson system' of radiotherapy." Fletcher's approach was ruthlessly scientific: "I have, through the years, relentlessly analyzed failures and complications as it is the only basis for assessment of existing policies and techniques of treatment and need for change. . . . Perseverance and continuity are essential for the accumulation of information."[11]

Finally, Fletcher's research established two founding principles in modern radiation oncology. He rejected the prevailing notion that large tumors, because of their mass, should receive lower doses and led the way in developing "shrinking field techniques designed to give the highest dose of radiation to the greatest mass of tumor." More than anything else, Fletcher refined the idea of subclinical disease, where cancer cells lurked beyond the reach of diagnosticians. Lower doses there, he demonstrated, were capable of containing microscopic spread. The concept revolutionized the treatment of breast cancer and tumors of the head and neck.[12]

As Fletcher's luster brightened, he yearned to split radiology and radiotherapy. Since the late 1950s, he had distanced himself from diagnostic radiologists performing no procedures on patients. Radiotherapists, on the other hand, worked directly on patients. Fletcher scuttled the word "radiologist" from his résumé and correspondence, referring to himself as a "radiotherapist," and he instructed Russell Cumley to use "radiotherapy," not "radiology," in the titles of all his publications. Fletcher wanted a department of his own.

The arrival of Robert D. Moreton strengthened his hand. Late in 1965, Clark hired Moreton as his assistant and head of the University Cancer Foundation. In Fort Worth, Moreton had been a prominent radiologist, an influential figure in the Texas Medical Association, and president of the Radiological Society of North America. His impeccable reputation, broad patient base, and community contacts increased Clark's ability to raise money throughout the state. Moreton was to Fort Worth what Bertner had been to Houston.

Moreton agreed that radiology and radiotherapy had evolved into distinct disciplines. In July 1966, Clark created a Department of Radiotherapy with Gilbert Fletcher in charge and a Department of Diagnostic Radiology, to be headed by Gerald Dodd, a graduate of Jefferson Medical College. After a six-year stint at M. D. Anderson, Dodd had moved to Jefferson in 1961. He had no sooner left than Clark began to woo him back. Dodd foresaw the looming technological revolution in radiology, and Clark called him every five to six weeks. The more Dodd talked about the future, the more Clark wanted him. Dodd returned on Clark's promise that the department would enjoy sufficient resources to purchase the latest technologies.[13]

M. D. Anderson had one of the best radiotherapy programs in the country, but in the early 1960s, chemotherapy began to hog the headlines of cancer medicine, and the best research was occurring elsewhere. Beginning in the 1950s, scientists at a handful of laboratories slowly revealed some of the mechanisms of cancer cell proliferation and demonstrated how chemotherapy works.

At the Southern Research Institute in Birmingham, Alabama, biochemists Frank Schabel and Howard Skipper unraveled the mathematics of survival and death. Thousands of mouse autopsies demonstrated that leukemia cells penetrated every organ and tissue. Leukemia cells proliferate at exponential rates, and Schabel and Skipper showed how the growth curves, when combined with the killing effects of the drugs, could determine how many malignant cells remained after treatment. Mouse leukemia cells double in number every day. By the fifteenth day, the malignant load will reach one billion cells, and the animal dies. Schabel and Skipper concluded that if an animal survives more than fifteen days, the number of cells killed by the treatment can be calculated. "The inescapable implication of these experiments," Skipper said, "is that in order to cure this mouse leukemia, we must kill every living leukemia cell in the mouse no matter where it is in the body, or else it will grow back to kill the mouse when it divides up to the magic number of about one billion cells . . . fifteen days later." Drugs used to treat leukemia, therefore, had to penetrate every tissue in the body, kill leukemia cells faster than they were being replaced, and be able to do this quickly. It takes one trillion cells to kill a child, not the mouse's one billion, but when exponential rates are at work, the distance between one billion and one trillion cells is just a matter of weeks. Equally significant, each dose of an effective chemotherapy drug kills the same percentage of cells, not the same number. If, for example, a mouse has one billion leukemia cells and an effective drug kills 99.9 percent of them, the surviving 999,000 or so cells will proliferate and make short work of the animal. The only way to save the mouse is to deliver a new dose just when the maximum number of cells have been killed but before the survivors begin to divide. "We had to find schedules of drugs to kill cells faster than they were able to grow back," Skipper said. "Only in that way could we ever hope to get the number of leukemia cells down to zero, and thereby cure the animal."[14]

While Schabel and Skipper were fashioning a pharmacological assault on leukemia, other scientists were addressing closely related issues. At St. Jude Children's Research Hospital in Memphis, Donald Pinkel tackled the tendency for errant leukemia cells to hide in the central nervous system. Leukemia cells floating in spinal fluid enjoy some protection from anticancer drugs, as do those in the brain, where the blood-brain barrier can filter out the medicines. Even one cell in the brain would, after it proliferated, kill a child in a matter of weeks. Pinkel injected chemotherapy drugs directly into the spinal canal to kill leukemia cells there and directed external beam radiation at those in the brain. At Memorial Sloan-Kettering, Joseph Burchenal and David Karnofsky tested many drugs and learned that in addition to the antimetabolites and aminopterin that Sidney Farber effectively employed, such new drugs as 2,6-diaminopurine and 6-mercaptopurine (6-MP) attack leukemia cells. They also employed methotrexate.

The hub of the wheel for chemotherapy research, however, was the National Cancer Institute, where Gordon Zubrod supervised brilliant clinical scientists, none more gifted than Emil Frei III and Emil J Freireich. Emil Frei was born in St. Louis to a devout Roman Catholic family. In 1943, the U.S. Navy sent Frei to Colgate for an accelerated premed curriculum, and without ever finishing a bachelor's degree, he matriculated to Yale School of Medicine and graduated in 1944. He interned at St. Louis University; between 1950 and 1953, he served with the Navy Medical Corps in Korea. In January 1953, Frei took a research residency at Washington University. Just as he arrived, so did Zubrod from Johns Hopkins. Zubrod immediately liked Frei.[15]

In 1953, Zubrod moved to NCI to head up its new clinical research program, and he brought Frei along. "Gordon was not recruited to NCI because he was

an oncologist," in Frei's recollection. "God knows there were virtually no oncologists at the time. . . . Gordon told me this was an opportunity to enter a
new era—to be on the ground floor of understanding cancer and treating cancer with chemotherapy. . . . There were few successes in the treatment of disseminated cancer in 1955. It was usually a matter of watching the tumor get bigger, and the patient progressively smaller." Zubrod's optimism ran counter to
the prevailing gloom, and Frei "became . . . committed to the position that science could solve the cancer problem."

Within two months of his appointment, Frei headed NCI's new Leukemia
Service. Although not a pediatrician, he treated children on a daily basis and
saw each one die. Nothing was more miserable than the hemorrhaging. Most
of the children endured multiple-site bleeding—from the nose, eyes, ears, bow-
els, and skin. After a few years, Frei made his first major contribution, that bleeding corresponded with a serious decline in the number of blood platelets, a condition hematologists label thrombocytopenia. Without the blood clotting that
platelets provide, leukemia patients end up anemic, bruised, bleeding, and
swollen. Frei recommended transfusions of blood platelets.[16]

NCI hematologists at first stonewalled. They worried that the platelets
would soon lose effectiveness as patients became immunized to them. In the
midst of the debate, Emil J Freireich, or "Jay," joined the Leukemia Service.
Raised on the mean streets of Chicago by a widowed, Hungarian immigrant
mother, Freireich displayed an urban ghetto instinct for survival—a combat-
ive, confrontational style, a bravado and certainty about his convictions, and
a spontaneous generosity toward those in need. "It has always been a short trip
between the top of [Freireich's] mind and the tip of his tongue," said a colleague,
"and in some ways the interaction between his personality and the application
of science has been his greatest asset and his greatest liability." Freireich possessed a capacity for seeing in patients what lesser minds never noticed, and the
courage to play his hunches.[17]

With twenty-five dollars from a Christian Scientist neighbor, Freireich enrolled at the University of Illinois when he was just sixteen. Six years later, he
graduated from the University of Illinois College of Medicine and interned at
Cook County hospital. He soon transferred to Rush Presbyterian; Abraham
Flexner's revolution had transformed Rush into an enterprise where academic
research rivaled in importance the training of physicians. Freireich's future un-
folded there: "So all of that converted me from a Cook County hospital atmosphere where you just put on Band-Aids and Scotch tape, to an academic

environment where the guys discovered things." When assigned a research project on the life cycle of red blood cells, he argued against the consensus that low serum iron explained anemia in patients with inflammatory disorders. His paper was published in the prestigious *Journal of Clinical Investigation.*

Personal loss drew him to hematology. At Cook County, Freireich fell in love with a young woman who soon died of ALL. He accepted a hematology fellowship at Massachusetts General Hospital and began to work with leukemia patients, a task few physicians relished, for such patients faced dire futures. Freireich joined NCI in July 1955 and was assigned an office next to Emil Frei. The nameplates in front of the offices—Dr. Emil Frei and Dr. Emil Freireich—intrigued both men, who struck up an immediate friendship. "To get a job here," Frei joked to Freireich, "you don't have to assume that name." The two men displayed different temperaments. Frei was calm and in control. Freireich was blunt and impulsive. Freireich had a knack for manufacturing ideas, Frei for converting them into plans. Frei was a peacemaker and Freireich a soldier.[18]

In 1955, the treatment for ALL rested on the clinical experience of a few physicians. The drugs methotrexate and 6-MP exhibited some effectiveness, but in an era before clinical trials and quantitative data, all assumptions were qualitative and anecdotal. Standards for measuring regression in tumors were imprecise, and definitions of complete remission did not exist. Insisting on prospectively designed, quantitative trials, Frei and Freireich helped give birth to medical oncology. Each trial began with a clearly defined objective, a cohort of patients carefully selected for similarities in disease, and standards for measuring tumor response. The two also defined "complete remission" as the absence of measurable cancer cells in the patient.

Frei and Freireich revolutionized the treatment of childhood leukemia. The 1955 trial constituted the first ever quantitative, prospective, experimental design for cancer treatment, demonstrating that complete remission after treatment with 6-MP or methotrexate most reliably prognosticated survival. Achieving complete remission quickly became the initial goal. Patients were treated even after that point because of the recognition that it was not possible to be certain that every leukemia cell had been killed. In fact, almost surely they had not. To address the problem of patient relapse and drug resistance, Frei and Freireich two years later launched the modern era of combination chemotherapy in a clinical trial treating patients by one drug first—methotrexate or 6-MP—and then by the other. For achieving complete remission, the combined use proved more effective than either drug used alone. That same year, the pair overcame considerable opposition within NCI and demonstrated the ability of blood platelet transfusions to stop hemorrhaging in leukemia patients. In 1959, their clinical trial using prednisone to achieve complete remission followed by 6-MP established the principle of continuing adjuvant chemotherapy even in the presence of complete remission. They had an especially fertile year in 1961, proving that full doses of combined cyclophosphamide (CPA), 6-MP, and methotrexate after complete remission with prednisone were far superior to half doses of each in sustaining complete remission. They proved that the best results depend on fine-tuning delivery schedules, and they achieved their best results with complete remission induced by vincristine and prednisone followed up by adjuvant treatment with 6-MP and methotrexate.

In 1962, Freireich suggested on a creative hunch a four-drug regimen that soon acquired the acronym VAMP—vincristine, amethopterin (methotrexate), 6-MP, and prednisone—each administered in a full dose. Freireich guessed that since each drug had a different method of killing leukemia cells, maximum doses could be administered without increasing toxicity. "At first I opposed it," confessed Gerald Bodey, a specialist in infectious disease who worked with Frei and Freireich. "Giving these kids four drugs all at once! As a Christian, I thought it was immoral because if they relapsed we would have no fallback. I thought Freireich was crazy." The results stunned Bodey. "We had sixteen patients, and eleven were cured. It was astonishing. I kept track of one of the children for decades until she died of breast cancer." Frei was awed, "To be able to do this with a lovely child . . . whom you think of almost as your own, is truly an extraordinary experience. . . . I had the evidence that you could succeed."

Frei then turned to Hodgkin's disease, employing a philosophy of clinical research "born out of scientific insights, biochemical insights, insights from cancer chemotherapy, and experimental models." Building on his success in treating leukemia, he jumped straight to four drugs in what became known as the MOPP regimen—mechlorethamine (nitrogen mustard), vincristine (Oncovin), prednisone, and procarbazine—administered in maximum doses with treatments continuing after the achievement of complete remission. And as was the case among his patients suffering with leukemia, Frei soon had some whose periods of complete remission stretched into months and years, to the point that he considered them cured.[19]

By the early 1960s, the internists delivering chemotherapy functioned in a disciplinary no-man's-land, caught between the basic scientists active in such groups as the American Association for Cancer Research (AACR) and the surgeons and, to a lesser extent, radiotherapists who dominated clinical care. Chemotherapists soon proposed a professional society. "No single organization," remarked the Youngstown, Ohio, internist Arnoldus Goudsmit, "has undertaken to represent this growing group of cancer clinicians, their areas of special concern, their body of knowledge, their opportunities for service, and their needs for communication." In 1964 at the annual meetings of the AACR, Goudsmit and six others founded the American Society of Clinical Oncology (ASCO). Harry Bisel, ASCO's first president, expected that the group would "provide clinical oncologists . . . a podium in order to make their feelings felt about the new directions the government is taking and an opportunity to share in the formative phases of that program." For Emil Frei, "It is vitally important that we make every effort to establish a subspeciality of medical oncology within the American Board [of Internal Medicine]." ASCO began to define medical oncology, and Frei took the lead. Prerequisites had to be determined, standards written, training programs defined, and examination questions developed.[20]

As Frei's reputation in clinical oncology magnified, Lee Clark targeted him for M. D. Anderson. Over the course of a generation, Gilbert Fletcher had built M. D. Anderson's reputation in radiotherapy, but Clark would not wait that long for prominence in medical oncology. Except for the work of Wataru Sutow, most of the medical oncology at M. D. Anderson was pedantic and derivative. With the skill of an admiring suitor, Clark began the courtship. He praised Frei for developing protocols that would save tens of thousands of lives and promised him hundreds of research patients in M. D. Anderson clinics, chairmanship of an academic department independent of the department of medicine, and a higher salary. Frei resisted for several years. A death in the family then left him with nieces and nephews to raise in addition to his own children. His NCI salary paled against Clark's offer. Clark made his pitch again and Frei accepted.[21]

Clark's success in recruiting Frei almost broke the heart of NCI director Kenneth M. Endicott. "Emil Frei has told me of your offer," Endicott wrote Clark, "and his decision to accept it. This is a serious loss to the National Cancer Institute and there is no prospect for replacing him with a man of equal accomplishments and capabilities, though we do have some younger men who show great promise. . . . [He] is the top man in his field. In addition, he is a splendid physician and a delightful person to have around. We will miss him more than you will ever know." Even worse was Endicott's worry that other members of Frei's team would follow him. "Now that he is going (and I understand he will be taking a substantial number of other investigators with him, which will add substantially to your already impressive capabilities in the field) . . . I am thinking of M. D. Anderson as the headquarters institution from the trans-Appalachian region and I hope that you will give this matter some thought." In a few years, a grateful Clark would nominate Endicott to lead the National Institutes of Health.[22]

Endicott's fears were well grounded. Emil Freireich and others, including infectious disease specialist Gerald Bodey and biostatistician Edward Gehan, a leader in designing NCI clinical trials, left for M. D. Anderson. Gehan had just published in *Biometrika* an article destined to become enormously influential in clinical cancer research—"A Generalized Wilcoxon Test for Comparing Arbitrarily Singly Censored Samples," which provided a statistical test for determining "the significance of the difference between two survival curves . . . when two different treatments are being compared for their influence on survival time."[23]

Clark forged a thorough administrative reorganization. He established the Department of Developmental Therapeutics, with Frei as chair, Freireich as head of the section of research hematology, and Myron Karon, another NCI transplant, as head of the section of applied molecular biology. Because childhood cancers constituted the most exciting frontier in medical oncology, Clark transferred the section of pediatrics from the Department of Medicine to the Department of Developmental Therapeutics early in 1966. Sutow's research into vincristine for the treatment of Wilms' tumor would soon constitute a major contribution to pediatric oncology. Wilms' tumor, a cancer of the kidney in children, had been treated with surgery and radiation since the late 1930s, and in the 1950s, Sidney Farber added the drug actinomycin-D to the regimen. Survival rates improved, and in some cases cures were achieved. In 1959, with the discovery of the drug vincristine in the Madagascar periwinkle plant, a new weapon was added to the arsenal. In binding to spindle proteins, vincristine disturbs cellular cytoskeletons, inhibits RNA synthesis, and disrupts cell division. Its discovery electrified oncologists at NCI, which launched a plant collection program that eventually tested thirty-five thousand botanicals for anticancer properties, the vast majority of them to no avail. Vincristine was one exception. Sutow added it to the existing actinomycin-D protocol for Wilms' tumor, and together with fine-tuning dosages and delivery schedules, he greatly improved survival rates.

One case stood out for Sutow. In 1966, a World War II veteran brought his ten-year-old son for treatment of Wilms' tumor. A former marine wounded at

Okinawa, the father loathed all Japanese. When Sutow entered the room to examine the boy, the father stiffened. Sutow, more adept than Winnie the Pooh at winning over children, ignored the father at first and slowly approached the boy, smiling and gently touching his hair. Sutow then turned to the father, bowed his head ever so slightly in an acknowledgment of respect, and assured him that the boy's outlook was promising. He explained the concept of adjuvant therapy, that even after surgery had removed the tumor, extensive chemotherapy would follow up to handle any residual, errant tumor cells. The father calmed down and stowed his racism. The boy survived, and his father forever remained an admirer of Wataru Sutow.[24]

Freireich's towering presence asserted itself immediately. Tall and stout, with a crop of dark hair and a smile permanently fixed to his face, Freireich, with both hands in the pocket of his long white lab coat, looked like a congenial emperor penguin. Back at NCI, Freireich had worked with an IBM engineer, whose son was Freireich's patient, on a device to separate blood into three components—plasma, erythrocytes (red blood cells), and leukocytes (white blood cells). A spinning centrifuge separated out the three kinds of cells according to densities. Whole blood from a donor was mixed with anticoagulant and pumped into a plastic bag and from the bag to the centrifuge. The needed cells were retrieved from the centrifuge. The remaining blood was then recombined, heated to body temperature, and returned to the donor. White cells from healthy donors could thereupon be combined and transfused into the leukemia patient, boosting damaged immune systems. Freireich's blood cell separator answered multiple questions. In acute lymphocytic leukemia, the number of healthy, oxygen-transporting red blood cells declines as malignant white blood cells crowd them out, leaving patients anemic. Anemia could be fought by collecting and combining erythrocytes from healthy people and giving them to leukemia patients. Normal white blood cells fight bacteria, viruses, and foreign matter, but they are replaced in leukemia patients by abnormal white blood cells without such properties. By infusing white blood cells from healthy people, physicians bolster the immune systems of leukemia patients. Rapidly proliferating malignant white blood cells also invade blood-forming tissues in the bone marrow, further disrupting the production of normal blood cells, and accumulate in the lymph nodes, spleen, liver, and kidneys. From sick patients, the blood cell separator removed the excess white cells. Finally, the blood cell separator reduced the time required to collect platelets, which enhanced the ability to fight hemorrhaging.[25]

Infection often killed leukemia patients before VAMP had time to wipe out the disease. Controlling infection in patients became another objective of treatment. In 1966, NCI funded the Life Island program at M. D. Anderson. The drugs and the disease could seriously compromise a patient's immune system, causing septicemia, neutropenia, and a variety of bacterial, viral, and fungal infections. The logic of "life islands" was to isolate patients from such germs. Each patient bed was shrouded in a plastic bubble, through which sterilized food, medicine, and other items were passed. Reducing infections increased the number of leukemia patients able to receive a full course of chemotherapy.

The increasing number of clinical trials at M. D. Anderson elevated the importance of informed consent and institutional oversight. Clark had approved the establishment of the Surveillance Committee. Clifton Howe, head of internal medicine, chaired the committee, which also included Gerald D. Dodd, chair of the Department of Radiology; Edgar C. White, chair of the Department of Surgery; and hematologist C. C. Shullenberger. They met for the first time on November 1, 1966, and approved seven projects, deferred one, and began to assemble the administrative machinery for ethical clinical trials.[26]

With M. D. Anderson now engaged in radiotherapy and medical oncology, Clark turned to cancer rehabilitation. Until then, interest in cancer rehabilitation had lagged. In June 1966, the New York Academy of Sciences invited Clark to develop and preside over a three-year series of symposia on cancer rehabilitation. The administrators wanted to convene in Princeton so that patients from Memorial would be available for case presentations. Clark confessed that to "date, very little has been done in cancer rehabilitation at M. D. Anderson Hospital, the main emphasis being on diagnosis and treatment." The only exception was UT Dental Branch, where Joe Drane and others worked to fit maxillofacial prostheses soon after head and neck surgery. To Robert Moreton, Clark delegated responsibility for managing the symposia. John E. Healey, a specialist at M. D. Anderson in physical medicine, assisted Moreton.

At the time, hospitals and public health agencies delayed cancer rehabilitation for two years, in some cases making it a matter of policy not to start until former patients had beat the odds for their particular disease. Why waste resources on people who would soon be dead? Once survivors had crossed the requisite statistical threshold, access improved. But delays reduced the chances for success. The sooner laryngectomy patients received speech therapy, for example, the more likely they were to recover some communication skills. Delays often led to long-term unemployment and self-imposed social isolation. Among breast cancer survivors with lymph node dissections, delays in exercise therapy could aggravate lymphedema and muscle atrophy. For amputees, early prosthetic fittings improved muscle feedback, the final shape of stumps, and postoperative coordination.

The sessions opened eyes. Clark covered rehabilitation from social, physical, and economic perspectives. Given his own skepticism about psychologists, he paid scant attention to emotional rehabilitation and ignored sexual rehabilitation for patients with urogenital cancers. The symposia—held in 1966, 1967, and 1968—put seventy-five influential oncologists under the tutelage of M. D. Anderson, giving the hospital an instant footing in cancer rehabilitation. In 1969, Clark elevated to departmental status the section on physical medicine and rehabilitation, with John Healey as chair. As bed shortages became acute, Clark needed to relieve the pressure and give rehabilitation patients a home. The Southern Pacific Railroad donated its hospital in north Houston for the new "University of Texas M.D. Anderson Hospital—The Annex and Rehabilitation Center." After extensive remodeling, the center would open in October 1972.[27]

───◁▧◖▧▷───

In the late 1960s, M. D. Anderson experienced rapid growth, with the 60,000th patient registering in 1965 and in 1970 the 70,000th. The number of employees rose from 1,385 in 1967 to more than 2,000 in 1970. Population growth, increases in the incidence of cancer, and more and more cancer survivors intensified the pressure. Between 1960 and 1970, as the petroleum and petrochemical economy boomed, the Harris County population increased from 1,364,569 to 1,903,191. Texas swelled from 9,579,677 people to 11,903,191. Along with population growth came an escalation in the incidence rate of cancer. In 1960, 125 per 100,000 people in the United States were diagnosed with cancer, a figure that reached 130 in 1970. The combined effect of more people and more cancer meant more patients. The multiplication of treatments afforded by surgery, radiotherapy, medical oncology, and rehabilitation added to the squeeze. When surgery alone constituted cancer treatment, patients underwent their operations and most left the hospital in ten to fourteen days. When radiotherapy was added to surgery, with fractionated doses administered daily for up to six weeks, a patient in a given year might be at M. D. Anderson a total of fifty days. Medical oncology exacerbated demand for beds and labs.[28]

The increase in patients, survivors, and treatments screamed for short-term as well as long-term expansion, particularly if the move toward ambulatory cancer care were to continue. Clark in 1966 leased 21,000 square feet of space from Center Pavilion, a fifteen-story hospital and residential facility two blocks from M. D. Anderson. There he installed the Domiciliary Care Unit, a complex of eight efficiency apartments for leukemia patients and family members. Clark also moved to Center Pavilion the offices of the Southwest Oncology Group, the Department of Developmental Therapeutics, the Department of Virology, and the sections of nuclear medicine, human genetics, and experimental surgery.[29]

A committee Clark had formed back in 1962 to assess future physical plant needs had concluded that the existing 350,000 square feet were inadequate. The board of visitors set about raising the money. Several federal agencies, including NCI, pledged $4.3 million, and the M. D. Anderson Foundation kicked in $1.5 million. From the estate of Mose Gimbel came $850,000, and a variety of other gifts generated nearly $1.4 million. In October 1964, the UT Board of Regents had signed a $7.06 million contract to add 230,000 square feet to M. D. Anderson. Into that space would go new programs in immunochemistry, clinical physiology, experimental pharmacology, and therapeutics; existing programs in immunology, virology, nuclear medicine, and biomathematics; and a new diagnostic clinic. The additions were dedicated on June 21, 1969. Physicians Referral Service, the financial pool that received income from M. D. Anderson physician fees, purchased the Mayfair, a fifteen-story apartment building across Holcombe. Offering 142 rooms and suites as a temporary home for some M. D. Anderson patients and as residential apartments for some physicians, it became known as the Anderson Mayfair.[30]

The rapid growth prompted administrative changes. In 1968, Robert Hickey was named deputy director of M. D. Anderson. His duties were ill defined but quickly became whatever Clark wanted. Adored by his wife and children and described by one associate as a "surgeon with the soul of an internist," Hickey soon turned into a combination of surgeon, gopher, and hit man. The loyal Hickey chafed at the chore. "He came to hate doing Dr. Clark's dirty work," one physician observed. "He was a nice man with a very difficult job." The board of regents in 1968 also changed Clark's title from director to president.[31]

━━━◉━━━

Lee Clark had strong programs in radiotherapy, medical oncology, and cancer rehabilitation, but racial segregation stood in the way of greatness until the mid-

1960s. When physicians and scientists from the Northeast, Midwest, and Pacific Coast visited M. D. Anderson, they hated the signs Colored and White adorning the entrances to bathrooms, waiting rooms, and cafeterias, and the separate wing for black patients. Lee Clark was polite, genteel, and paternalistic, but he was also a child of West Texas, where whites and blacks knew their place and trembled at the suggestion of change. Clark was not unlike the West Texas writer Larry L. King who confessed that his own racism came as "naturally as breathing. . . . Never would it have crossed my vacant mind that blacks might someday attend schools north of the tracks." Grant Taylor bragged that children had never been segregated in pediatrics, but on antebellum cotton plantations, the children of slaves and the children of whites had not been segregated either, at least not until the onset of puberty stimulated southern sexual obsessions and dictated separation. Some black nurses were confined to 6E during daylight hours or had to work the graveyard shift, with strict instructions to be out of the hospital before dawn.[32]

Ethel Fleming, the second black professional nurse hired at M. D. Anderson, had joined the staff in 1951. She quickly came to love the institution. But she suffered indignities every day. At lunchtime, she descended to the segregated basement cafeteria along with other black employees; they received their food trays through a slot and then tried to eat in air sometimes heated to 100 degrees by the adjacent hospital laundry. She changed clothes every day in a segregated locker room, went to the toilet in a "colored" bathroom, and often drew the nastiest assignments. Fleming spent one night in pediatrics on 6-West with a hemorrhaging leukemia patient. By morning, her crisp white uniform was speckled and smeared red with the child's blood. As daylight approached, Fleming was late in getting to the locker room, late in discarding her uniform, and late getting out the door. She took an elevator down to the basement, and changed to street clothes in a colored bathroom. Acutely aware of regulations requiring black personnel to be out of the hospital before dawn, Fleming had a black maid keep the uniform overnight and then exited through a door far from the parking lot.

Although black patients and employees endured the Jim Crow culture, Lee Clark insisted that all patients—rich, poor, white, black, Hispanic, out-of-state, and foreign—possessed a right to every treatment that M. D. Anderson offered. Once a patient gained access to M. D. Anderson, every technology in the arsenal of cancer medicine was available. The wealthiest white patients might have enjoyed VIP treatment in plush 2-West rooms and expeditious access to laboratory tests, x-rays, and clinics, but when the time came to go under the surgeon's scalpel or the radiotherapist's beam, the richest white patient and the poorest black or Hispanic all ended up in the same surgical suites, under the same cobalt-60 or betatron, and getting the same chemotherapy drugs at the same schedules with the same supportive care.

Desegregation came to M. D. Anderson in 1965 as a fait accompli, the result of President Lyndon Johnson's Great Society and the demands of medical economics. In 1964 and 1965, Johnson steered through Congress two pieces of legislation that doomed segregation. The Civil Rights Act of 1964 prohibited discrimination on the basis of race, religion, color, national origin, and gender and promised to cut federal funding to businesses along with state and local government agencies that segregated employees. In 1965, Congress passed legislation implementing Medicare and Medicaid, which provided federal government health insurance to the elderly and to the poor. Together, the programs promised hospitals enough cash to significantly reduce the financial burden of indigent care. In 1965, Clark received a letter from Philip R. Lee, deputy assistant secretary of the U.S. Department of Health, Education, and Welfare, warning him to desegregate or make do without federal funding. At its August 11, 1965, meeting, the M. D. Anderson Administrative Committee decided to integrate.[33]

Desegregation commenced immediately at M. D. Anderson and at hospitals and medical schools throughout the South. The Colored and White signs came down, and the cafeterias and locker rooms were gradually opened to everyone. Ed Munger noticed the timidity of black employees. "Back in those days, we had a black cafeteria and a white cafeteria. When the memo came to merge them, some of my black employees were afraid to go into the line with whites, especially white doctors. I just told them to get in line and get their food. And nothing happened."[34]

One visible M. D. Anderson program did not survive desegregation. Ever since 1945, the Order of the Eastern Star (OES), the women's auxiliary of the Masonic Order, had offered a handicrafts program for M. D. Anderson patients. The OES financed the program with an annual banquet and bazaar, and OES women became fixtures at M. D. Anderson. Some OES leaders, however, hated desegregation. As white patients were introduced to the 6 East Nursing Unit and assigned beds next to black patients, OES absenteeism increased. Marion Kelly, the OES leader, refused to budge. She stated that "until the [OES] board meets and reconsiders the situation in light of white patients being admitted there, it will be impossible to provide handicraft services to the 6 East Nursing

Unit." Clark tried to be patient, but the problems persisted, and in August 1966, Arthur Kleifgen wrote a blunt letter. "Communications had broken down and little had been done to improve communications between the OES workers and the Director of Volunteer Services. It was our conclusion that the program was being conducted for the benefit of the members of the OES rather than for the benefit of the patients." On November 1, 1966, Clark told OES leaders that the handicrafts program "is being terminated" and placed under an occupational therapist within the new section of Physical Medicine and Rehabilitation.[35]

When Lurleen Wallace, governor of Alabama and wife of former Alabama governor George Wallace, checked into M. D. Anderson with metastatic ovarian cancer in 1967, the new order in race relations was well established. Four years earlier, Lurleen's husband had fulfilled a campaign promise by barring the entrance of two black students in a symbolic attempt to prevent integration at the University of Alabama. "Segregation now, segregation tomorrow, segregation forever," he had proclaimed during his first inaugural address. His carefully orchestrated stand in the schoolhouse door was a media moment, and he became an icon for bigots. Ethel Fleming had a chance to put Wallace in his place.

She had grown up in Houston near the Baker estate and as a girl climbed through the hedges to play on the grounds. After graduating from high school, Fleming moved to New York City to attend the Harlem Hospital School of Nursing. In November 1951, she became an Anderson nurse. With Renilda Hilkemeyer as a mentor, Fleming developed into a nurse without peer—tough, fair, and perceptive. She went out of her way to touch her patients—a grasp of the hand, a squeeze of the arm, a touch on the head—even those patients with foul, fungating tumors. She often called patients' families after hours to report on their condition, and when patient relatives could not afford to stay overnight in town, she fed and boarded them for a few days. When frustrated, racist, and pain-wracked patients lashed out and called her "nigger," Fleming turned the other cheek. On one occasion, a white woman suffering intractable pain screamed "niggers" at several black nurses trying to bathe her and change the sheets. She ordered them out of the room, and they complied, but her offense enraged them. A few minutes later, when the patient began to whimper in pain and ask for assistance, they responded slowly. Fleming heard about their revenge and ordered them back. Patients always came first, even the most troublesome and the most racist. Hilkemeyer had carefully selected Fleming as the first black nursing supervisor, much like Branch Rickey of the Brooklyn Dodgers picked Jackie Robinson in 1947 to integrate major league baseball.

Rickey needed a man with athletic talent and a steady temperament. Hilkemeyer needed a black nurse who could withstand racist pressures and earn the respect, and obedience, of black and white nurses.

When she got into the hospital bed, Lurleen Wallace came under Fleming's care. In a few minutes, the governor showed up and took immediate offense, insisting that Fleming leave the room and bring in the nursing supervisor. "I am the supervisor, Governor," Fleming quietly replied. Wallace stomped out of the room and headed for the seventh floor, hoping to get satisfaction from Clark. He got none. The South was changing.[36]

On January 19, 1966, Kenneth Endicott lectured in Houston on "the role of the University of Texas M. D. Anderson Hospital and Tumor Institute in the national cancer plan," describing M. D. Anderson, Roswell Park, and Memorial Sloan-Kettering as coequals. No longer embarrassed by the burden of Jim Crow, Clark resumed his quest to bring the Tenth International Cancer Congress to M. D. Anderson. The timing was perfect. In 1964, Lyndon Johnson had appointed Clark, along with the Houston heart surgeon Michael DeBakey, to the President's Commission on Heart Disease, Cancer, and Stroke, with Clark chairing the cancer subcommittee. Emil Frei III, a founding father of the American Society of Clinical Oncology, was charged with developing the board certification exam in medical oncology. One year later, NCI asked Clark to chair its Clinical Cancer Training Grants program. Gilbert Fletcher was now president-elect of the American Society for Therapeutic Radiologists; Robert Moreton had just completed his term as president of the Radiological Society of North America; Robert Shalek would soon be elected president of the American Association of Physicists in Medicine; the Ewing Society had awarded William MacComb the Janeway Medal for his pioneering efforts in surgical oncology; Robert Hickey had just completed simultaneous terms as the vice president of the Radium Society and the James Ewing Society; and Murray Copeland had just ended his term as president of the American Cancer Society.

Copeland became Clark's ambassador to the Ninth International Cancer Congress in Tokyo, with a mission from Clark to bring the next congress to Houston. Copeland had fertilized the ground four years earlier at the Eighth International Congress in London, where many Soviet delegates concluded that he presided over cancer medicine in the United States. In 1966, before the final vote to name the site of the next Congress, Copeland worked the crowd like the floor manager at a presidential nominating convention. The American and Soviet delegates voted as a bloc. The Soviet oncologist Nikolai N. Blokhin was voted president of the Congress, Murray Copeland as vice president for North America, and Houston as the host city. "I was able," Copeland wrote Clark, "to bring pressure to bear . . . for the Tenth International Cancer Congress. Both Professor [Nikolai] Blokhin and [I. A.] Rakov were enthusiastic about having the Tenth Congress in Houston, and were of significant help when the voting began." The delegates also selected Copeland as secretary-general of the International Union against Cancer. Copeland returned to M. D. Anderson like an Olympic athlete bearing a gold medal. In four years, the world of cancer medicine would descend on M. D. Anderson.[37]

7

The Summit,
1970–1971

*If Congress would appropriate a billion dollars a year for ten years,
we could lick cancer.*

—R. LEE CLARK, 1969

On May 22, 1970, Lee Clark arose earlier than usual. His wife, Bert, got up too and baked a batch of biscuits, a delicacy the health-conscious couple rarely indulged in. That day, just a few hours short of the opening of the Tenth International Cancer Congress, both needed extra energy. For Houston, being chosen as the host city was like winning a bid for the Olympic Games. The congress would bring to the city five or six thousand visitors, filling up hotels, bars, restaurants, and taxis for a week. "Man through medicine has thrown down the gauntlet, challenging the disease cancer," Clark declared on the eve of the event. "He will conquer and achieve the victory over cancer for all time." Much of the burden for planning the conference fell on Murray Copeland, whose calm demeanor inspired confidence. "He could sleep through an earthquake," William Russell once said of Copeland. "I'm constitutionally built to withstand pressures," Copeland agreed. "I always sleep well at night, no matter what."[1]

Clifton Howe felt ill at ease. At the end of March 1970, barely six weeks before the congress, Howe as manager of local arrangements fretted over every detail, smoking more heavily than usual. The last few years had exacted a toll.

The congress would be his last major assignment at M. D. Anderson. After the arrival of Frei and Freireich, Howe and the Department of Medicine lost influence to Developmental Therapeutics. C. C. Shullenberger would soon replace him as chairman. And although Howe still considered Clark a close friend, their relationship seemed more distant, the camaraderie less spontaneous.[2]

Gilbert Fletcher trembled in nervous anticipation. During the past two months, he had grown increasingly annoyed with delays in delivery of a $525,000 linear accelerator, the first of its kind in the United States designed to treat cancer. With the capacity to generate electron and photon beams, to switch back and forth between them, and to treat deep-seated tumors, the linear accelerator represented the latest technology. Fletcher wanted its 32-million electron volts on display. Finally, on May 14, 1970, it arrived. Fletcher behaved like a new father in a maternity waiting room. In his long, white coat, he skittered about the parking lot while cranes lowered the accelerator into a giant hole near the recently completed Gimbel Building. Once the machine came to rest on wheeled dollies at the bottom of the pit, the baby was born. Fletcher gave it a couple of gentle pats.[3]

In the spring of 1970, as talk of cancer and conferences buzzed throughout the Texas Medical Center, a combination of science, politics, and popular culture expanded the nation's awareness of the disease.

A symbolic expression of it lay in a widely distributed work of fiction and drama. *Love Story*, a bubblegum Romeo and Juliet romance novel, features a young woman who dies of cancer. Published in March 1970, *Love Story* was an instant best seller and spawned an equally popular movie. On her deathbed, the heroine is beautiful and feisty, even sexy, her brown eyes as sparkly as the sheen on her long black hair. The film bears no hint of intractable pain, of weight loss and cachexia, of incontinence, catheters, and morphine drips. Viewers might have concluded that dying of cancer was not so bad.

The Tenth International Cancer Congress, a joint effort of M. D. Anderson and the National Academy of Sciences, offered panels and a total of 130 papers focusing on chemical carcinogenesis. But M. D. Anderson scientists were conspicuously absent from the program. Houston anchored the world's oil, natural gas, and petrochemical industries, and large corporations paid huge sums to purchase influence in Austin and Washington. Wherever hydrocarbons were pumped, refined, and shipped, Lee Clark had prominent oil and gas men and their wives serving on Anderson fund-raising committees. "In Texas politics," a longtime associate said, "these were the last people Dr. Clark wanted to offend[;] he knew . . . the cancer incidence maps in Texas better than anybody." But becoming known as an environmental activist would have cost him influence in Austin. His politically strategic response was to turn from the public glare of the big city and the sight of the ship channel and retreat to the more rural environs of Smithville, Texas, where he planned to build an environmental research campus.

On April 22, 1970, one month before the opening of the cancer congress, Senator Gaylord Nelson of Wisconsin, Representative Paul McCloskey of California, and environmental activist Paul Hayes sponsored a series of Earth Day rallies throughout the nation, each dedicated to reducing pollutants in the water and the air.[4]

Gaylord Nelson had also been in the thick of the controversy over artificial sweeteners. In 1969, he raised the alarm about cyclamate, an artificial sweetener shown to induce bladder cancer in rats. No proof existed that it was carcinogenic in human beings, but Nelson called on the FDA to ban the chemical. Coca Cola and Pepsi protested, as did Abbott Laboratories, the primary manufacturer of cyclamates. They denied any carcinogenic effects of the sweetener.

The FDA reacted quickly. The thalidomide disaster of the 1950s had not faded from public memory. In 1957 Grünenthal, a West German pharmaceutical company, sold thalidomide throughout Europe. Besides reducing nausea in pregnant women, the drug had a sedative effect for anxiety and insomnia. The FDA had not approved the drug for sale in the United States because Grünenthal did not satisfy safety testing requirements. In 1961, Grünenthal pulled the drug after learning that it caused stillbirths as well as phocomelia, a birth defect in which babies are born with short, stumplike arms and legs and useless hands and feet. Between 1958 and 1962, more than ten thousand babies around the world were born with phocomelia, while the FDA ban kept down to twelve the number in the United States, cementing the FDA's image as protector of American consumers. In October 1969, the FDA banned cyclamate.[5]

Soft drink manufacturers responded immediately. Long anticipating the possibility of a ban, they had developed new artificial sweeteners, most notably saccharin. Nelson urged an FDA ban on saccharin as well. The House Committee on Government Operations investigated other artificial sweeteners. On the commodity markets, sugar futures fluctuated wildly. Millions of Americans wondered whether Coke killed.[6]

The issue of pharmaceutical carcinogenesis was not limited to soft drinks. In 1969, Felix Rutledge and Taylor Wharton in the Department of Gynecology worried about an extremely rare form of vaginal adenocarcinoma appearing in adolescent girls. At Massachusetts General Hospital, Arthur Herbst and Robert Scully identified diethylstilbestrol as the culprit. After World War II, gynecologists had prescribed the drug to assist women with a history of miscarriages. It helped carry babies to full term but was gestationally carcinogenic, leaving the infant girls with a susceptibility to cancer of the vagina during their teens and early adulthood.[7]

The cancer congress was to be in the Albert Thomas Convention and Exhibits Center. Last-minute snafus drove Clifton Howe to drink. On Friday afternoon, he panicked upon learning that extra telephone lines had not yet been installed at the convention center, nor was there a separate press room for journalists. The tenth congress, having registered more than six thousand physicians and scientists and credentialing eighty-five journalists, would soon be inundated by thousands of busy people unable to connect with the outside world. Howe also learned that one of the center's air-conditioning units was kicking on and off without regard to the thermostats. He called Clark. Clark barked four words: "Take *care* of it."

While Bert whipped up the biscuits, Clark drove in the dark to the convention center. Every light seemed to be on. Just outside the entrance, *The Crab*, a large, metal sculpture by Alexander Calder, adorned the sidewalk. Selected by Clark and Copeland as the symbol for the tenth congress, it had been relocated from the Houston Museum of Fine Arts. After more than two thousand years, however, the crab was losing its impact as a cancer icon. For Houston psychiatrist Harry Ricketts, the image of "millions of little crabs circulating throughout my body" simply did not give the right effect. But the sculpture did not bother Ricketts, who could not see a crab in Calder's design anyway. "Most people wouldn't have known what it was unless you put a sign in front of it saying, 'This is a giant crab.'" As a faculty member from Purdue noted, "Comparing cancer to a ten-foot-tall crab reminds me of the monster bug, B-movies of the 1950s. Can you imagine M. D. Anderson today putting a giant crab in its lobby and telling patients, 'This is what's inside you.'" But to Lee

Clark on May 22, 1970, the Calder crab was beautiful. Everything seemed to be in order. He drove home to the biscuits.[8]

———⬛〰⬛———

International events beginning a few weeks before the opening of the congress threatened its success. The U.S. invasion of Cambodia intensified the anti–Vietnam War movement and increased hostility toward the Nixon administration. Had Vice President Spiro Agnew not been scheduled to offer the keynote speech at the congress, the invasion would probably have caused minimal commotion in Houston, but since Nixon's January 1969 inauguration, Agnew had become a lightning rod for criticism. His presence in Houston might make the conference a target of the antiwar movement; a few scientists threatened to boycott the congress unless Clark removed Agnew from the program. Renato Dulbecco of the Salk Institute, for example, was scheduled to cochair a panel discussion and deliver a paper. He threatened to lead scientists in a boycott unless Clark dropped Agnew. Clark did not even bother to respond to Dulbecco, and the boycott fizzled.[9]

Clark worried more about the invasion's impact outside the United States. The Soviet Union had vehemently protested the invasion, and Clark fretted that the hardline Soviet leader Leonid Brezhnev might make it more difficult for Soviet scientists to travel to the United States. A total of 142 papers were scheduled to be delivered by scientists from Soviet bloc countries, and if Moscow pulled their travel visas, the presenters would be trapped behind the Iron Curtain.

More than 1,400 presentations were scheduled. The Union Internationale Contre Cancer had always considered itself immune to cold war politics. East and West shared an equal stake in curing cancer. Even more serious than the potential Soviet boycott was the possibility that Nikolai N. Blokhin, the prominent Soviet oncologist, would not be able to attend. At the Eighth International Cancer Congress in Moscow, he had helped Murray Copeland secure M. D. Anderson's bid. Blokhin wanted to be there. "This type of international congress is really important for me and other scientists," he said. "Face to face discussions [provide] a much better forum for understanding what is new, different and promising than the impersonal publications in journals."

Security for Agnew's visit was tight. The 1963 assassination of President John F. Kennedy in Dallas lingered in the minds of law enforcement officials. On the day of Agnew's speech, snipers, undercover officers, and squad cars were placed along the route from Ellington Air Force Base to downtown. Two hours before Agnew's arrival, feisty demonstrators gathered outside the Sam Houston Coliseum, where the opening speeches were to be held. Agnew's limousine drove straight into the coliseum garage and he was escorted to the platform. Of the ten thousand people watching Agnew ascend the platform, most erupted in cheers. The University of Texas Longhorn Marching Band, dressed in burnt orange uniforms, paraded into the coliseum playing "The Eyes of Texas" full blast.

Clark welcomed the crowd to the opening session and gratefully acknowledged the presence of Nikolai Blokhin, awarding him a size seven cowboy hat that perched precariously on the Russian's size eight head. "Blokhin looked like one of the bobbing-head dashboard ornaments popular with teenage drivers," as one registrant remembered. Blokhin's smile, however, was genuine. Clark greeted a few more dignitaries, including GOP congressman George Bush, who had lost his daughter Robin to leukemia in 1953. Bush introduced the vice president. A few minutes into Agnew's keynote speech, shouts of "Peace Now" cascaded down from the bleachers. For many delegates, such political antics were unprecedented. "Politics," commented one German scientist, "is a problem in all our countries, but I suspect a lot of delegates really don't know what the fuss is all about." Police hustled the protestors away. Agnew's speech consisted of several minutes of platitudes; he then returned to the limousine. After Agnew's remarks, the pianist James Dick performed Tchaikovsky's Piano Concerto No. 1 in B-flat minor with the Houston Symphony orchestra. The orchestra concluded with the overture to Reznicek's *Dona Diana*.[10]

In his opening address, Clark noted the virtual revolution taking place in biology and criticized radical surgery and the dominant role of general surgeons. "Heretofore," he said, "the attitude has been that a Board-qualified general surgeon had knowledge of all problems and, therefore, of cancer problems also. This assumption has proven to be one of the greatest drawbacks to the betterment of the care of the cancer patient. Too often the opportunity to cure has been lost through inept surgical therapy or by aggressive operations applied with no real understanding of the specific and unique factors encountered with cancer." One Houston physician remembered sitting in the audience with several surgeon friends. "When Clark said that they turned silent and sullen. You could've heard a mouse squeak." Clark continued, warning that too many diagnostic radiologists performed radiotherapy "without the special knowledge required in the application of radiation energy." Too few pathologists had a "primary interest in cytology . . . backed up by ultrafine structure interpretation." And too few internists had specialized in oncology. "In the past ten years the role of the cancer chemotherapist and hematologist [has] evolved as a medical specialty." Finally, Clark pitched multidisciplinary teamwork: "There is no longer a place in medicine for the single physician except in cancer detection. The oncologic team is the only answer to better treatment."[11]

—◦◦◦◦◦—

As the general surgeons in the audience grumbled, Clark and Copeland left for a weekend of activities. For most of the foreign delegates, Texas had an exotic appeal, the stuff of John Wayne movies. They expected a taste of Texas, and Clark gave them a mouthful. At a ranch south of Houston, he treated 4,685 registrants to a Texas barbecue, complete with pork and beef ribs, chicken, brisket, sausage, beans, potato salad, and coleslaw. The guests attended a rodeo that featured bucking broncos, bull riding, steer wrestling, calf roping, quarter horse racing, and country western musical performances. To demonstrate that Texas did everything in a big way, Clark scheduled tours of the Astrodome, to its admirers the Eighth Wonder of the World, an indoor, air-conditioned stadium with forty-two thousand seats. To dispel stereotypes about Texas, Clark treated the delegates to performances of the Houston Symphony, the Houston Opera, and the Alley Theater.

On Monday morning, the first day for panel discussions and the presentation of scientific papers, Clark and Copeland returned to the Albert Thomas Convention Center. On the streets of downtown Houston, the delegates stood out like tourists of an unusual breed, with an uncommonly high percentage of bow ties and a penchant for talking. In place of sports, politics, or the economy, they wandered in the more esoteric worlds of radiation dosimetry, cell kinetics, and viral loads. They clutched their programs like bettors at the horse races, scheduling their days around windy presentations in hope of picking up some clues to cancer. Murray Copeland looked as though he were at his high school prom, greeting everyone by first name.

By far the most popular exhibit was NASA's Apollo spacecraft. In July 1969, probably the vast majority of the registrants had huddled around television sets watching Neil Armstrong plant the Stars and Stripes on the moon's surface. On April 11, 1970, NASA had launched Apollo 13, its third lunar landing mission. The drama that followed, when the failed craft managed to return to earth, neither crashing, burning up from the friction with the atmosphere, nor shooting past the planet carrying its occupants hopelessly into space, gained NASA further fame. Now the delegates could stand in the presence of history. They peered through the windows, examined the location of the heat shield, and marveled at this testament to humanity's scientific ability. Tour buses shuttled delegates back and forth on the sixty-mile round-trip from the convention center to NASA's Mission Control Center. The Apollo spacecraft and Mission Control symbolized the full potential of science and technology. Might it not, then, bespeak the possibility of a triumph over cancer?[12]

⚬⚬⚬

If not quite omnipotent, Lee Clark seemed omnipresent at the congress. He enjoyed attention and turned up his magnetic personal charm. Blessed with an uncanny talent for speaking comfortably about a wide range of topics with a variety of people, Clark could effortlessly intuit the feelings of others. He could break off a conversation without offending, his gentle smile revealing that he had to be somewhere else soon, squeezing a hand or gracefully touching an arm and departing with an "it's good to see you" that left guests satisfied. One foreign journalist found, "your Dr. Clark . . . a most amazing fellow . . . everywhere at once, yet so thoughtful and so knowledgeable."[13]

Clark paid special attention to the presentations. Collectively, they represented the state of oncology in 1970. The sessions marked the point at which medical oncology began its rise to parity with surgery and radiation oncology, and general surgery surrendered ground to surgical oncology. As proposals to deliver papers arrived, Clark had tracked them carefully. When the final count was tallied, M. D. Anderson offered seventy-three presentations, more than Memorial Sloan-Kettering, Roswell Park, Ellis Fischel, Harvard, and NCI combined. The program reinforced Clark's announcement that the imperial reign of general surgeons over oncologists was ending. Only a handful of papers and panel discussions touted radical surgery.

Clark did prominently display one surgeon. John S. Stehlin reported on what Clark considered to be the most important clinical research development to come out of the Department of Surgery in the 1950s and 1960s. In developing his perfusion procedures—surgically isolating the blood supply to a tumor and then pumping drugs to the tumor site, not to the whole body—Stehlin allowed for dramatic increases in doses that upped the odds of achieving "tumoricidal" doses and reducing the frequency of amputations. Stehlin had worn the cloak of multidisciplinary care in his collaboration with medical oncology.[14]

The program also reflected NCI's new focus on medical oncology and molecular biology. In 1968, NCI had established its Human Tumor Cell Biology Branch, and ninety-nine papers explored the importance of tumor biology and demonstrated how research at the molecular level had blurred the once firm boundaries separating biology, pathology, genetics, and biochemistry. Sixty-eight papers involved the new field of cancer immunology. More than two hundred papers reported on experimental and clinical chemotherapy protocols. Only ten papers handled genetic issues, one by David Anderson, the young geneticist whose research on "cancer eye" had so endeared him to Clark. Anderson had since extended his research into human melanomas.[15]

Clark also featured mammography, certainly M. D. Anderson's greatest contribution to date in diagnostic radiology. Robert Egan had already left for Emory, but Gerald Dodd and others continued the research. Although mammography had already changed the medical landscape in diagnosing breast cancer, the technology still displayed nagging shortcomings, none more serious than its propensity for false positives and unnecessary biopsies. Refining mammography, through better physics and more sophisticated x-rays, continued to be an important goal. Dodd and physicist Al Zermeno described how new technologies offered better spatial resolution and led to fewer false positives.[16]

Parading through the convention of leaders in medical oncology, Clark had Emil Frei, Emil Freireich, and Gerald Bodey describe the Life Islands and laminair air flow room and the prophylactic use of antibiotics to protect patients undergoing chemotherapy. Frei reported promising results in the treatment of melanoma with multidrug therapy, and in a panel discussion he mused on the importance of incorporating research on cell kinetics into the development of chemotherapy drugs. With Gerald Bodey, Freireich expounded on the superiority of COAP (Cytoxan [cyclophosphamide], Oncovin [vincristine], Ara-C [cytarabine], and prednisone) over simple Ara-C in treating acute myelocytic leukemia. With Evan Hersh, Freireich described how the blood-cell separator transferred peripheral blood leukocytes. Hersh reported on immunosuppression. Edmund Gehan spoke to the critical role of biostatisticians in designing research protocols and interpreting results.[17]

By 1970, Emil Frei, Emil Freireich, and Gehan had planted M. D. Anderson firmly in medical oncology, but radiotherapy and medical physics had given the institution its first truly national reputation.[18]

Gilbert Fletcher held court at the congress. He glided from panel to panel and paper to paper, relentlessly driving home his conviction that radiotherapy had bankrupted radical surgery, except in isolated instances, and that radiation alone was curative for some cancers. Felix Rutledge agreed in a panel discussion that radical and extended radical surgery for gynecological tumors ought to be superseded by conservative surgery combined with radiotherapy. M. D. Anderson surgeon Richard Jesse and radiotherapist Robert Lindberg, along with Joe Drane of the UT Dental Branch, discussed progress in head and neck tumors, particularly how radiotherapy lessened the number of radical operations. At a panel discussion on bone and soft tissue sarcomas, Richard Martin explained how multidisciplinary care was reducing the need for limb amputations. Eleanor Montague criticized the ongoing use of the radical mastectomy.

Physicians approached Fletcher in the hallways. Younger oncologists recognized him by his name tag and maneuvered for an introduction, waiting for him to sit down for coffee or get into an elevator or head for a bathroom. At receptions, they patrolled the room, drink in hand, eyes fixed on Fletcher, waiting for a chance to approach him. Some just ogled and whispered his name. At the tenth congress, he was M. D. Anderson's certifiable celebrity.

Although Developmental Therapeutics garnered much of the limelight, medical oncologists labored in the Department Medicine as well, and several presented papers. Among them was Raymond Alexanian. A thin wisp of a man with a résumé bearing a heavy Ivy League imprint—an undergraduate degree from Dartmouth and an M.D. from Harvard—Alexanian was no stranger to suffering. With one leg shorter than another, he had hobbled through life with a five-inch heel. In 1964, after teaching for a year at the University of Washington, Alexanian came to M. D. Anderson. While still a fellow at Christie Hospital and the Holt Institute, he had published an article on stem cell kinetics in the *International Journal of Radiation Biology*. More articles followed, including a landmark paper on treating multiple myeloma with melphalan. In 1962, oncologist Daniel Bergsagel of M. D. Anderson had first reported on the efficacy of melphalan in treating multiple myeloma. Alexanian scored in 1969 with another article on melphalan, this time in JAMA (the *Journal of the American Medical Association*), which elaborated on attacking multiple myeloma with a combination of melphalan and prednisone.

At the tenth congress, Alexanian presented a paper on the treatment of plasma cell disorders, a cluster of related cancers rooted in antibody-producing white blood cells, including multiple myeloma and Waldenström's macroglobulinemia. Alexanian specialized in multiple myeloma, a disease of B lymphocytes, in which the cancer seemed to be monoclonal, or descended from a single cell.

At the time, multiple myeloma brought an average survival time of less than two years and could be as nasty as childhood leukemia. Many patients experienced severe bone pain, kidney failure, hypercalcemia, osteoporosis, suppression of the immune system, and bone resorption, which eroded bone density and increased the likelihood of multiple fractures. Until the mid-1960s, physicians had no treatment for the disease.

Melvin Samuels discussed the first real success in the treatment of embryonal testicular cancer through the use of vinblastine alone and vinblastine in combination with the L-phenylalanine and reported dramatic responses, one patient enjoying a remission that lasted for nearly six years, astonishing for a disease that typically killed young men in a matter of months. It was also clear, however, that testicular cancers vary greatly according to tissue type. Tumors identified as teratomas, teratocarcinomas, and embryonal carcinomas were the most responsive; seminomas were more resistant.[19]

With sixty-eight papers, viral carcinogenesis attracted journalists like June bugs to light, especially after Clark and Blokhin prompted a feeding frenzy by mentioning the possibility that the viral origins of some human cancers would soon be uncovered. Clark had planned to give prominence at the congress to M. D. Anderson's Department of Virology.

M. D. Anderson's brightest star in the search for the viral origins of human cancers was Leon Dmochowski, who chaired the Department of Virology. At the congress, Clark pushed forward what he considered to be the seminal work of Dmochowski, who delivered eight papers, more than any other participant.

Late in the afternoon on the final day of the congress, after 1,342 scientific papers and eighty-five panel discussions, a handful of scientists remained. As taxis carried registrants to the airport and janitors started folding up tables and stacking chairs, the last session featured a presentation on the replication and persistence of RNA oncogenic viruses. Howard Temin, a young virologist at the McArdle Laboratory in Wisconsin, discussed his discovery of reverse transcriptase, an enzyme employed by all retroviruses to transcribe genetic information from the virus for RNA to DNA, which can then integrate into the host genome. A fundamental tenet of molecular biology held that genetic information always moved from DNA to RNA. Temin demonstrated that because of reverse transcriptase, genetic information could pass from RNA to DNA, disorient the mechanisms of cellular reproduction, and generate more cancer cells. The claim stunned every virologist still in the room, heralded the death of classical virology, and put Temin on the fast track for Stockholm and the Nobel Prize. Leon Dmochowski's entire career of chasing viral markers would soon be eclipsed by genetic manipulation. Reuters picked up the story and the news spread. Within days, Clark was talking about changing the name of the Department of Virology to the Department of Molecular Carcinogenesis. Dmochowski, the star of the Tenth International Cancer Congress, was about to learn how fickle science can be. At the beginning of the week, journalists had sought him out. At the end of the week, his star had been eclipsed.[20]

—————

Prematurely bald, with a well-trimmed goatee hugging his fleshy face, Jeffrey Gottlieb came to M. D. Anderson in 1970. Born in New York City in 1940, Gottlieb was the kind of child parents like to put on display, dazzling friends with an early fund of knowledge. He attended Amherst and Harvard Medical School. In 1967 and 1968, while in the midst of a residency at Boston Children's Hospital, Gottlieb met Sidney Farber, who recounted for him the history of chemotherapy in treating childhood leukemia and his confidence that other cancers would soon succumb; Nobel Prizes awaited the clinical scientists involved. Children's Hospital was a perfect apprenticeship, but oncologists anxious to acquire the real scientific union card needed to be at the National Cancer Chemotherapy Service (NCCS), the site where Hodgkin's disease, choriocarcinomas, and childhood leukemia began their surrender to science. For two years, beginning in 1968, Gottlieb trained at the NCCS and exhibited precocious talent.[21]

During his residency, however, Gottlieb had to trade the long white coat of a physician for the cotton gown of a cancer patient. A hard lump appeared in one of his testicles. At the time, physicians feared testicular cancers almost as much as childhood leukemia. It killed adolescent boys and young men, sparing only those whose disease had not cropped up elsewhere in the body. Once pathologists determined that Gottlieb's lump was malignant, surgeons cut out the testicle. Testicular cancers have a penchant for metastasis. They seem to lurk in the bodies of boys until a full rush of testosterone at puberty prompts rapid proliferation. More than most patients, of course, Gottlieb understood his predicament.

In 1970, Emil Frei and Emil Freireich accepted Gottlieb as a senior fellow and assigned him a number of investigational drugs. One year earlier, a Japanese pharmaceutical company had begun production of bleomycin, an antibiotic with anticancer properties. NCI started to investigate it just as Gottlieb arrived. The drug affected Hodgkin's disease, squamous cell carcinoma, and testicular cancers. Soon after its discovery, Italian scientists synthesized doxorubicin (Adriamycin), which Gottlieb also tested. It exhibited positive effect against a variety of cancers, although one side effect—cardiac toxicity—had to be monitored carefully.

On a less personal level, Melvin Samuels shared Gottlieb's interest in testicular cancer. Samuels, an oncologist in the Department of Medicine, had in 1960 settled into M. D. Anderson as section chief for general medicine, "a boring place," he remembered. "We did all the dirty work—medical histories, presurgical physicals, and treatment of the nonmalignant diseases of many patients. All the excitement was over in Developmental Therapeutics." Anxious to get in on the action, Samuels took a one-year leave of absence and worked with Frei and Freireich. He then returned to Medicine as head of genitourinary oncology. In 1969, Samuels began to treat testicular cancer patients with the drug Velban (vinblastine) and noticed positive responses and short-term remissions. "All of those young men," he sighed. "It broke your heart to see them die so quickly." To achieve longer remissions, Samuels upped the doses, pushing patients to the very limits of toxicity and acquiring the sobriquet "Megadose Sam." He had little respect for oncologists who "used itsy-bitsy doses. They may be popular for a while with parents and nurses, but their patients die." Administering huge doses of Velban, Samuels cured up to 25 percent of his patients with early stage disease. "Tumors learn quickly; you must get it all," he concluded, which required combination chemotherapy. When Samuels went to combination chemotherapy, adding bleomycin to Velban, his cure rates for early stage disease jumped to 75 percent. In just a few years, embryonal testicular cancer had gone from a diagnosis of doom to one with a potentially good outcome. Other varieties of the disease, however, defied the regimen.[22]

Chemotherapy was entering a splendid age, and many oncologists actually began to tender the notion that cancer might soon be consigned to the past. Chemotherapy treatments of choriocarcinomas, childhood leukemia, and now testicular cancer—each a death sentence just a few years earlier—burnished the hopes of scientists and physicians who, like R. Lee Clark, intended to make the conquest of cancer an official national priority.

At the Tenth International Cancer Congress, Clark basked in success. Nothing his grandfather and father had accomplished could compare. He had danced their dance and won the prize. Many of the world's most brilliant scientists had gathered around him, broadcasting his praises and M. D. Anderson's and fixing on both the attention they deserved.

——◁◁◁◁◁ ◁◁◁——

Clark hoped to follow one triumph with another. Among his politically powerful friends was Senator Ralph Yarborough, a liberal Democrat with a conservative Texas constituency. Yarborough backed President Lyndon B. Johnson's civil rights, health, education, and antipoverty programs. First elected to the Senate in 1957 to fill the seat vacated by the resignation of Price Daniel, Yarborough voted for the Civil Rights Act that year and still eked out reelection in 1958. Even his subsequent call for an end to the poll tax and his backing of the Civil Rights Act of 1960 failed to turn voters against him. In 1964, when Johnson won in a landslide, Yarborough let the president's coattails pull him back to the Senate.[23]

In the Senate, he sponsored bill after bill for the president and dutifully complied with Johnson's every suggestion. A vocal proponent of national health insurance, Yarborough worked tirelessly for Medicare and Medicaid in 1965 and earned the enmity of the American Medical Association. He also surfaced on the radar scope of Mary Lasker, head of the American Cancer Society, who saw in him a politician to advance her agenda in congress. She had turned the ACS into a cash machine. Her husband's death from stomach cancer in 1952 then transformed passion into obsession. Lasker was also on a first-name basis with leading Democratic women including Eleanor Roosevelt, Bess Truman, Jacqueline Kennedy, and Lady Bird Johnson. In 1964, she urged Johnson to establish the President's Commission on Heart Disease, Cancer, and Stroke to recommend appropriate federal action. Johnson agreed with "the nice little lady who had become friends with Lady Bird," and appointed seven people to the commission, including R. Lee Clark, heart surgeon Michael DeBakey of the Baylor College of Medicine, and oncologist Sidney Farber. She also suggested to Johnson that curing cancer might become an enduring legacy. In 1965, Congress amended the Public Health Service Act, providing $50 million in 1966, $90 million in 1967, and $200 million in 1968 to establish regional research medical programs throughout the nation for research into heart disease, cancer, and stroke.[24]

In mid-1968, Lasker read Solomon Garb's *Cure for Cancer: A National Goal.* The director of science for the American Medical Center in Denver, Garb promoted a "national commitment to make the cure . . . of cancer a national goal." He attributed NASA's success to the billions invested in its mission and to the privilege the agency enjoyed of reporting directly to the president, without having to fiddle with multiple levels of federal bureaucracy. Completely taken with Garb's logic, Lasker distributed copies of *Cure for Cancer* to friends and political associates. She invited Garb to dine with her in New York. Garb convinced her that curing cancer was only a matter of money and political resolve. On July 20, 1969, she sat glued to her television set in New York City as Neil Armstrong walked on the moon. With billions of dollars in funding, a specific scientific objective, and a presidential mandate, NASA had demonstrated, as had the Manhattan Project, the federal government's ability to successfully stage a partnership with scientists. Lasker saw no reason why the United States could not do for cancer what it had done for atomic bombs and space exploration. Discovering a cure was a seductive illusion, but the scientifically naive Lasker formed the Citizens Committee for the Conquest of Cancer. On December 9, 1969, she took out a full page ad in the *New York Times*: "This year, Mr. President, you have it in your power to begin to end this curse. . . . We beg you to remember those 318,000 Americans who died of cancer in 1968. . . . America can do this. . . . There is no doubt in the minds of our top cancer researchers that the final answer to cancer can be found. . . . Why don't we try to conquer cancer by America's 200th birthday?"[25]

Curing cancer in seven years was as unlikely as the resurrection of Lasker's husband, but the Laskers had made their fortune in advertising, and better than most Mary Lasker understood that in politics, image and slogans can trump substance. Lasker left millions of Americans with the impression that a cure for cancer was imminent. Although Lee Clark knew better than anybody that the notion of curing cancer by 1976 was ludicrous, he nevertheless fawned over Lasker, telling the press that "if Congress would appropriate a billion dollars a year for ten years we could lick cancer." Sidney Farber agreed: "We are so close to a cure for cancer. We lack only the will and the kind of money and comprehensive planning that went into putting a man on the moon. Why don't we try to conquer cancer by America's 200th birthday?" Both men in their well-intentioned deception set a dangerous precedent later to befall many cancer crusaders—overselling the cause and trafficking in the rhetoric of imminent cure.[26]

Informed scientists were not fooled. The *Wall Street Journal* excoriated the delusion. "At the time the moon effort was begun, there was no question that it could be done. The technology was available, and no new scientific discoveries were required. . . . By contrast, little is known about cancer. . . . Although to a trained pathologist, the appearance and behavior of a cancer cell differ from those of a normal cell, the exact differences . . . are largely a mystery. Without . . . basic knowledge, researchers are more or less stabbing in the dark in developing new methods of treatment. Such stabs are sometimes successful, often not, and it is impossible to be certain in advance which effort will work and which won't."[27]

Cancer, however, was a cause whose time had come. Yarborough, moreover, needed an issue. His standing in the party depended largely on the quality of his relationship with the president, who still had great influence on Texas politics. As long as Johnson considered Yarborough an ally, conservative Democrats kept quiet. In 1967, however, Yarborough publicly questioned the wisdom of the war in Vietnam. After the Tet Offensive in February 1968 and the president's subsequent decision not to seek reelection, Yarborough endorsed the candidacy of Senator Robert F. Kennedy of New York. Johnson seethed in rage. So did former governor John Connally and other conservative Democrats, including the former Democratic congressman Lloyd Bentsen. The politically harried Yarborough hoped to ride a cancer wave to victory.

On March 25, 1970, with the primary just six weeks away, Yarborough sponsored Senate Resolution 376 calling for a panel of consultants on the conquest of cancer. It moved quickly through committee and won unanimous approval. Congressman John J. Rooney, Democrat from New York, sponsored House Concurrent Resolution 675, which proclaimed that it "is the sense of the Congress that the conquest of cancer is a national crusade to be accomplished as an appropriate commemoration of the two hundredth anniversary of the independence of our country. . . . Congress [should] appropriate the funds necessary for a massive program of cancer research . . . so that the citizens of this land and of all other lands may be delivered from the greatest scourge in history." The panel consisted of twenty-six people—scientists and nonscientists—including oncologists Joseph Burchenal and Sidney Farber; I. W. Abel, head of the United Steelworkers of America; Laurance Rockefeller, the venture capitalist and patron of Memorial Sloan-Kettering; Benno Schmidt, a graduate of the University of Texas Law School, close friend of John Connally, partner in the New York investment firm J. H. Whitney and Company, and chairman of the board of Memorial Sloan-Kettering; and R. Lee Clark. Schmidt, Rockefeller, Burchenal, Farber, and Clark were all among Lasker's confidants.

The measure did not bring Yarborough what he needed. Few Texans sympathized with the antiwar movement. Lloyd Bentsen challenged Yarborough in the primary and concocted a smear campaign that linked him to radical students, black power advocates, and antiwar demonstrators. On May 2, 1970, patriotism derailed Yarborough. Bentsen won easily.[28]

Clark had rarely brooded about election outcomes. He spent his political capital promoting M. D. Anderson and carefully avoided taking political sides, but Yarborough was a great loss. Together with Benno Schmidt, Clark speeded the work of the panel of consultants. Yarborough would leave office in January 1971, and Clark intended to have the panel's recommendations finished before then. Like the surgeon general's committee on smoking in 1964, they met in secret, hoping to avoid media scrutiny.

For the scientific community, the secret meetings smacked of conspiracy. "The future of all biomedical research . . . is perhaps being decided," editorialized *Drug Researcher Reporter*, "or at least prejudiced, by a congressional commission that is operating in a framework of 'closed-door meetings' and 'secret hearings' . . . The[ir] 'instant reporting' apparatus bypasses the traditional method of publishing findings in established scientific journals where they will be subjected to critical review, not only by the editorial board but by scientific peers."[29]

Basic scientists tried to block the maneuver. Seymour Cohen of the University of Pennsylvania, at loggerheads with Clark and Lasker, convened several meetings of NIH and medical school scientists. In August, Cohen caustically declared that "the best researchers are in universities—not in cancer institutes. . . . Some cancer institutes have good people . . . but they are not the really first rate people. We must mobilize the first rate talent in this country toward the cancer effort." Harry Rubin of UC Berkeley expressed doubt about curing cancer in short order. "The question itself," he insisted, "implies a Faustian attitude about 'conquering' cancer, about which I entertain some skepticism. . . . I question the appropriateness of the contract, Manhattan project–type approach. Its effectiveness depends on a confident understanding of basic mechanisms, which we clearly do not have." Fundamental research, explained the biochemist Heinz Fraenkel-Conrat, "is the key to all and any long-range scientific problem, and . . . it would be shortsighted to support only that research which directly and immediately pertains to the cancer problem. . . . I should strongly favor . . . increased support to basic biological researches relevant to cancer." Robert Sinsheimer of Cornell echoed him.[30]

When the panel of consultants officially reported on December 4, 1970, the fingerprints of Lee Clark and Mary Lasker could be seen on every page. The body called on Congress to replace NCI with a new National Cancer Authority (NCA) independent of NIH and reporting directly to the president. Clark had always credited M. D. Anderson's spectacular growth to his original success in securing freedom from the dean of the medical school. "Having its own budget and the right to defend it as one of the free standing units of the University of Texas System," he insisted, "has been [M. D. Anderson's] greatest prerogative and perhaps the basic reason for its singular growth." A similar independence would liberate the NCA from the choking bureaucratic entanglements of NIH, the surgeon general, and the Department of Health, Education, and Welfare. Clark also envisioned a network of cancer centers throughout the nation—like M. D. Anderson, Roswell Park, and Memorial Sloan-Kettering—each free of medical schools and connected to the National Cancer Authority.[31]

The consultants recommended more money for research and took sides in the intense debate about the relative importance of basic against targeted, categorical research such as went on at M. D. Anderson. The culture of NIH emphasized the fundamental research popular in medical schools, and NIH often cited the discovery in 1928 of penicillin by Alexander Fleming to illustrate how basic research could produce new ways to treat disease. The advocates of targeted research retorted that since 1937, NCI had gone through more than $1 billion with little to show for it.

Kenneth Endicott, as chief of NCI, cited the successes of the institute's chemotherapy program. As head of a categorical institution, Lee Clark defended M. D. Anderson's narrow mission. A. Clark Griffin had revealed much about carcinogenesis and Leon Dmochowski about viruses, but Christmas cards from grateful survivors did not arrive at their offices. They cured nobody. Gilbert Fletcher in radiotherapy, Emil Frei and Emil Freireich in Developmental Therapeutics, Richard Martin in surgery, Wataru Sutow in pediatrics, and Melvin Samuels in Medicine cured patients through clinical research. Clark hoped to accelerate the translation of basic research into clinical settings. He called for ongoing contact between "the investigator, as the originator of new information, and the physician, as the converter and purveyor of that information." At what Clark increasingly called "comprehensive cancer centers," basic researchers and clinical investigators needed to be on the same team.

At the crux of the controversy, each methodology dictated a particular form of designing and managing research. At NIH, decisions about the distribution of resources depended on peer review. Experts rendered judgment about the merits of a proposal, and NIH followed their recommendations religiously, supplying research money in the form of grants in which the investigators enjoyed considerable intellectual freedom in generating data and reaching conclusions. Targeted research relied on the contract rather than the grant. During the 1950s, Kenneth Endicott had pioneered the use of research contracts, in which his scientists planned the proposal and closely monitored its execution, leaving the investigator little freedom to adapt the program. Although Lee Clark for years had acknowledged the importance of basic research in biochemistry, genetics, and molecular biology, M. D. Anderson thrived on targeted clinical research.

The panel of consultants, much to Clark's liking, called for the establishment of sixteen new clinical cancer centers, the creation of an international cancer data base, and the establishment of a National Cancer Advisory Board (NCAB) to monitor federal cancer control programs. Lee Clark, Mary Lasker, and Benno Schmidt, like most Americans, were in a hurry. Yarborough put it best: "We can no longer afford half-hearted efforts in the field of cancer research. . . . We . . . need to get busy with the job of saving American lives. A board of advisors that included not only scientists relishing the abstract elegance of laboratory research but citizens wanting the quick results the public craved would be sure to endorse that goal."[32]

When Ralph Yarborough left the Senate, Senator Edward Kennedy took the lead. Since 1962, Kennedy had promoted various federal initiatives to cure heart disease, cancer, and stroke, and the gathering political momentum for a government assault on cancer offered him a stage from which he could make a run for the White House. President Richard Nixon kept a close eye on Kennedy. In anticipation of the 1972 election, when he would come up for a second term, Nixon worried that Kennedy might make political capital out of cancer. So he highjacked the issue. His State of the Union message on January 22, 1971, called for a war on cancer, telling the nation, "The time has come when the same kind of concentrated effort that split the atom and took man to the moon should be turned toward conquering this dread disease" and suggesting an initial appropriation of $100 million. The speech precipitated a flurry of partisan posturing. Three days later, Senator Kennedy and Republican senator Jacob Javits of New York introduced S 34, Conquest of Cancer Act, to the Senate. S 34 adhered to the elements outlined in the report from the panel of consultants—an independent National Cancer Authority, an emphasis on both clinical and basic research, and research funding based on contracts as well as grants.

On March 9 and 10, the Senate Subcommittee on Health opened hearings and learned that the business of curing cancer would not be akin to endorsing motherhood. S 34 attracted intense opposition from scientists. Among academicians, giving the new National Cancer Authority independence from the NIH and a direct pipeline to the president aroused the most hostility. The Association of Professors of Medicine, the American Medical Association, and the American College of Surgeons condemned the notion, as did the Federation of American Societies for Experimental Biology and the Association of American Medical Colleges, which represented the interests of medical school faculty. They lacerated the proposal for an independent National Cancer Authority that would emphasize clinical over basic research and contracts over grants. In their view, it would make the ferreting out of fundamental cancer secrets less likely. James Shannon, former director of the NIH, took aim at Mary Lasker, testifying that S 34 represented the work of "uncritical zealots, experts in advertising and public relations and rapacious 'empire builders.'" Officials from the American Cancer Society, the American Association for Cancer Research, and members of the panel of consultants opposed Shannon. "It is important," testified Lee Clark, "to get this program out from under the six tiers of bureaucracy . . . and have an Administrator responsible for cancer who is not subordinate to those responsible for eleven other health institutes."

To sharpen his own profile, Nixon opposed S 34. A newspaper column by Ann Landers on April 20, 1971, demonstrated the depth of public support. "If the United States can place a man on the moon," she wrote, "surely we can find the money and technology to cure cancer." The column prompted an avalanche of letters to Congress. On May 11, Nixon announced his Cancer Cure Program and Republican Senator Peter Dominick of Colorado introduced S 1828, an alternative to S 34 that called for $100 million in money for cancer research but left the administration of the program with NCI, which would remain within NIH. The Senate Subcommittee on Health convened hearings on June 10. Several weeks later, the Senate passed S 1828, 85 to 0.

The debate engaged basic scientists, clinical scientists, physicians, government officials, and a host of advocacy groups. Nearly 300,000 Americans died of cancer in 1969, and once Congress passed appropriate legislation, NCI would become the mother lode, a font of hundreds of millions of dollars annually for universities, hospitals, and state and local public health departments. The disease marched on inexorably. Epidemiologists expected cancer in 1972 to strike 650,000 Americans. An article in *Newsweek* warned, "Of the 200-odd million Americans now alive and well, fully 25 percent, or some 50 million, will one day hear their doctors announce the dread diagnosis. Of these, 34 million will die. . . . Other thousands will linger on, for months, for years, some for quite a few years. Perhaps the cruelest truth of all is that of those doomed to die this year of cancer, 4,000 are children."[33]

In September, Representative Paul G. Rogers, a Florida Democrat, introduced HR 10681, the House counterpart to S 1828, which expanded the reach of the legislation to include heart disease and stroke. Rogers had strong feelings about the matter. A new agency, he contended, would only duplicate existing agencies, multiply administrative costs, and reduce the total amount of money available for research. Rogers wanted to keep NCI inside NIH. The director of NCI would become an associate director of NIH, as would the directors of the National Heart and Lung Institute and the National Institute of Neurological Diseases and Blindness. HR 10681 required NIH peer review of all research grants. The NCI director could establish other review groups only after securing permission from the NIH director. Lasker marshaled the American Cancer Society against Rogers. Hearings opened on September 15, 1971, and Lasker attended. The panel of consultants made its case. Clark testified that creating a separate National Cancer Authority would not separate cancer research from the other biomedical research programs of NIH, subordinate research grants to contracts, elevate clinical research above basic research, or weaken the medical schools. But the committee sided with Rogers. Lasker was wired into the Democratic political establishment, but in 1971, a Republican sat in the Oval Office.[34]

——◦◦◦◦◦——

In the eye of the research storm stood NCI's contract-based Special Virus Cancer Program (SVCP). Most biomedical scientists at NIH and the medical schools held the SVCP in contempt. Presided over by Robert Huebner, the SVCP dispensed tens of millions of dollars in the form of contracts without real peer review. Many investigators receiving money sat in judgment of their own research. "The SVCP," complained one scientist, "is a masquerade. They make continuous proclamations of progress to justify the vast amounts of money being spent. But the nature of the program is that it excludes people who are highly critical. It has created a kind of stampede in which everyone rushes lemming-like in the same direction, and critical discussion, points of obvious contradiction, are ignored." The writer indicted the contract system. "With the motto 'Nothing too stupid to test' . . . the chemotherapy program has handed out some $330 million since 1955 in search of the magic bullet against cancer, yet has managed to miss discovering many of the more useful anticancer agents in current use." A virologist complained that "most of the people making the top decisions at the SVCP are not top scientists. They are allocating enormous amounts of money on the basis of relatively little knowledge. . . . The reason why the peer review system grew up is that no one individual can make these decisions intelligently."[35]

In the midst of the debate, Clark sought to embarrass critics of the Panel of Consultants. In the early 1960s, classical virology had begun the shift to onco-genic virology. At NIH, the virologists directed the business, and NCI's SVCP consumed an inordinate share of the research budget. Laboratories around the world competed to find the viral origins of cancer in humans. It was a dog-eat-dog competition with huge amounts of money at stake. Clark wanted M. D. Anderson to solve the riddle. With three NCI contracts totaling $1.5 million, Clark gave Leon Dmochowski carte blanche to hire fifteen new Ph.D.s and lab-oratory space at Center Pavilion. Dmochowski in turn allocated to virologist Elizabeth S. Priori a huge lab. Center Pavilion, a dreadful facility suffering from mold and air-conditioning problems, sparkled with anticipation. Priori soon told Dmochowski that she had isolated a human cancer virus, a C-type virus replicating in the pleural effusion cells of a five-year-old boy with Burkitt's lym-phoma. Doubtless eager for scientific immortality, she created an acronym based on her own initials and designated the culture ESP-1. Dmochowski had been working with electron microscopes for years hunting for viral particles in can-cer cells, and he always seemed just one step short of greatness. Dmochowski called Clark on the telephone and shouted, "We've done it!" Clark and Dmo-chowski flew to Washington, D.C., to reveal their success to NIH and NCI vi-rologists. The news earned headlines around the world. On July 14, 1971, Dmo-chowski and Priori reported the discovery in *Nature.* For President Richard Nixon, the news was a stroke of political good fortune, sure to promote his war on cancer, and he sent Priori a letter of congratulations.[36]

Clark and Dmochowski both had something to prove. Over the years, Dmochowski had engaged in an ongoing feud with Robert Huebner, who con-sidered him a mediocre scientist. Having trained as a pathologist, not a virol-ogist, Dmochowski possessed modest scientific credentials and an abrasive personality. Huebner also accused him of the cardinal sin—overinterpreting data and seeing what nobody else could see. NIH scientists accused M. D. An-derson virologists of practicing old science. Virologists like Robert Gallo were hunting human cancer viruses via reverse transcriptase, not Dmochowski's stodgy tumor markers. Dmochowski yearned for scientific stardom, and Clark clamored to show that a categorical institution with a contract-based program could conduct good science.[37]

Then disaster struck. NIH virologists asked for samples of the tissue culture. Vials of ESP-1 also went to other laboratories in Europe, California, and New York. NIH virologists gloated when they could not replicate Priori's findings. Dmochowski and Priori defended themselves. "We are aware of the controversy," Priori told reporters. "We do not believe that this virus is from a mouse." In a courtesy call, Robert Gallo informed Clark of NIH's findings. In mid-September 1971, NIH announced that ESP-1 was a mouse virus, not a human virus, present through an accident of laboratory contamination. *Medical World News* reported that rather than carefully corroborating the research, "M. D. Anderson went for the headlines." Huebner wrote that "certain obviously essential steps in addition to those taken should have been required before announcing a human cancer virus to the world." Newspapers around the globe reported the mistake. Clark was furious, and he demanded that Priori and Dmochowski draft a memo of explanation in case members of the state legislature and board of regents probed the mess. A September 10, 1971, editorial in *Nature* entitled "Of Mice and Men" labeled the so-called discovery a "red herring." Clark did not like being embarrassed, and Dmochowski and Priori would soon pay for their sin.[38]

The gaffe played into the hands of critics in the NIH, the medical schools, and the larger biomedical community. On September 6, the *Washington Post* published a letter signed by ten prominent scientists, including four Nobel laureates in medicine. They opposed establishment of an independent cancer agency. So did Democratic congressman Claude Pepper of Florida, a close friend of Mary Lasker and cosponsor of the National Cancer Act of 1937. Pepper's opposition was crucial. Clark, Lasker, and the panel of consultants were finished. In October, the House Subcommittee on Public Health unanimously approved HR 11302. The House Commerce Committee reported the bill to the floor of the House for a roll call vote. On November 15, 1971, by a vote of 350 to 5, the House adopted HR 11302. A conference committee worked out the differences between HR 11302 and S 1828. Clark, Benno Schmidt, and Mary Lasker all remained in Washington, D.C., during the deliberations, taking every opportunity to influence the outcome. On December 9 and 10, both houses of Congress adopted the conference committee report. Two weeks later, President Nixon signed the National Cancer Act of 1971, offering it as a "Christmas gift to the American people."[39]

The legislation represented a political compromise. NCI, not a new National

Cancer Authority, would direct the war on cancer, and NCI would remain within NIH, subject to scientific peer review panels that gave equal status to basic and clinical research. The NCI annual budget, however, would be submitted directly to the president, bypassing the controls that NIH and HEW exercised over other health institutes. The measure gave the president the authority to appoint the director of NCI and to establish an eighteen-member National Cancer Advisory Board. The new three-person President's Cancer Panel would monitor the national cancer program, hold periodic public hearings, and submit an annual progress report to the president. In consultation with the NCAB, NCI was to establish new cancer centers and manpower training programs, award contracts and grants for research, conduct cancer control activities, and create an international cancer research data bank. The panel of consultants had obtained less than they had hoped, but Nixon soon buoyed their spirits. Early in 1972, he appointed Lee Clark and Benno Schmidt to the president's cancer panel, along with Robert A. Good, a pediatrician at the University of Minnesota Medical School. The bill allotted $400 million for research in 1972, $500 million in 1973, and $600 million in 1974. Clark delighted in the canonization of M. D. Anderson, along with Memorial Sloan-Kettering and Roswell Park, as "comprehensive cancer centers" with fully developed programs in research, education, and patient care. He returned to Houston savoring victory. An all-out war on cancer loomed, and M. D. Anderson would be in the vanguard.[40]

The war on cancer was under way for America, but not for Brian Piccolo, the workhorse fullback of the Chicago Bears. Throughout 1971, the film *Brian's Song* packed theaters. First shown as a television movie, *Brian's Song* produced such high Nielsen ratings that the producers arranged for the film's theatrical release. Starring James Caan as Piccolo and Billy Dee Williams as his teammate Gale Sayers, the film portrayed the friendship between the two men, one black and the other white, and Piccolo's death from cancer in 1970. The football player had complained of trouble breathing, and coaches noticed that his speed had diminished. Tests revealed a melon-size tumor next to his heart, with embryonal testicular origins. Had Piccolo been treated by Melvin Samuels and Jeffrey Gottlieb at M. D. Anderson in an earlier stage of the disease, he might have survived, but he was not in the right place at the right time. Few were. So in 1971, while Congress debated the National Cancer Act, Brian Piccolo embodied the disease.[41]

8

Waging War
and Fading Away,
1971–1977

I don't want to leave my institution after blowing it up.
—R. LEE CLARK, 1977

B rush and hair baubles in hand, Frances Cording sat behind her ten-year-old daughter Susan for the morning ritual. It was August 1970, and the family had just arrived in Huntsville, Texas, where Dick Cording had accepted a faculty position in the Department of Philosophy at Sam Houston State University. Frances parted the hair in the back, pulled together the fine brown strands, and braided them tightly. Blessed with a keen mind and graceful athleticism, Susan Cording faced a bright future in the land of sunshine and ice cream, where all dreams are supposed to have happy endings. But as Frances's nimble fingers danced around the fifth grader's neck, they detected a swelling on the left side. Susan had noticed the mass too. When the swelling persisted for several days, and Frances remembered that Susan had seemed unusually lethargic all summer, she took her to a local physician. Even after two full rounds of antibiotics, the mass grew. A subsequent needle biopsy revealed Hodgkin's disease—stage 2A. The doctor referred the Cordings to M. D. Anderson. Two pediatric oncologists, veteran Margaret Sullivan and newcomer Jordan Wilbur, assumed responsibility for Susan's care. Wilbur had trained at Stanford under Harry S. Kaplan; under the tutelage of Gilbert Fletcher, Sullivan had aggressively treated Hodgkin's disease with radiotherapy.

Sullivan and Wilbur made sure that the Cordings fully grasped their predicament. Until recently the disease had been mercilessly unforgiving. When Dick asked whether Susan would have a normal life, the two physicians responded with brutal honesty, "We're not talking here about a full life. We're talking about survival and existence." Stunned into numbing bewilderment, their world suddenly askew, the Cordings managed to drive home. "We'd never heard of Hodgkin's disease. We cried and cried all the way [home]."

Lee Clark would leave a legacy of aggressive, multidisciplinary treatment, and Susan Cording was about to inherit it. To reduce the total number of Hodgkin's cells in her body—and to improve the odds of survival so that radiotherapy and chemotherapy could kill every last one—the pediatricians recommended surgery as a first line of treatment. Surgeons opened Susan from neck to groin, scrutinized the lymph nodes and internal organs, and excised every tumor they could see. The abdomen appeared free of active disease, but clusters of malignant cells studded her neck and chest. Because Hodgkin's cells often congregate in the spleen, the surgeons also performed a splenectomy, removing the organ in a procedure that the M. D. Anderson surgeon Charles McBride had only recently perfected. Experience showed that Hodgkin's disease spreads early to regional lymph nodes, so in a peremptory strike, with daily doses over the course of six weeks, radiotherapists delivered nearly five thousand rads of radiation to Susan's neck, chest, abdomen, and groin. Under what is surely the world's ultimate tanning salon—the bed under a megavoltage radiotherapy contraption—the targeted areas of Susan's body acquired a deep, reddish hue. She teetered precariously on a physiological high wire separating the therapeutic from the lethal. Sullivan and Wilbur also considered chemotherapy but changed their minds after one dose, deciding to reserve the drugs as a fallback in the event of recurrence.

Several weeks after the final treatment, pathologists searched Susan's bloodstream for Hodgkin's disease, with its telltale Reed-Sternberg cells. When none materialized, they pronounced her "in remission," hardly a cure but certainly evidence that the treatments had extended her life. The Cordings then entered the ethereal world of cancer survivors, waiting nervously to see whether luck holds, whether the remission stretches into months and the months into years.

They waited and waited. Susan's disease remained at bay. She had fallen ill far enough ahead of the scientific curve to survive a disease once universally fatal. Sports, school, music, and the concerns of a teenager rushed back into her life.

Cancer treatments, however, can exact an expensive toll. Susan's ovaries, lungs, kidneys, bones, thyroid gland, and heart had absorbed massive amounts of radiation. Sometimes the cancer clock ticks slowly. The Cordings had not seen the last of M. D. Anderson.[1]

—⁕—

The ESP-1 debacle left Clark embarrassed. In the hallways of the NIH, NCI, and medical schools throughout the country, Dmochowski was the target of jokes, an albatross—and critics gloated. Priori and Dmochowski yearned for exculpation. But the NCI had ordered Clark back to Bethesda, where officials announced that research contracts with the Department of Virology were canceled. Smoldering beneath a frigid exterior, Clark threatened a nasty fight and then managed to renegotiate with NCI. The Department of Virology would eventually be reduced to a section in the Department of Biology, and a new Department of Molecular Carcinogenesis would be created. With the assistance of Felix Haas, chair of the Department of Biology, Clark worked to rebuild M. D. Anderson's reputation at NCI. In 1969, biochemist Ralph Arlinghaus had arrived at M. D. Anderson from the U.S. Department of Agriculture's animal research laboratory on Plum Island to strengthen virology. Virologists were losing power at NIH, and biochemists exploring chemical carcinogenesis were seizing it. Haas and Clark touted Arlinghaus as a scientist worthy of NCI trust. The ESP-1 debacle began to fade.

Arlinghaus soon lived up to the sales pitch Clark had delivered to NCI. From his laboratory came papers published in such prestigious journals as *Virology*, *Journal of Virology*, *Archives of Virology*, and *Cell*, in which he established, using the Rauscher murine leukemia virus, the mechanism of murine leukemia retrovirus protein formation in viral-infected cells.[2]

When the funding to the Department of Virology evaporated, so did the jobs of seven scientists at M. D. Anderson, including Dmochowski and Priori, as well as twelve support personnel. Clark secured for Priori a one-year appointment with NCI and managed to extend it into a second year, but her tenure at M. D. Anderson was finished. Over the years, Dmochowski had been fond of saying, "Being on the cutting edge of science is often like walking barefoot on a freshly stropped razor's edge." He was about to experience the truth of that.[3]

Clark stewed over the looming personnel changes. He hated ugly confrontations. Robert Hickey usually did the dirty work of firing people, but Hickey was unavailable. So Clark turned to Elmer Gilley, M. D. Anderson's director of finance and administration, a man devoted to the hospital but with little power beyond the front door of his own office. Clark told Gilley to fire Dmochowski, and Gilley dutifully obeyed, delivering the news at 6:15 one evening in the virologist's laboratory. Dmochowski had few friends at M. D. Anderson. His formality had precluded intimacy. The next day Clark offered the interim chairmanship of the as yet unformed department of molecular carcinogenesis to the virologist James Bowen with instructions to break the bad news to those losing their jobs. "That's your first job," Clark said. "If you do it, you'll have a future at M. D. Anderson." Bowen, wanting such a future, swept out the department and dismantled classical virology at M. D. Anderson.

The 1972 Bertner Award confirmed the demise of classical virology. The award went to Howard Temin, the codiscoverer of reverse transcriptase, an enzyme employed by all retroviruses to transcribe genetic information from the virus for RNA to DNA, which can then integrate into the host genome, disorient the mechanisms of cellular reproduction, and generate more cancer cells. At M. D. Anderson and around the nation, investigators largely abandoned classical virology, with its emphasis on viral markers, in an effort to study the activities and implications of reverse transcriptase.[4]

———※———

Clark could not afford to spend too much time on ESP-1 damage control. For decades he had been harnessed to his career. Whenever he found a momentary vacuum in his personal life, M. D. Anderson soon filled it. Early in the 1970s, he implemented a series of plans that some would later call the "metastasis of M. D. Anderson"—its expansion beyond the Pink Palace. The original bill creating M. D. Anderson authorized the board of regents to establish substations of the hospital without legislative approval; enormous growth combined with the complexities of cancer research had created unprecedented pressure for more hospital rooms, clinic space, and laboratories.

In April 1971, the UT board approved Clark's request for a campus devoted to environmental carcinogenesis, an idea that had first surfaced in 1962 when he read *Silent Spring*. The legislature appropriated a hundred thousand dollars in planning money. Political prudence, Clark realized, dictated a rural location because hypersensitive petrochemical corporate executives resented allegations that their products or production methods caused cancer. Clark had a spot in mind. Over the years, traveling back and forth between Houston and his ranch in Bastrop County, he had enjoyed the ambience of Buescher State Park and had noticed Camp Swift, a former military training base and prison for German POWs. Overcoming opposition from UT at Austin, which was wary of a state-supported basic research facility in its backyard, Clark raised the money for an environmental science park—seven hundred acres to be carved out of Buescher State Park for an ecology study area and an animal study compound at Camp Swift. Construction was completed in 1977, and with the Environmental Science Park, Clark stood poised to exploit NCI's new infatuation with chemical carcinogenesis.[5]

At about the same time, M. D. Anderson opened its first treatment substation, taking over the Rio Grande Radiation Treatment Center in McAllen. Cancer patients in South Texas had long been forced to seek help in Houston, a journey many could not afford. The residents of South Texas wanted M. D. Anderson radiotherapists to deliver the treatments locally. In September 1977, Governor Dolph Briscoe dedicated the treatment center.

The need for more space knew no bounds. "University administration is somewhat like a rabbit hutch," Clark once told Richard Martin. "Every time you hire two new people, a new litter of employees soon appears." By the mid-1970s, M. D. Anderson's nearly four thousand employees needed more space, and across the street and down a block stood the Prudential Building, a twenty-story limestone edifice. The Prudential Life Insurance Company wanted to relocate, and the board of regents purchased the building and renamed it the Houston Main Building to absorb the rapidly increasing employee base at M. D. Anderson and at the University of Texas Health Science Center at Houston.

When the real estate transaction, the construction, and the personnel transfers were complete, M. D. Anderson was a much enlarged institution made up of the main campus on Holcombe Boulevard, the Houston Main Building, the Annex and Rehabilitation Center, the radiation treatment center in McAllen, and the Environmental Science Park.[6]

The expansion, however, failed to keep up with growth. In 1945, M. D. Anderson had treated only 135 individuals. By 1965, the total number of individual patients treated reached 57,971 and a decade later 112,971. Between 1944 and 1965, M. D. Anderson had registered more than 60,000 *new* patients, and during the next decade, another 40,000. Clark added employees steadily, from 1,485 in 1963 to 2,500 in 1970, and 3,800 in 1975. In 1976, the hospital was operating at 96 percent capacity. "Sometimes I feel like Sisyphus," Clark once remarked to Martin. "I no sooner push the boulder up to the top of the mountain than it rolls back down and I start all over again."[7]

The future of ambulatory cancer care at M. D. Anderson was also at stake. If growth continued unabated, and the goals were to be met of diagnosing and treating as many patients as possible in clinics, rather than through overnight hospital stays, Clark needed far more space.

In late 1971, Clark announced the most audacious building program in M. D. Anderson history. The expansion was financed by Marshall G. Johnson, a devout Lutheran and self-made millionaire rancher from Wharton, Texas, and his wife, Lillie. A heart attack in 1955 had left Johnson eager to invest in a good cause. The Johnsons established the Johnson Foundation, endowed it with assets exceeding $6.63 million, plus fifty-one thousand acres of land in Florida, and vowed to build a Lutheran hospital in the Texas Medical Center. Between 1959 and 1961, several institutions pitched ideas, but the Johnsons found most persuasive the presentations of Grant Taylor and Lee Clark. Cure cancer. Over the next several years, the general plans distilled into architectural drawings for a 14-floor, 330-bed, 270,500-square-foot Lutheran Hospital Pavilion and a 12-story, 345,500-square-foot Clinic Building, all attached to the existing M. D. Anderson Hospital and Tumor Institute. The Johnsons wanted a nondenominational chapel included.[8]

The widening of the Vietnam War, however, ballooned the federal deficit and delayed the project. Hill-Burton funds for hospital construction disappeared. The plans gathered dust until late in 1971, when only thirty-five thousand U.S. troops remained in Vietnam. Hill-Burton funds materialized. In September 1972, the UT Board of Regents announced the expansion.[9]

The fund-raising campaign sought $39.2 million. Hope ran high. Much of the money was already on hand. The Johnson Foundation's Florida real estate was now valued at $17 million. To raise the balance, the board of visitors organized local fund-raising committees throughout Texas and solicited donations from many well-endowed former patients. Dozens of Lutheran churches donated. Clark hosted dinners seemingly everywhere, always reiterating the somewhat misleading statistic that in thirty years, the five-year survival rates for cancer had climbed from one in five patients to one in three.[10]

It was a promising time to raise money for cancer. Passage of the National Cancer Act teased Americans with the expectation that a cure merely awaited more effort. And prominent Americans kept the disease on the front pages.

Just when Clark announced the fund-raising campaign, Senator Birch Bayh abandoned his campaign for the Democratic presidential nomination to care for his wife, Marvella, then recuperating from breast cancer surgery. Shirley Temple Black, the former child actress, informed the media of her recent treatment for breast cancer. First Lady Betty Ford revealed in 1974 that she had just undergone a mastectomy. A few weeks later, Happy Rockefeller, the wife of Vice President Nelson Rockefeller, spoke of her double mastectomy. In 1973, Senator Ted Kennedy of Massachusetts announced that physicians had amputated the right leg of his twelve-year-old son Patrick to treat an osteosarcoma. The boy was treated at the Dana-Farber Cancer Center in Boston, where Emil Frei now presided.[11]

Emil Frei had left M. D. Anderson to become physician in chief at Dana-Farber. Patrick Kennedy was assigned to Norman Jaffe, a pediatric oncologist from South Africa destined for tenure at M. D. Anderson. At Patrick's first appointment, an Irish governess barked to the boy, "Behave yourself" in Gaelic. Jaffe had learned to speak some Gaelic in South Africa, and he responded in Gaelic, much to her delight. In front of the senator, Jaffe placed an IV in Patrick with little trouble, sticking the vein on the first try. The day before, nurses and phlebotomists had repeatedly failed to insert the IV, leaving the boy's nerves raw and his arms resembling pin cushions. At the moment Jaffe hit the vein, he became a friend of the Kennedy family. "The Senator treated me royally," Jaffe remembers today. The chemotherapy protocol worked. Osteosarcoma cells never colonized the boy's lungs. That Patrick Kennedy survived his disease, as did Shirley Temple Black, Betty Ford, and Happy Rockefeller, provided healthy faces for the war on cancer.[12]

The National Cancer Act of 1971 had rendered Lee Clark one of the most influential people in cancer medicine. The bill appropriated $400 million for research in 1972, $500 million in 1973, and $600 million in 1974, and the lion's share of the money would flow to NCI-recognized "comprehensive cancer centers," with fully developed programs in research, education, and patient care. M. D. Anderson, Memorial Sloan-Kettering, and Roswell Park were the templates for those to come. Applications for recognition inundated NCI. As a member of the President's Cancer Panel, Clark enjoyed great influence in the recognition process, and he joined several site teams to investigate qualifications. NCI fielded dozens of applications in 1972, rejecting the vast majority because they could not demonstrate serious programs in all three areas. Comprehensive cancer center status that year went to the Albert Einstein Cancer Research Center of the Albert Einstein College of Medicine in New York; the Dana-Farber Cancer Institute in Boston; the Duke Comprehensive Cancer Center in Durham, North Carolina; the Lombardi Comprehensive Cancer Center at Georgetown University Medical Center in Washington, D.C.; and the Mayo Clinic. In 1973, the network of comprehensive cancer centers was expanded to include the Abramson Cancer Center of the University of Pennsylvania in Philadelphia; the Sidney Kimmel Comprehensive Cancer Center at Johns Hopkins University in Baltimore; and the Paul P. Carbone Comprehensive Cancer Center at the University of Wisconsin in Madison.[13]

The process of awarding comprehensive cancer center status became a political football, with constituency-conscious congressmen throwing around political weight in an effort to bring more federal money to their districts and states. Clark worked diligently to hold the line, insisting that comprehensive cancer centers needed to be just that, endowed with the infrastructure and personnel to treat a wide variety of cancers, conduct both basic and clinical research, and engage in a variety of cancer education programs. Many of the institutions applying for comprehensive cancer care status simply did not possess the resources. In 1974, for example, the Cancer Research Center of Columbia, Missouri, failed in its bid. John S. Pratt, its director, protested to Clark, "We do comply and we are simply asking [for] official cognizance of the fact that this is so. Such recognition would be extremely helpful to us in increasing our level of federal support."

St. Jude Children's Research Hospital in Memphis, Tennessee, provides another example. St. Jude confined its research and treatment efforts to pediatric malignancies and by mission therefore could not be considered comprehensive. At the same time, however, St. Jude, because of its research effort, was far more than a local practice in oncology, as was the Cancer Research Center in Missouri. As a compromise, the status of NCI-recognized Cancer Center emerged, acknowledging institutions whose research and treatment missions set them apart from small hospitals and local oncology practices but that did not enjoy the diversity of programs needed to qualify for comprehensive cancer status. In 1974, for example, UCLA enjoyed recognition as a cancer center associated with NCI but not as a comprehensive cancer center. By the end of the decade, NCI had recognized sixteen comprehensive cancer centers.[14]

—⁂—

Almost as soon as President Nixon signed the National Cancer Act of 1971, the rancor flared. The legislation faced periodic renewal, and the partisans of biomedical research still fumed. In September 1975, Michael DeBakey, representing the NIH, testified before the panel, claiming that NCI's recently acquired quasi-independent status fragmented NIH research efforts and gave to cancer research an inappropriate share of federal money. But President Gerald Ford had bigger items on his agenda. He was still taking heat for pardoning Richard Nixon in the Watergate scandal; the Arab oil boycott of 1973 had created gasoline shortages and inflation; and in April 1975, North Vietnamese troops overran Saigon and proclaimed the Socialist Republic of Vietnam. Overwhelmed by serious domestic and foreign challenges, Ford knew that any attempt to change the legislation would prompt Democrats to portray him as a friend of cancer. DeBakey's testimony worried Clark, but Benno Schmidt assured him, "I think it is too early to reach any conclusions about the potential influence of the President's Biomedical Panel on categorical research support. However, I do not personally anticipate that the Panel will take any action which will be detrimental to the cancer program."[15]

Clark was as optimistic as ever, although he traded the rhetoric of "cure" for the promise of "control." The nation's bicentennial had come and gone and critics were beginning to label the war on cancer a failure. In 1972, Clark told *Houston Post* reporter Mary Jane Schier that he was "excited because I believe— have believed for a long time, in fact—that a truly coordinated national effort can bring many forms of cancer under control in the foreseeable future," and he set the tone for the next generation of oncologists. The combination of early detection and aggressive treatment promised to reduce the incidence and mortality of cancer.[16]

While the NCI budget swelled from $378 million in 1972 to $815 million in 1977, Clark and Schmidt as members of the President's Cancer Panel carefully monitored the war on cancer. As the billions piled up, so did the animosities. Professor Arthur B. Pardee of Dana-Farber captured the sentiments of many scientists: "The truly basic research is not the only activity that the NCI must support, but I do believe it is the most important . . . I am not at all persuaded that the best expenditure of American taxpayers' dollars is for a small number of second-rate [institutions] that employ numerous third-rate investigators and use contract materials made by an army of technicians. These 'ideas' are the warmed-over generalizations of committees or managers who are often inadequate scientists at best. The tone-deaf do not select music; why should the idea-deaf select research?"[17]

The biomedical community branded the NCI organ-site research program, which designated certain cancer centers to focus on cancer in specific organs, a colossal waste. M. D. Anderson, for example, became home to the large bowel program. Saul Roseman, director of the McCollum-Pratt Institute at Johns Hopkins, sarcastically promised failure. "I not only believe that such programs as yours will prove ineffective, but [think that they will] probably delay finding a cure for cancer in general (if this is possible), and for targeted tissues in particular. . . . NASA's marvelous accomplishments are essentially engineering accomplishments, achieved by applying basic and fundamental laws derived over the course of centuries." NCI, he insisted, "must never be diverted from making its *major* investment in the area most likely to provide the final pay-off, basic research in cell and molecular biology." Roseman said sarcastically, "[I know] where your program ends, but am not quite certain where it starts. Do you support work on the small as well as the large bowel? How far up the G.I. tract do you go, and does some other group pick it up from there?" Jonathan Rhoads, a surgeon at the University of Pennsylvania, also harbored serious misgivings. To Murray Copeland he confessed "grave doubts as to the purpose these organ site programs serve and the necessity for having them."[18]

The debate intensified the rivalry between M. D. Anderson and the medical schools. In promoting the strengths of categorical institutions, the panel of consultants had sometimes taken careful aim. "Medical schools," the panel said at one point, "have been traditionally organized on a functional basis with very little cross-fertilization among specialties. Multi-discipline programs are rare. As a consequence, the present medical research community is neither trained nor organized to undertake a large-scale, multi-discipline, nationally coordinated research effort against cancer."[19]

In June 1972, the board of regents recognized M. D. Anderson's preeminence in cancer medicine with a name change to the University of Texas System Cancer Center, with its headquarters in Houston. Clark and others successfully campaigned to use uppercase letters for THE UNIVERSITY OF TEXAS SYSTEM CANCER CENTER, with "The University of Texas M. D. Anderson Hospital and Tumor Institute" in lowercase letters.

The change rubbed raw the descendants of M. D. Anderson. When Thomas Anderson protested, Frank Erwin of the UT Board of Regents countered that Marshall and Lillie Johnson had contributed millions without insisting that their name be in the title. Anderson replied that demoting "M. D. Anderson" from the institutional title sacrificed the best-known brand name in oncology and would confuse many physicians, patients, and federal agencies. He protested in vain.[20]

Clark meanwhile undertook for NCI the task of collecting, analyzing, and disseminating worldwide all data related to the prevention, diagnosis, and treatment of cancer. The NCI had awarded a contract to Informatics, a private company specializing in management information systems, to develop an international cancer research data bank, and Informatics executives brought M. D. Anderson in on the contract.

For assistance, Clark turned to biomathematician Stuart Zimmerman. Whenever Clark needed advice about computers and applications of mathematics to biological problems, he approached Zimmerman, whose Ph.D. in biomathematics from the University of Chicago and keen analytical mind equipped him to fathom and reduce very complex problems into recommendations on which Clark could act. Late in 1968, Clark had explored the possibility of working with California-based Rand Corporation on the development of a new data management system. In August 1970, Rand officials visited Houston. Zimmerman grew impatient. The system included neither time sharing arrangements for multiple users nor simultaneous access to the data from multiple users, and could only work with IBM hardware. Zimmerman predicted serious bottlenecks. Zimmerman warned Clark that Rand seemed "inordinately eager... to obtain the magnetic tapes from the Epidemiology Registry.... All these records are privileged information as they involve patient care." Zimmerman's suspicions convinced Clark to back away, and the project fizzled.[21]

Another challenge needed attention. By the mid-1970s, the hospital's relations with referring physicians had soured to the point that it was spoiling the quality of patient care. Bureaucracy and inefficiency were primary factors. As patient loads accumulated, so did complaints from referring physicians. Few M. D. Anderson physicians had ever spent much time in private practice. Clark had originally recruited young internists and surgeons straight out of the military, and in the 1960s and early 1970s, he invigorated the medical staff with brilliant clinicians, oriented toward clinical research, drawing them from NCI and the medical schools. In 1977, when Joseph T. Ainsworth came to M. D. Anderson to direct the employee health service, he was "astonished at how little experience the hospital's most influential clinicians had in private practice.... Most of the staff did not have a clue about the challenges of practicing medicine in the real world, about being on call day after day, covering the overhead, and treating patients." When busy, tired, and harassed local physicians received casual responses from M. D. Anderson staff, they complained loudly.[22]

Clark helped to widen the split. Having spent his career dealing with physicians in private practice, his patience had worn thin. Every M. D. Anderson initiative that ever even hinted at removing a nickel from the pockets of private physicians generated shrill protests. Accusations that Clark practiced "socialized medicine" abounded. He had repeatedly explained that the salaries of his physicians did not come from state coffers. All fees for service went to the Physicians Referral Service, which then paid the medical staff fixed salaries. But the explanation satisfied few critics. Clark sometimes took wicked pleasure in their discomfort: "I like them shaking and quaking and crapping in their pants." He also knew, however, that fences needed mending. In 1974, he hired Joseph T. Painter as assistant director of M. D. Anderson for extramural affairs. A cardiologist and the son of the former UT president and geneticist Theophilus Painter, Painter was widely known in the Texas medical community.[23]

Clark also had to implement new federal rules on human subjects. The National Cancer Act of 1971 had promised to open the taps of research money, but investigators would also find themselves subject to external scrutiny as never before. In 1966, James Shannon, director of the NIH, revealed two egregious incidents involving human subjects. Using NIH grant money, Tulane University scientists had transplanted a chimpanzee kidney into a human being. Although they secured the consent of the human subject, the project seemed more whimsy than experimental. Shannon also revealed that Chester M. South, a highly respected physician at Memorial Sloan-Kettering, had injected live cancer cells into elderly patients at the Brooklyn Jewish Chronic Disease Hospital without securing their consent. Emmanuel E. Mandel, the medical director at Memorial, had approved the experiment. NCI investigated; Senator Estes Kefauver poked around; and the New York media attacked. Research centers across the nation reexamined surveillance procedures. The Public Health Service tightened the definition of "informed consent." At M. D. Anderson, the Surveillance Committee informed all clinical investigators of the need for accurate and detailed records of research protocols involving human subjects. When M. D. Anderson was slow to submit "a general assurance of compliance," NIH promptly issued a "final reminder."[24]

These incidents pale against the Tuskegee syphilis experiment. Informed by a source inside the U.S. Public Health Service (PHS), the *Washington Star* broke the story on July 25, 1972. Not since the Nuremberg Trials in 1949, with their grisly portrayals of Nazi medical experimentation, had the issues of medical ethics, informed consent, and clinical research garnered so much attention.

In 1932, the PHS had enrolled 399 African American men in a long-term study of end-stage syphilis. Most were poor, illiterate sharecroppers. Over the course of the next forty years, researchers informed the subjects that they would receive free treatment for their "bad blood"—a euphemism for syphilis—and annual physical examinations at no cost. The original PHS objective was to track each man until his death, noting all along the way each new deterioration in condition. At first, the men received bismuth, mercury, and neoarsphenamine, the standard medications for syphilis, but in nontherapeutic doses. The PHS wanted to study the natural course of *untreated* syphilis. Over time, however, the PHS scuttled even that facade, switching to a daily dose of the "pink pill," which they assured the subjects was a new treatment for "bad blood." Actually, the pill was an aspirin. Even the development of injectable penicillin in the early 1940s did not disturb the project. "As I see it," wrote one PHS physician, "we have no further interest in the patients until they die." At the end of World War II, one PHS staffer proudly reported, "So far we are keeping the known positive patients from receiving treatment." By 1972, twenty-eight of the subjects had succumbed to tertiary syphilis and another hundred to complications related to the illness. Forty of their wives contracted the disease, and nineteen children were born with congenital syphilis.

The PHS insisted that the men had volunteered and had never complained about treatment, which only made the agency seem more obtuse and the issue of informed consent more essential. An Alabama state health officer accused the media of "trying to make a mountain out of a mole hill." ABC Evening News anchor Harry Reasoner, however, saw a mountain and described the study as an experiment that "used human beings as laboratory animals in a long and inefficient study of how long it takes syphilis to kill someone." Congress passed the Research Act of 1974 requiring all entities experimenting with human subjects to establish an institutional review board to approve research projects.[25]

Many clinical investigators chafed under the regulations and choked under what they considered an avalanche of paperwork. The maze through the federal bureaucracy, they argued, discouraged innovation and slowed the development of new treatments. In 1976, Emil J Freireich testified before the President's Cancer Panel, "There is now, particularly in our country, a new very important, significant, impediment to progress to therapeutic research and that is regulation under existing law. This particular impediment to research is very significant because it is unambiguously malignant." Freireich explained that minor officials in the FDA had authority to stop research projects. "I can discover no benefit from this dramatic change in direction . . . of the Food and Drug Administration. . . . To frustrate these patients [who are without hope] with these unnecessary regulations is an outrage." One FDA official even objected to the use of the words "therapy," "treatment," "chemotherapy," or "immunotherapy" in any research proposal because such words carry "a connotation of assured benefit to the patient and as such tend to be coercive."[26]

M. D. Anderson investigators felt the squeeze. Each proposed project involving human subjects first went to the Office of Research, where its scientific merits were evaluated. The proposal then moved on to the Surveillance Committee for an assessment. Any project entailing substantial risks to human subjects needed the consent of the Executive Committee of the Medical Staff. If research involved the use of investigational drugs, a Surveillance Committee quorum had to include at least two members licensed to administer drugs and one member not so licensed. M. D. Anderson also broadened the criteria of informed consent. Prospective subjects had to be "so situated and informed as to be able to exercise choice without elements of force, fraud, deceit, duress, or other forms of constraint or coercion." Before an individual could make an affirmative decision, the investigators had to demonstrate that he or she understood "the nature, duration, and purpose of the experiment[;] . . . the method and means by which it is to be conducted; [and] all inconveniences and hazards to be expected within reason." All unanticipated adverse reactions had to be reported to the Surveillance Committee and then to the Department of Health, Education, and Welfare (HEW). Use of any investigational drug required FDA approval, and then could be used only as part of an approved research protocol. Many older medical oncologists remembered nostalgically the era when they had enjoyed enormous latitude in treating patients with experimental drugs. "Looking back on it now," Norman Jaffe remembered, "it's hard to believe how much freedom we had."[27]

Other federal agencies also crowded into research projects. The AEC and the Occupational Safety and Health Administration (OSHA) set new standards to protect physicians, nurses, and technicians working with radioactive materials. The Environmental Protection Agency (EPA) worried about the disposal of radioactive materials used in diagnosis and treatment. The Office of Research Safety at HEW monitored compliance. The new field of recombinant DNA research, in which molecular biologists tinkered with the genes of living organisms, was also a cause for concern. Scientists and the general public worried that careless investigators might inadvertently unleash manufactured pathogens on an unprepared world. To prevent such a nightmare, the Public Health Service issued its "NIH Guidelines for Research Involving Recombinant DNA Molecules." At M. D. Anderson, Clark established the DNA Review Subcommittee and the Chemical Carcinogenesis Review Subcommittee.

Investigators using warm-blooded animals in experimental research also faced new scrutiny. In response to rising public concern, the NIH released its first "Guide for Laboratory Animal Facilities" in 1963, and three years later Congress passed the Animal Welfare Act, requiring humane treatment for animals in transportation for commerce and involved in medical research. The legislation gave the Department of Agriculture authority to monitor compliance. Subsequent amendments strengthened the law.[28]

In the 1960s and 1970s, Congress required that institutions receiving federal money for the use of animals in research be accredited by the American Association for Accreditation of Laboratory Animal Care, and that a scientific peer review panel within the institution screen all research proposals involving the use of animals. The research must have the potential to yield results beneficial for society and be attainable through no other means. All unnecessary injury and suffering to animals must be avoided and all experiments terminated when they cause unnecessary injury; anesthesia is required; animals must be killed in such a way as to cause immediate death; and postexperimental animals have to be cared for properly. The sheer volume of animal experiments turned a few heads. In 1976, the incinerator within the Section of Experimental Animals at M. D. Anderson was burning nearly four tons of animal and anatomical waste materials each month. The new animal facilities at Camp Swift had been constructed strictly according to the new codes.[29]

Basic and clinical researchers now confronted an alphabet soup of internal and external reviews—the Office of Research, the Surveillance Committee, the Biohazards Committee, the Isotope Committee, the Executive Committee of the Medical Staff, the NCI, the NIH, OSHA, HEW, the Department of Agriculture, the AEC, the EPA, the PHS, and the FDA. To all these were added state and local public health and safety regulations; a bewildering array of professional accreditation standards for hospitals and for medical and scientific disciplines; the scrutiny of more and more activist groups, such as the Coalition for Responsible Genetic Research and the Medical Committee on Human Rights; and a legion of politicians eager to protect the public and to get their names in the newspaper.

Although Clark was busier and his job more complex than ever, retirement loomed. He had groomed no successor. "Perhaps he thought he would never be succeeded," recalls virologist James Bowen. In many ways, the timing seemed perfect. In the summer of 1977, Clark's terms expired as president of the American Cancer Society and as a member of the President's Cancer Panel. A. M. Aikin, the state senator from Paris, Texas, who had fought many battles for Clark, announced that he would not seek reelection. Clark seemed smaller in his suits and a bit more vulnerable. A few years before, he had pretended to attend a professional meeting in Los Angeles in order to surreptitiously visit a plastic surgeon for a facelift. And over the years, in a hopeless quest to mask male-pattern baldness, he had taken to strategically combing his remaining hair. He looked his age, and his health was far from good. In April 1977, Clark turned down an invitation from Mary Lasker to visit in New York. "I am scheduled to have a little plumbing done on my heart on Friday, April 22," he wrote, "and will not have an opportunity to see you until later on this summer." Associates worried about his health. Joyce Alt, director of nursing, began accompanying Clark to social events "to keep an eye on him."[30]

Bert Clark hungered for her husband's retirement. His hectic schedule for more than thirty years had tired her. The prospects of Lee's continuing to work until he died held no appeal. Lee was approaching seventy, then the mandatory retirement age for UT administrators. Board of regents regulations provided for exceptions, but Allan Shivers now sat on the board, and he believed that M. D. Anderson needed an overhaul. With nearly five thousand employees and an upcoming budget of $124.6 million, the cancer center had outgrown Clark's leadership style. A micromanager, Clark found it increasingly difficult to oversee the institution's complicated bureaucracy. At a meeting in 1977 that Clark did not attend, a physician could not find a pencil for taking notes. "How many pencils do we have around here?" he asked a colleague. "Wait, I'll call Dr. Clark," cracked the other physician. "He'll know." He had created, according to one surgeon, "an atmosphere that's almost spiritual in its nature," but the time for change had arrived.[31]

Shivers knew that the rivalry between the Department of Developmental Therapeutics (DT) and the Department of Medicine needed resolution. When Emil Frei left for Boston in 1972, Emil Freireich had taken over DT. A seminal figure in the history of oncology, and a physician fiercely committed to his patients,

Freireich won friends and made enemies with equal efficiency, and some physicians in Medicine resented his success. The conflict-averse Clark, however, did little, letting competition turn into rivalry, rivalry into resentment, and resentment into raw confrontation. In 1973, Cliff Howe threatened Jeane P. Hester, Ken McCredie, and Jeffrey Gottlieb with the loss of medical privileges unless they obeyed his new admission policies. Howe's low opinion of chemotherapy antagonized the medical oncologists. "Despite heated denials of a few and well publicized zealots, most all physicians . . . would agree," he wrote in 1973, "that progress in chemotherapy is slow. The reason for this, of course, is its total irrationality . . . the administration of cytotoxins which destroy or damage both cancer cells and the normal cells indiscriminately." He questioned the use of combination chemotherapy, laminar air flow rooms, and perfusion therapy.[32]

The basic sciences needed reconstruction. Clark bragged that external funding for M. D. Anderson research had leaped from $8.2 million in 1973 to $26 million in 1978 and boasted of the recent advance in cytogenetics among M. D. Anderson biologists under the direction of T. C. Hsu. They developed what became known as the C-banding technique to reveal heterochromatin in human cells. But even T. C. Hsu candidly admitted, "I have an impression that the scientific community in the country does not think our biology department is first-class." Hsu agreed with critics that securing a worldwide reputation for scientific research was difficult in a "mission-oriented institution," but he also blamed leadership in the department for a "lack of ambition and vision." Felix Haas, head of the Department of Biology and the target of Hsu's comment, even confessed "that the bulk of effort on cancer research over the past fifteen years to finding a *cancer cure* has been *premature* and even detrimental to this objective." In 1975, a visiting team from NCI concluded that although the hospital was justifiably renowned for the quality of its patient care, its research had real deficiencies. "Some of the projects are very good, some moderately so, and others are obviously poor." Clark knew that the basic sciences needed reconstruction, but he could not bring himself to do it. "I don't want to leave my institution after blowing it up," he told Shivers.[33]

Clark took comfort in leaving his successor with a physical plant that was up to the emerging challenge of ambulatory care. The newest expansions had added 717,000 square feet to the previous 580,000 square feet. Included was a 330-bed hospital, to be known as the Lutheran Pavilion, clinic expansions sufficient to handle 1,200 patients per day, and better radiation treatment and research facilities. On October 2, 1976, the dedication ceremony attracted twenty-three hundred people, and the guest list read like an honor role of cancer medicine. First Lady Betty Ford, perhaps the best-known cancer survivor in the world, delivered the dedicatory speech.[34] After thirty years fighting the war on cancer, Clark still suffered no battle fatigue. He enjoyed going to work every day and had never felt more excited about cancer medicine in general, M. D. Anderson in particular, and the prospects of gaining ground on the disease.[35]

And never had cancer been more fascinating or more mysterious. With every page turned in a new journal article, cancer humbled those trying to fathom it. Clark marveled at the biological complexity of the disease. In the late 1940s, he had referred to cancer as more than sixty diseases; by the 1960s he was talking of a hundred different forms of cancer. In 1975, the M. D. Anderson surgeon Marvin Romsdahl, who was conducting research on melanomas, observed to Clark that among "the most important things we have been able to determine [about melanoma] is [that] the tumor is made up of a multitude of individual cell populations, each with its own genetic, biochemical, and physical characteristics." Clark had agreed in 1972 when Howard Skipper told him, "Cancer is a hundred or more diseases relatable to their normal cells of origin. . . . Some grow rapidly; some grow slowly. . . . Some types of cancer are almost always widely disseminated when detected."

Even at the fundamental level of treatment, an intellectual revolution was shaking oncology. In 1946, when Clark became director of M. D. Anderson, the fifty-year-old logic of William Stewart Halsted had still governed cancer medicine—that except for the leukemias and lymphomas, cancer at first was essentially a local disease, confined to the tumor site, contiguous tissues, and neighboring lymph nodes, and that the lymph nodes formed a temporary barrier to cancer cells. In the late 1960s and the 1970s, University of Pittsburgh breast cancer surgeon Bernard Fisher argued that breast cancer, and by logic other solid tumors, were systemic from the beginning, and that patient survival demanded systemic treatments. Most surgeons and radiotherapists at M. D. Anderson had come by the mid-1970s to accept this logic. Patients needed the local treatments of surgery and radiotherapy and the systemic treatments of combination chemotherapy.[36]

Unlike sufferers a generation before, patients afflicted with acute lymphocytic leukemia, choriocarcinoma, early stage embryonal testicular carcinoma, and early stage Hodgkin's disease could now reasonably anticipate a cure. In 1976, M. D. Anderson oncologists were employing thirty-three individual drugs and ten combination regimens. With combined melphalan-prednisone therapy, hematologist Raymond Alexanian was extending survival times for multiple myeloma patients. Although metastatic solid tumors exhibited powerful survival instincts, several seemed ready to surrender. Early stage Wilms' tumor was now curable; the surgical perfusion of chemotherapy drugs had enjoyed considerable success against melanoma; and a number of sarcomas wilted in the presence of new combination chemotherapy regimens. A few other solid tumors had even brighter forecasts. M. D. Anderson medical oncologists Jordan Gutterman, George Blumenschein, and Aman U. Buzdar had doubled survival time in certain breast cancers with a combination drug protocol of 5-fluorouracil, Adriamycin, cyclophosphamide—the FAC protocol.[37]

Emil Freireich, the 1972 winner of the prestigious Lasker Award and chair of DT, presided over a department that now handled 60 new leukemia patients a year, 150 patients with metastatic melanoma, 30 with metastatic breast carcinoma, 60 with soft tissue sarcomas, 70 with lymphomas, 70 with bronchogenic carcinoma, and 100 with other neoplasms. In 1975, Freireich had started a bone marrow transplant program to repopulate the blood-forming systems of leukemia patients. Hematologists Karel Dicke and Kenneth B. McCredie collected bone marrow from patients in remission, filtered out the leukemia cells, and stored the marrow for reintroduction into patients at the time of recurrence. The Clinical Research Center had forty-two beds—twenty on the twelfth floor and twenty-two on the eleventh. In June 1977, the first 5 patients were placed in germ-free laminar air flow rooms, where giant filters cleaned incoming air that otherwise might have carried dangerous pathogens and threatened patients with compromised immune systems.[38]

———

For all of its successes, however, M.D. Anderson could not save Jeffrey Gottlieb. His testicular cancer surged back and defied treatment. At the time of his death on July 1, 1975, Gottlieb headed M. D. Anderson's Chemotherapy Service. Freireich was heartbroken. "That we should lose him is nothing short of tragic. Every day we can cure more and more people of cancer. But there is always that small percentage for whom it seems nothing we can do really helps. The day we can stop every single case, is the day we are working for. It is the goal toward which Dr. Gottlieb was working." To honor him, M. D. Anderson established the Jeffrey Gottlieb Memorial Award, which recognizes leading medical oncologists.[39]

Clark held his surgeons in high esteem. The sharp animosities with radiotherapists had eased somewhat, and the virtues of cooperation had never been more evident. Helmuth Goepfert, a head and neck surgeon born in Chile and addicted to motorcycles, in the 1970s had fashioned a voice conservation protocol for patients with cancer of the larynx. Goepfert combined radiotherapy and surgery, and with technical innovations, he made most radical laryngectomies obsolete. For many patients, Goepfert saved not only their lives but also their voices. Ed White retired in 1978 after presiding over the department for thirty years. Veteran surgeon Richard Martin replaced him. White had never quite fulfilled Clark's academic expectations. Clark had asked him repeatedly, for example, to write a comprehensive textbook on cancer surgery, but White preferred scalpels and surgical suites to pencils and medical libraries. Still, during his watch, M. D. Anderson became a leader in emphasizing conservative surgical procedures. The logic of conservative surgery combined with radiotherapy revolved around the objective of preserving or even improving existing survival rates while doing less damage, especially to patients with early stage disease. Radical mastectomies for breast cancer became less common, as did amputations for soft-tissue sarcomas, pelvic exenterations for uterine and cervical cancer, extensive procedures for head and neck cancers, and removal of the rectum and anus for rectal cancer. In the new Department of Urology, Douglas Johnson was developing alternatives to the radical prostatectomy, a procedure that often left patients incontinent, impotent, or both.[40]

Clark advertised his new Department of Dental Oncology, the first of its kind in the world. For years, M. D. Anderson physicians and Dental Branch dentists had consulted regularly, usually in the treatment of head and neck cancers. The new department formalized the relationship. In addition to working with head and neck surgeons to preserve dental function and to construct maxillofacial prosthetic devices, its staff developed a sodium fluoride gel that decreased the severity of dental cavities in radiotherapy patients; reduced jawbone atrophy among radiated head and neck patients; and established joint programs for cancer prevention and cancer education.[41]

Clark would leave behind one of the best radiotherapy programs in the nation. More than two thousand cobalt-60 devices were at work around the world, and cancer treatment centers sought radiotherapists trained at M. D. Anderson, who now numbered in the hundreds. The 32–million electron volt linear accelerator was the largest of its kind in clinical use, and treatment of M. D. Anderson patients with the fast neutron therapy of the cyclotron at Texas A&M University would soon commence.[42]

Diagnostic of Radiology compared to Radiotherapy in preeminence. A decade before, when Clark brought Gerald Dodd back to Houston, he had promised all the money needed to build an advanced radiological laboratory, and Dodd certified that "he kept his promise." Radiology was undergoing an expensive technological revolution. The advent of computerized tomography imaging (CAT scan) in the early 1970s offered images of high quality at very low doses. Magnetic resonance imaging (MRI) ensued, providing radiologists with high soft-tissue resolution and discrimination on any imaging plane, using less dangerous, nonionizing radiation. Dodd had pioneered the technology of lymphangiography, which injected radioactive substances that then collected in denser tumor tissues, making it possible to identify some lymphatic metastases without surgery.

M. D. Anderson had also ventured into therapeutic radiology, in which radiologists finally crossed the threshold from diagnosis to treatment. Under Dodd's guiding hand, Sidney Wallace led the way. By the mid-1970s, he was introducing catheters into the vascular system and guiding them to a tumor site, where chemotherapy could be administered or where blood vessels could be occluded with gels and metal coils. They could be used to stop hemorrhaging or starve some tumors of their blood supplies.[43]

Genetics still held Clark tightly in its grip. In the late 1950s and early 1960s, M. D. Anderson geneticist David Anderson had shifted his research from cancer eye in cattle to melanoma in human beings. He identified the nevoid basal cell carcinoma syndrome, an inherited proclivity for basal cell carcinomas, medulloblastomas, and ovarian fibroid disease. In 1970, Louise C. Strong, a physician, came to M. D. Anderson and immersed herself in cancer epidemiology. Both Anderson and Strong focused on population genetics and clinical epidemiology rather than molecular genetics. Strong's investigation of sarcoma patients confirmed the existence of disease aggregates in some families. She also demonstrated that the etiology of many cancers could be found in complex interactions between genetic and environmental factors. Lee Clark found in her a natural choice to lead genetics.[44]

Cancer genetics had never been more promising. At the University of California, San Francisco, School of Medicine, biologists Michael Bishop and Harold E. Varmus had just reported their discovery of oncogenes, which regulate cellular growth and differentiation but can go awry under certain circumstances. When infected with some viruses or affected by certain chemical carcinogens, oncogenes can stimulate unregulated cell proliferation. The discovery, which would earn Varmus and Bishop the 1989 Nobel Prize in Medicine, displaced the prevailing wisdom that cancer was caused by viral genes, distinct from a cell's normal genetic composition, which exist indolently within the body's cells until triggered by carcinogens. Although molecular genetic research at M. D. Anderson was not nearly so sophisticated, Clark anticipated that genetic treatments for cancer would someday materialize.

Clark had also inserted M. D. Anderson into immunotherapy, with the young physicians Evan Hersh and Jordan Gutterman at its forefront. In 1972, NCI decided to fund immunotherapy programs on a contract basis, particularly to study the antitumor properties of bacillus Calmette-Guérin (BCG), an attenuated variety of mycobacterium bacillus long used in treating tuberculosis. In studies of mouse leukemia, BCG had stimulated the animals' immune systems, which identify the cancer cells as foreign and attack them. Clark handed the $350,000 NCI contract over to DT.

The NCI tumor immunology program doubled in size by 1973 and doubled again in 1974. Hersh and Gutterman conducted a variety of clinical research projects in immunology and medical oncology, testing a host of drugs, among them melphalan, Adriamycin, BCG, and arabinosyl-6-mercaptopurine on a host of cancers, including melanoma, leukemia, lymphomas, fibrosarcomnas, and breast and colorectal carcinomas. Between 1972 and 1977, they wrote or helped write more than ninety peer-reviewed articles. In Jordan Gutterman and Evan Hersh, Clark thought that he saw the future outlines of oncology.

On August 20, 1977, the UT Board of Regents announced Clark's retirement. They praised a career that had elevated cancer awareness throughout the nation, marshaled private and government resources against the disease, and transformed M. D. Anderson from a sleepy medical clinic to one of the preeminent medical institutions in the world. But the cure evaded him. The disease, or better yet the diseases, had defied his willpower, scorning the billions of dollars thrown at it. Actually, the intricate mix of tissue origin, tumor biology, and the immunological status of the patient produces a complex calculus of probabilities. In the year of Clark's retirement, more than 700,000 Americans were diagnosed with cancer, and 300,000 died of it. The identification of some universal principles governing cancer cells still appeared distant, but better control was in the offing.[45]

When he began his career, Clark liked to say, one of five cancer patients survived at least five years. In 1978, one in three did. Some attributed it to earlier diagnoses, not progress in treatment. Clark measured progress in the visage of survivors. A few days after his retirement was announced, former patient Gloria (Belsha) Robertson showed up for a friendly visit. In 1962, as a twelve-year-old, she had had an osteosarcoma with lung metastases. Surgeons amputated the leg and removed lesions from both lungs. A medical oncologist then tried phenylalanine mustard (PAM). The cancer never returned. In 1977, Gloria's husband and two-year-old son accompanied her to Clark's office. Clark bounced the little boy on his knee.[46]

Monroe Dunaway Anderson, a Tennessee banker who became a successful Houston cotton merchant, decided to create a charitable foundation in 1936, three years before he died. In 1942, trustees of the M. D. Anderson Foundation provided five hundred thousand dollars, an interim site, and land on which to build what would become the University of Texas M. D. Anderson Cancer Center.

Part of the former stables and carriage house on the six-acre Baker estate was converted into a research laboratory where scientific studies were conducted while the hospital was being planned. Shortly before Christmas 1942, four scientists began biochemistry and biology projects that would lead to one of the world's most productive cancer research programs.

Twelve surplus army barracks from World War II purchased by the M. D. Anderson Foundation were moved to the Baker estate in the late 1940s and converted to inpatient wards and an operating room. Behind the barracks is the Baker family's main house, which was used as administrative offices by M. D. Anderson Hospital for more than a decade.

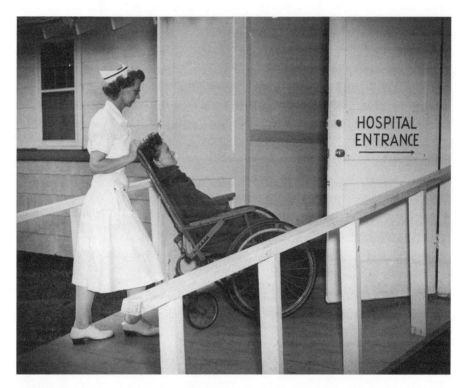

Ornell Balzer, R.N., one of the first nurses hired to care for hospital patients at the Baker estate, wheels a patient into one of the converted barracks in 1951.

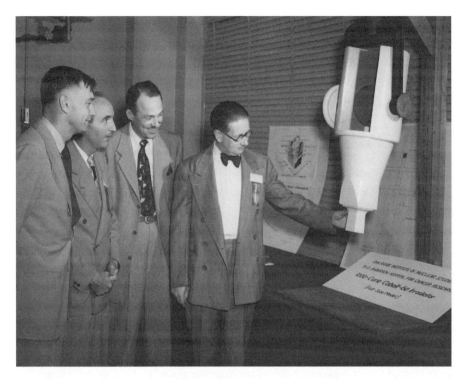

The cobalt-60 machine was designed on the Baker estate by Leonard G. Grimmett, Ph.D., an English physicist, and Gilbert H. Fletcher, M.D., a French-born pioneer in radiotherapy. It revolutionized the delivery of radiation beams to treat cancer. The machine was chosen in 1950 over eleven other proposals for development by the U.S. Atomic Energy Commission. Examining a model are (*left to right*) Marshall Brucer, Ph.D., of the Oak Ridge Institute of Nuclear Studies; Dr. Fletcher; R. Lee Clark, M.D., who was M. D. Anderson's first full-time director and president from 1946 to 1978, and Dr. Grimmett.

The first newly constructed M. D. Anderson Hospital was built in the early 1950s on a heavily wooded site donated by the M. D. Anderson Foundation. The hospital, in the newly created Texas Medical Center, was described by *Time* magazine as the "Pink Palace of Healing" because of the shining Georgia Etowah pink marble exterior, which Dr. Clark thought represented optimism for controlling cancer.

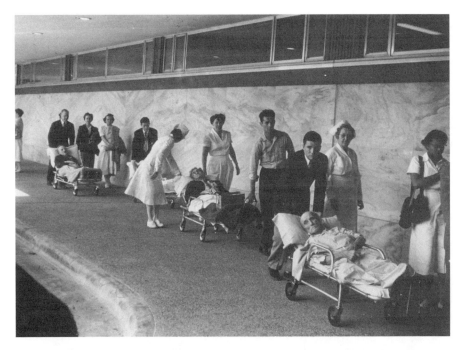

More than thirteen thousand patients had been treated on the Baker estate by the time a caravan of ambulances moved forty-six patients to the new M. D. Anderson Hospital on March 19, 1954.

Soon after becoming director of nursing in 1955, Renilda Hilkemeyer, R.N. (*third from right*), started the Foreign Exchange Nurse Visitor Program to help nurses from many countries learn how to provide cancer nursing care. She is generally regarded as the architect of oncology nursing philosophy and practice. During her twenty-nine years at M. D. Anderson, she trained two generations of nurses to mix and administer many new anti-cancer drugs on inpatient units and helped open the first outpatient clinic for patients to receive chemotherapy.

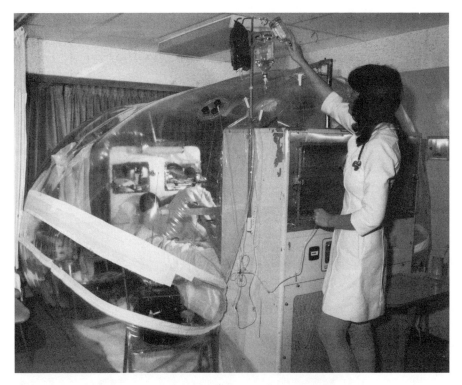

In 1966, M. D. Anderson was the nation's first cancer hospital to install "life islands" for patients whose immune systems were severely depressed. During the eleven years they were in use, the germ-free plastic isolation units helped reduce life-threatening infections for leukemia patients.

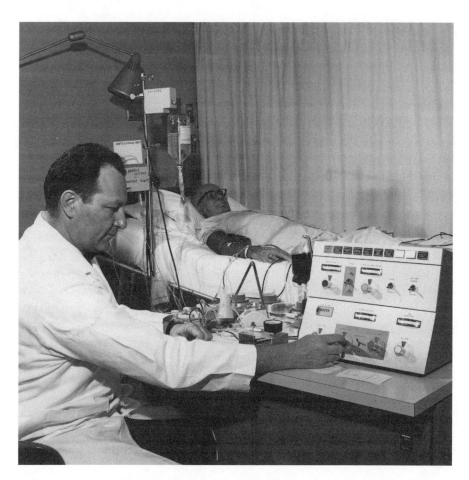

Leukemia specialist Emil J Freireich, M.D., who joined the M. D. Anderson faculty in 1965, helped design the continuous-flow blood cell separator now used throughout the world. The machine collects whole blood and separates it into cellular components that can be transfused to patients to combat infections, control hemorrhaging, and help manage other complications of cancer and its treatment.

T. C. Hsu, Ph.D., led the team of M. D. Anderson scientists who in 1970 developed a technique known as C-banding, which made it possible to pinpoint the location of genes on chromosomes. Years earlier, he had demonstrated that normal cells have forty-six chromosomes, which are arranged in twenty-three matched pairs. Dr. Hsu, often called the father of modern cytogenetics, continued to teach, write, and work in his laboratory until shortly before he died in 2003.

Under terms of the National Cancer Act of 1971, M. D. Anderson was named one of the first three Comprehensive Cancer Centers. The legislation created a three-member President's Cancer Panel to oversee federal progress against cancer. The first panel, shown here from *left to right,* included philanthropist Benno C. Schmidt, President Richard Nixon, M. D. Anderson president R. Lee Clark, M.D., and immunologist Robert A. Good, M.D.

Left to right: Charles A. LeMaistre, M.D., president from 1978 to 1996, Houston banker Ben Love, and San Antonio businessman Red McCombs—the last two, former chairs of M. D. Anderson's Board of Visitors—discuss plans to expand the M. D. Anderson complex. In 1957, Love's wife, Margaret, was treated for the first of three successful bouts with cancer, and over the years he led several major capital campaigns for the cancer center. In 1995, the Margaret and Ben Love Clinic was dedicated in the couple's honor. The generosity of McCombs and his family, who donated $30 million to support bold new research programs, was recognized in 2005 when the Red and Charline McCombs Institute for the Early Detection and Treatment of Cancer was announced.

All M. D. Anderson employees were encouraged to wear T-shirts with the slogan "Fighting Cancer. Now *That's* a Job!" during a National Hospital Week celebration in the early 1980s. Setting an example are key management members (*left to right*) Robert D. Moreton, M.D., vice president for patient affairs; Charles A. LeMaistre, M.D., president; E. R. Gilley, executive vice president for administration and finance; and Robert C. Hickey, M.D., executive vice president and head of the Division of Surgery.

M. D. Anderson's Science Park–Research Division is located in the Lost Pines region of Central Texas near Smithville, where teams of scientists at the Virginia Harris Cockrell Cancer Research Center study the environmental causes of cancer. Their landmark discoveries include demonstrating the first direct link between a chemical carcinogen in tobacco smoke and human lung cancer.

During her thirty years as a radiotherapist at M. D. Anderson, Eleanor D. Montague, M.D., was instrumental in showing that many women with breast cancer could be treated successfully without undergoing the once-routine radical mastectomy. Her milestone reports published in the 1970s and 1980s demonstrated that lumpectomy (a breast-sparing procedure in which only the tumor is removed) combined with radiation therapy could be as effective as radical mastectomy for certain breast cancer patients.

John Mendelsohn, M.D., a prominent physician-scientist, became M. D. Anderson's third president in 1996. He has pioneered laboratory and preclinical research to prevent cancer cell proliferation by blocking epidermal growth factor receptors on cell surfaces. He developed the monoclonal antibody 225 that led to creation of the anticancer agent cetuximab (Erbitux). Dr. Mendelsohn (*left*) discusses related translational research with Rakesh Kumar, Ph.D., professor of molecular and cellular oncology, and Zhen Fan, M.D., associate professor of experimental therapeutics. Photo by Beryl Striewski.

In 2001, Margaret L. Kripke, Ph.D. (*left*), then executive vice president and chief academic officer, presented the first bachelor's degree awarded by M. D. Anderson to Christina Rollins, who received her degree in cytogenetic technology from the School of Allied Sciences. Since graduating, Rollins has worked in the cancer center's Department of Cytogenetics, where she performs staining techniques on chromosomes extracted from patients' blood, bone marrow, and tissue samples. These tests help physicians make definitive diagnoses and monitor treatments. Photo by F. Carter Smith.

In 2004, former president George H. W. Bush and his wife, Barbara, learned that the Robin Bush Child and Adolescent Center would be named for their young daughter who died from leukemia in 1953. Three pediatric cancer patients brought the Bushes a present announcing the naming of the center. At the event, which marked the June birthdays of both Mr. and Mrs. Bush, the couple also learned that more than $50 million had been raised for the George and Barbara Bush Endowment for Innovative Cancer Research at M. D. Anderson. Photo by Pete Baatz.

Cyclist Lance Armstrong (*left*), who won seven Tour de France races after treatment for metastatic testicular cancer, brought the 2005 Bristol-Myers Squibb Tour of Hope to M. D. Anderson, where President John Mendelsohn, M.D., welcomed him for a survivors' rally and seminar about the importance of clinical trials. Photo by F. Carter Smith.

This aerial view includes most of M. D. Anderson's main campus, where patients and visitors enter through the R. Lee Clark Clinic (*center*). Outpatient clinical buildings (*right*) are named for philanthropists Margaret and Ben Love and former M. D. Anderson president Charles A. LeMaistre, M.D. The Lutheran Hospital Pavilion is left of the main entrance; directly behind it are the Albert B. and Margaret M. Alkek Hospital, the Clinical Research Building, and the George and Cynthia Mitchell Basic Sciences Research Building. Skybridges connect these facilities to the Lowry and Peggy Mays Clinic, the Cancer Prevention Building, the Faculty Center, the T. Boone Pickens Academic Tower, and the Jesse H. Jones Rotary House International, which are across the street. Photo by Jim Olive.

In 1997, Bill Schultz, a long-term pancreatic cancer survivor, became a volunteer at M. D. Anderson so he could comfort other patients confronting cancer. In just over ten years, he had logged almost sixty-five hundred hours in the Volunteer Services' Gift Shops and the Anderson Network's Hospitality Center, as well as with other programs that support patients and their families. In 2008, he began volunteering one day a week at M. D. Anderson's satellite Radiation Treatment Center in Fort Bend County near his home in Rosenberg, Texas. One of more than fifteen hundred hospital-based volunteers, Schultz says, "[I am] grateful that I can give hope to others," especially patients with pancreatic cancer. Photo by Wyatt McSpadden.

The Red and Charline McCombs Institute for the Early Detection and Treatment of Cancer provides new opportunities for teams of scientists and clinicians from many different departments to collaborate on research problems of common interest. International leaders in six fields direct programs incorporating the latest research in molecular-based approaches to cancer diagnosis and treatment. Shown here with M. D. Anderson president John Mendelsohn, M.D. (*center front*), they are Gordon B. Mills, M.D., Ph.D., director of the Robert J. Kleberg Jr. and Helen C. Kleberg Center for Molecular Markers (*left front*), and Isaiah J. Fidler, D.V.M., Ph.D., director of the Cancer Metastasis Research Center (*right front*); (*standing left to right*) James D. Cox, M.D., director of the Proton Therapy Center; Garth Powis, D.Phil., director of the Center for Targeted Therapy; Yong-Jun Liu, M.D., Ph.D., director of the Center for Cancer Immunology Research; and Juri G. Gelovani, M.D., Ph.D., director of the Center for Advanced Biomedical Imaging Research. Photo by John Everett.

9

Charles A. LeMaistre and the Consolidation of Excellence, 1978–1983

They were just so sick and there was so little we could do.
—PETER W. MANSELL, 2005

In 1978, Charles A. LeMaistre arrived at M. D. Anderson. So did Melissa Gallardo. Like Freddie Steinmark nine years before, she had bone cancer. Just sixteen years old, Melissa had lost her little brothers Marcos and Leo to osteosarcomas, Marcos when he was five and Leo at eleven. Amputation and radiation did not save Marcos's leg or his life. He died nine months after diagnosis, and the osteosarcoma in Leo's hip took his life in just eleven months. The Gallardos came to M. D. Anderson. As expected, the pediatricians recommended amputation and adjuvant chemotherapy. Melissa blanched. She remembered Marcos limping and dying anyway. She insisted on keeping her leg.

For years, M. D. Anderson surgeons had worked toward "limb salvage" surgery—the treatment of soft tissue sarcomas and bone tumors without amputation. Long-term survival for many soft tissue sarcomas was the same with amputation as with excision surgery and radiation. Since Melissa recoiled at the thought of amputation, pediatricians decided to radiate the tumor site and then administer chemotherapy. After six months of unsuccessful treatment, they again recommended amputation and Melissa again refused. Pediatric oncologist Wataru Sutow then tried a different drug, which shrank the tumor substantially, enough for surgeons to remove the diseased bone and to install a stainless steel implant. Melissa returned to El Paso for eighteen months of chemotherapy, and rehabilitation specialists soon had her walking again.

Multidisciplinary care explained the turnaround. Raphael Pollock sat on the cusp of the change. In 1981, after a surgical residency at the University of Chicago, he arrived at M. D. Anderson for a fellowship. In one of his last procedures at Chicago, a patient presented with an osteosarcoma in the arm. Surgeons amputated the arm and the shoulder, a standard, radical approach to the lesion. The M. D. Anderson fellowship required six months of training in surgery, three months each for radiotherapy and chemotherapy, and a separate research project. When medical oncologist Robert Benjamin asked Pollock how to treat a young man from India with an identical osteosarcoma, Pollock recommended amputation. To Pollock's astonishment, Benjamin called for limb salvage surgery, chemotherapy, and radiation. "I have since played tennis with that young man," Pollock remembered.[1]

<hr/>

R. Lee Clark's successor at M. D. Anderson would inherit a culture nearly forty years in the making and marked by commitments to high-quality patient care, multidisciplinary cooperation, and treatments designed, whenever possible, to cure the disease, as long as the hospital possessed the resources for supportive management of side effects.

The quality of M. D. Anderson patient care, widely acknowledged as among the best in the world, revolved around a respect for patients permeating the entire staff. New employee orientations, annual evaluations of employee performance, and meetings in every department repeated the mantra of quality care through cleanliness, respect for patient opinions and anxieties, and helpfulness. Nothing could get an M. D. Anderson employee in trouble more quickly than rudeness to a patient. Between 1942 and 1978, Ernst Bertner and Lee Clark had approached patients with warmth and optimism, values that over time solidified into an institutional culture. The new president would do well to have a similar personality.[2]

To find such a person, former governor Allan Shivers and the board of regents created a search committee chaired by Charles A. LeMaistre, the chancellor of the University of Texas System. LeMaistre also chaired a second committee to select a president for the University of Texas Health Science Center at Houston. Shivers served on both committees. Except for LeMaistre and Shivers, the first committee consisted largely of M. D. Anderson people.

Robert Hickey was interested. His credentials were above reproach, and he was a decent man, but he carried too much baggage. Over the years, Clark had delegated to Hickey much of the dirty work—having him deliver the bad news of why budgets had to be cut, why promotions or pay raises could not be given, why laboratory space could not be increased—all the "why nots" of academic administration. Hickey had been forced on so many occasions to deliver bad news that some critics labeled him the "Abominable 'No' Man."

Joseph T. Painter was interested as well. Clark had encouraged him to apply, and Shivers respected him immensely. The Painter name still carried weight at the University of Texas. Judicious and diplomatic, Painter had the temperament for the job. A long career in private practice suited him as a liaison with the medical community, but he was not well known among the faculty at M. D. Anderson because most of his time had been spent on external activities working with physicians, medical societies, and professional organizations.

Emil J Freireich was one of the finest medical oncologists in the world. Throughout his career, he had refused to employ placebos in clinical research on the grounds that some of those patients would die. Freireich designed clinical trials in which all patients received protocols that offered at least a measure of hope. For all of his acumen, however, he was too blunt for academic administration where persuasion and patience are more likely to produce success.

The search committee gave close consideration to one outside candidate—Paul Marks, the highly regarded cell biologist and director of the Columbia University cancer center. Marks would go on one day to lead Memorial Sloan-Kettering, but in 1978 he had the wrong zip code. Shivers wanted new blood at M. D. Anderson, but not New York City blood. He preferred someone with Lone Star sensibilities. And Marks's appearance before the committee went poorly, with some members concerned about whether he would fit in at M. D. Anderson. Marks felt the same way; on the drive to the airport, he said as much to LeMaistre. As the interviews and discussions proceeded, it became clear that, in LeMaistre's words, "there was no 'new' Lee Clarks among the candidates."

Governor Shivers had several favorites, but the candidates either did not do well at interviews or were rejected by faculty members on both committees. After all of the candidates had been eliminated, Shivers asked LeMaistre if he wanted to be a candidate. If not, Shivers intended to reintroduce the names of those he had championed. LeMaistre then agreed to become a candidate for both positions, resigned the chairmanship of the search committees, and participated in interviews with the faculty of both institutions. When each committee returned a favorable vote, Shivers laughed and told him, "You will have to choose which one you will be unpopular with." Because of his interest in lung cancer and his relationship with Lee Clark, LeMaistre chose M. D. Anderson.[3]

———

Charles Aubrey LeMaistre was born in Lockhart, Alabama, on February 10, 1924. His father, John Wesley LeMaistre, had hired on in 1902 as a surveyor for the Jackson Lumber Company to help build the town of Lockhart. In 1908, he settled down, marrying Edith McLeod, a schoolteacher from Orion, Alabama. They became members of the First Baptist Church of neighboring Florala, Florida, and he joined the local Masonic Lodge. In 1918, voters put him on the school board and in 1923 sent him to the state legislature.

In 1923, LeMaistre invented the technique of end-matching white pine flooring, an innovation that eliminated vast amounts of wastage and improved the bottom line of every company in the industry. In 1926, he introduced and the legislature passed a reforestation bill, the first in the nation, requiring the planting of seedlings on harvested land. Aware that timber production in Covington County had peaked, LeMaistre set in motion the gradual deconstruction of the town, making sure that company employees could purchase their own homes, a decision that solidified his reputation as a "high and important man . . . widely esteemed by the meek and the lowly." His salary, investments, company stock, and directorship of the First National Bank made him a rich man.

Charles LeMaistre was the youngest of Edith and John LeMaistre's six children. He was nicknamed "Mickey." "My schoolteacher aunts referred to me as Charles," LeMaistre later recalled, "as did most other teachers and my professors. . . . But I always introduced myself as Mickey LeMaistre, and after a while no one tried to talk me out of using it. Frankly, I think my nickname has made me more approachable."[4]

In 1929, Mickey's world flipped. After several years of congestive heart failure, John Wesley LeMaistre died suddenly of a strep infection. Anxious to flee the plantation mentality in Lockhart, Edith moved the family to Tuscaloosa, home of the University of Alabama. When the "Scottsboro boys" hit the headlines, Edith gave her children an object lesson in racism. In 1931, eight black men between the ages of fifteen and twenty had been falsely accused of raping a white woman. A change of venue brought the trial to Tuscaloosa, where a spectacle ensued. It attracted the KKK, a brass band playing "Dixie," and thousands of jeering spectators. Edith LeMaistre thought the Scottsboro case was disgraceful. She put the children into the family car and drove through downtown on University Avenue so they could get a good look at the dark underbelly of American life. The streets were lined with rolled barbed wire and National Guardsmen. Mickey never forgot the lesson. He absorbed the polite gentility of southern life without its visceral hatreds.

From his mother, Mickey inherited a sense of destiny that John Calvin would have appreciated. At the age of ten, he suffered a ruptured appendix while walking in the woods, surviving only because his pet dog, Inga, raced home barking a distress call. To Edith, the entire episode seemed understandable only through a biblical sense of destiny. God had something special in store for Mickey. When he became president of M. D. Anderson, LeMaistre remarked, "I have believed for some years now that I was spared for a special reason."[5]

In 1940, Mickey graduated from high school. He tinkered with the idea of attending the U.S. Naval Academy, but a recent eye injury sidetracked him. Mickey went to the University of Alabama instead and enrolled in premed in the Army Specialized Training Program. The university had only a two-year medical curriculum, and in 1945 he transferred to Cornell University Medical College in New York City, finishing the curriculum in 1946. With an interim of approximately six months before his internship started, he was assigned to Halloran General Hospital on Staten Island as a medical corpsman. In 1947, with a freshly minted medical degree, he accepted an internship and then a residency in internal medicine at New York Hospital. LeMaistre followed up with postdoctoral fellowships there under physician Walsh McDermott, a leader in infectious diseases who was treating pulmonary tuberculosis with antibiotics. Impressed with the young fellow, McDermott recommended that David Barr, chairman of the Department of Medicine, hire LeMaistre as an instructor, and Mickey focused on diseases of the chest.

At the same time, LeMaistre trained at the Communicable Disease Center (CDC) in Atlanta and was in the first class of the Epidemic Intelligence Service (EIS), charged with hunting the earth for infectious microorganisms and North Korean and Soviet-funded germ warfare programs. Because EIS required a national security clearance, LeMaistre kept the appointment a secret from everybody, including Joyce Trapp, a University of Alabama student he had recently met. In January 1952, at the Navajo Indian Reservation in Tuba City, Arizona, an outbreak of infectious hepatitis had sickened 315 of 417 Navajo students at a government school. The Bureau of Indian Affairs contacted CDC director Alexander Langmuir, who dispatched LeMaistre on a top secret mission to Tuba City to investigate. Two children had died, and after examining others, LeMaistre identified the disease as hepatitis.

The challenge was twofold, first to care for the 313 Navajo children ill with hepatitis and second to address the problem of untreated tuberculosis among the Navajo. Because no hospital was available, the task involved converting the boarding school into a care facility and using the faculty and staff as nursing personnel. LeMaistre contacted Langmuir and requested air drops of hospital and medical supplies. Within thirty-six hours, LeMaistre had supplies for taking temperatures and for recording clinical findings. After training the well-educated boarding school staff to double as medical assistants, LeMaistre asked McDermott at New York Hospital to dispatch Mary Ann Payne, a specialist in hepatitis at Cornell. She arrived in Tuba City in short order. Payne completed the laboratory and clinical studies and for the next year treated the children. All of the children recovered.

With the nearby hospital for tuberculosis full of seriously ill patients, others were being turned away with no treatment at all. The reservation's contaminated water supplies, poor sanitation systems, and windowless mud and log houses were incubators for tuberculosis. With the permission of Langmuir, LeMaistre asked McDermott for assistance in controlling the serious problem of untreated tuberculosis on the Navajo Reservation. McDermott had a contract with Squibb Company to test isoniazid, a new drug with antitubercular properties that could be used on an ambulatory basis. Cultural and political roadblocks, however, stood in the way. Navajos trusted Arizona rattlesnakes more than white people and had little inclination to let government doctors experiment on them.

At LeMaistre's request, the Navajo Tribal Council invited McDermott to build an ambulatory health system on the reservation, dedicating ten thousand dollars of tribal funds to the cause. Aware that the project would cost far more, LeMaistre raised money privately. McDermott assigned Dr. Kurt Deuschle to develop the new program. The ambulatory care system he developed, which was known as "Many Farms," became the pattern for ambulatory care programs on Indian reservations throughout the nation. While Deuschle did his job, LeMaistre worked with Payne and executed other assignments for CDC. McDermott's persuasive manner assured Navajo leaders that they were receiving the best treatment available and that it had already been fully tested on whites. The drug rinsed TB bacillus from the blood of Navajo children.[6]

On June 3, 1952, in the middle of the Tuba City project, LeMaistre married Joyce Trapp, and they settled temporarily in Atlanta when LeMaistre was chosen to teach the second EIS class. With its combination of urban living and southern ambience, Atlanta suited the LeMaistres, but Mickey had unfinished business at Cornell, and when he completed the EIS tour, they moved to New York. The couple soon missed Atlanta, however, and they returned in 1954 when the Emory University School of Medicine offered Mickey an academic appointment. LeMaistre shone at Emory, exhibiting fine clinical teaching abilities and even better administrative skills. In the most high-pressure and chaotic situations, he always retained a clear head and emotional equilibrium. When staff members approached him with problems, he assumed the role of a psychologist, not a pulmonologist.

In 1957, Emory named him coordinator of clinics at Grady Memorial Hospital and chair of its new department of preventive medicine and community health. LeMaistre was a natural. In Tuba City, he had seen how Navajo poverty exacerbated disease and how a sustained community health campaign could contain it.

After five years in Atlanta, LeMaistre took a faculty position at the UT Southwestern Medical School in Dallas. Mickey and Joyce loaded three children in the family car and drove west. In August 1959, Dallas welcomed them with 108-degree temperatures. "I felt certain Joyce thought we'd made a mistake, but everything worked out well," he recalled. At Southwestern, LeMaistre supervised residents and fellows, treated patients at Parkland Hospital, and served as medical director of the chest division at Woodlawn Hospital. From that post in August 1962, he was appointed to the U.S. Surgeon General's Advisory Committee on Smoking and Health. Its report in 1964 set off a powerful public health wave, with LeMaistre at its crest.[7]

In 1965, he became associate dean of UT Southwestern. The position brought him to the attention of Harry Hunt Ransom, the visionary chancellor of the University of Texas. With the demographic shift to the Sunbelt well under way, Ransom asked LeMaistre to take a leave of absence and study the state's future health manpower needs. LeMaistre accepted, organized a committee, and produced a report that recommended two new medical schools, in San Antonio and Houston, five new nursing schools, five allied health sciences schools, a school of public health, and a doubling of enrollment at the existing medical schools.

Ransom found in LeMaistre a supple intellect and political savvy, and he plot-
ted a future for him in the UT System. In 1966, he convinced LeMaistre to become the system's vice chancellor for health affairs. It was not an easy sell. "I
really left Cornell and then Emory to avoid becoming an administrator,"
LeMaistre remembers, "[but] Harry Ransom was the most persuasive person
I'd ever met." The vice chancellor had no trouble delegating authority, wanted
people to succeed but had no need to please, and was unflappable. Ransom soon
promoted him to executive vice chancellor, and then in 1969 to deputy chancellor. In 1971 the board of regents appointed LeMaistre chancellor of the University of Texas System, the only physician to ever occupy the office. LeMaistre
had planned to spend only four years at the helm, but former governor Allan
Shivers, a recent appointee to the board, had conditioned his acceptance on
LeMaistre's staying. When the fourth year was about to pass, Shivers cajoled
LeMaistre into remaining at the helm. Under LeMaistre, the University of Texas
System added campuses throughout the state.

—⟶ଊୋୋ⟵—

LeMaistre had a full plate, and he first concentrated on working out the kinks in
the hospital's relationship with private practice physicians, the men and women
who referred patients to M. D. Anderson. By the end of Clark's tenure, the problem had become acute. Because all patients seeking treatment needed referrals
from a private physician, the issue spoke to the financial health of M. D. Anderson and to the well-being of its patients. Three of M. D. Anderson's leading physicians headed the effort: Joseph Painter, Joseph Ainsworth, and Robert Moreton.
    Painter was wired into the Texas medical establishment, a man personally
familiar with hundreds of physicians, known by thousands, and acutely sensitive to the emotional pressures on cancer patients. He strengthened relations
with medical and professional societies. At Painter's suggestion, M. D. Anderson organized the Texas Physicians Advisory Council in 1980. Composed of
twenty-one private physicians from throughout the state, the council suggested
that M. D. Anderson make a priority of sending patients home with information important to their own recovery and informing their physicians of the recommendations as well.
    Ainsworth focused on relationships with family physicians, drilling into M. D.
Anderson specialists the importance of keeping good records and maintaining contact with the referring doctors. Expeditiously returning telephone calls
to physicians concerned about their patients helped to ease frustrations, as did

a beefed-up staff to make sure that correspondence from private physicians was answered promptly.

Robert Moreton, whose gentle demeanor and reassuring voice could calm the beast in anybody, assumed the task of telephoning disgruntled doctors, apologizing for the actions that produced the complaints, promising remedial action, and then following up with patients. He invested herculean energy in the task, spending several hours a day soothing bruised feelings.[8]

The office of director of research was going to be elevated to a vice presidency, and LeMaistre considered the post a critical priority. He offered the position to Frederick Becker, a three-year veteran of M. D. Anderson who had replaced William Russell as chair of pathology. José Trujillo, a native of Panama and beloved chair of the Department of Laboratory Medicine at M. D. Anderson, had suggested Becker as Russell's replacement. Trujillo was widely known for demonstrating the clinical relevance of chromosomal abnormalities, and he eventually produced routine laboratory tests to reveal the subtypes of leukemia and recommendations for the optimal treatment of each. At the time, Becker was interim chairman of the department of pathology at NYU and director of pathology at Bellevue Hospital, and he enjoyed an international reputation in liver pathology. On Trujillo's recommendation, Clark had pursued Becker. Over dinner at the Marco Polo Club in New York during meetings of the American Cancer Society, Clark offered Becker the job. Becker accepted, though he had once "thought it was more likely that I would go to the moon than to Texas, and I was not sure that I wanted to live in a place where the only sin is shouting." He considered M. D. Anderson highly inbred, with a southern identity and a pathology department strong on diagnosis but weak on research. "Lee Clark, however, was very persuasive," Becker remembered. At Becker's first faculty meeting, when Clark bragged about his education at Columbia, NYU, and Harvard, somebody shouted, "And not a good football team among them."[9]

In recruiting Becker, Clark had bucked tradition. Throughout its history, M. D. Anderson had never appointed a Jewish department chair. The concerns about anti-Semitism bubbled at a subterranean level, but they were nevertheless real, at least in the memories of some Jewish faculty. Becker epitomized a Texan's stereotype of New York Jews—blunt, bright, and not easily intimidated. During the recruitment process, Clark sent Robert Hickey to New York to check out Becker, and Hickey came away from the meeting reassured that a New York Jew could bring esteem to the Department of Pathology at M. D. Anderson and elevate its commitment to research. Times were changing.

Diminutive in stature, endowed with an unusually absorbent mind, and given to pacing while he talked, Becker immediately established a presence at M. D. Anderson. One colleague likened his mind to a "galactic garbage pail. It collects everything and forgets nothing." Becker soon realized that his previous suspicions about M. D. Anderson had been justified. The institution was badly inbred and incapable of self-criticism. In 1978, before accepting the presidency of M. D. Anderson, LeMaistre visited with his friend Lewis Thomas, the president of Memorial Sloan-Kettering, who warned that anti-Semitism, racism, and sexism still lurked at M. D. Anderson. In LeMaistre's memory, "These were not big hurdles. . . . The institution was ready for change with no resistance with regard to race and very little if any anti-Semitism but was rather staunch in its feelings that women oncologists did not measure up to the standards of men." He remembered visiting the most recalcitrant department heads. "I gave them the opportunity to change the status or to see their successors do so." One year later faculty search committees were all given the charge to search for minority leaders in oncology. In his search for a vice president of research, LeMaistre did not look far. Becker accepted and became the first Jew in M. D. Anderson history to exercise such administrative power.[10]

Progress in hiring minority and women faculty members, however, would prove to be daunting. Nearly a decade later, a survey would be conducted by several women faculty, including experimental radiation oncologist Elizabeth Travis, immunologist Margaret Kripke, Lillian Fuller, epidemiologist Margaret Spitz, and Judy Watson in the president's office. News circulated quickly. Frequently the women would lunch together at a table in the dining room. On one occasion, a male faculty member would joke that the women were plotting the downfall of the institution. "No," they replied, "just the takeover."

The survey would reveal that of 127 full professors at M. D. Anderson, only 2 were women. There were 15 women among the 126 associate professors, and 35 among 153 assistant professors. In 1988, M. D. Anderson would count no black full professors, 3 black men and no black women as associate professors, and 1 black man and no black women as assistant professors. Hispanics were only marginally better off. In 1990, the women established the Women Faculty Organization.[11]

—⟨⟨⟨⟩⟩⟩—

As 1978 gave way to 1979, LeMaistre also christened cancer prevention. After the surgeon general's report, investigators throughout the nation hastened to further document the relationship between smoking and cancer, and by the early 1990s, more than sixty thousand studies would confirm tobacco use as a factor in the etiology of cancers of the lung, larynx, oral cavity, bladder, renal pelvis and ureter, breast, prostate, colon, esophagus, cervix, pancreas, stomach, penis, and rectum, as well as myeloid and lymphoid leukemia. The science spawned a powerful, grass-roots political movement. In the 1970s, such political action groups as Action on Smoking and Health and Americans for Nonsmokers' Rights sought bans at the local, state, and federal level on the free distribution of cigarettes, commercial advertisement of tobacco products, smoking on airlines, and restrictions on smoking in public venues. In 1978, the American Medical Association issued its report *Tobacco and Health*, indicting the tobacco industry as a purveyor of disease and death.

At approximately the same time, diet and nutrition appeared as the second early front in cancer prevention. Epidemiological studies revealed substantial ethnic differences in cancer rates according to the volume of animal fat in the typical diet. Nations and ethnic communities with low-fat diets had lower incidences of cancers of the breast, stomach, ovaries, prostate, colon, and endometrium, as did those with diets rich in fiber. In 1982, the National Academy of Sciences released its landmark study on diet, nutrition, and cancer.[12]

Cancer prevention at the time involved little more than acting on classical epidemiological data by reducing tobacco use and improving nutrition, but the field accelerated as other modalities seemed to stall. Critics sounded the alarm, wondering whether hundreds of millions of dollars had been spent poorly, rewarding bad science at the expense of prevention, and whether better-financed public health and education programs could have prevented more cases of cancer than surgeons, radiotherapists, and medical oncologists had cured. Some environmentalists insisted that government regulation of chemical companies could reduce the incidence of cancer, and they cited as proof the case of Love Canal, a residential neighborhood in Niagara Falls built atop a toxic waste site and afflicted with high rates of liver disorders, miscarriages, birth defects, and cancer. The ghost of Rachel Carson once again stalked the land.

In 1978, Samuel Epstein, head of the toxicology and carcinogenesis laboratories at Children's Cancer Research Foundation in Boston, insisted in his book *The Politics of Cancer* that increases in the incidence of cancer could be traced to industrial pollutants, a claim disturbing to corporate America, which found environmentalists guilty of carelessly applying the results of animal testing to humans and trafficking in scare tactics. In her books *Panic in the Pantry* (1975) and *Toxic Terror* (1985), Elizabeth Whalen, a Harvard-trained demographer and cofounder of the American Council on Science and Health, discounted industrial pollutants as a cause of human cancer. Biochemist Bruce Ames of UC Berkeley rejected chemical pollution as a significant cause of cancer and insisted that the natural environment wallowed in carcinogens posing a greater threat than chemical companies.[13]

Congress responded with the Community Mental Health Center Extension Act of 1978 and the Biomedical Research and Research Training Amendments of 1978, mandating an expanded research program for the prevention of cancers caused by occupational or environmental exposure to carcinogens and requiring that five of the appointees to the National Cancer Advisory Board be knowledgeable in environmental carcinogenesis.[14]

The data on chemical pollution and carcinogenesis were mixed, and LeMaistre was a measured man, inclined neither to quick conclusions nor rash statements. He would let the Environmental Science Park grapple with the issue. To determine whether America was drowning in chemical carcinogens required more research; in the meantime, he intended to have M. D. Anderson take the lead in cancer prevention. He established the Department of Cancer Prevention, the only one in the world, and to lead it, he hired Guy Newell, the deputy director of NCI. Highly respected for his research in cancer epidemiology, environmental carcinogenesis, and nutrition, Newell brought to M. D. Anderson's fledgling program a certain cachet. With Newell came Peter W. Mansell, a British-born surgeon and oncologist, as associate director.[15]

LeMaistre then set out to "cope with the burgeoning size of the institution" and to ensure that day-to-day decisions "were made at the point closest to the action." Under Clark, the administrative flow chart had come to contain more dotted than straight lines. An outside consultant had characterized M. D. Anderson as a group of silos. LeMaistre created several new vice presidencies, including the vice presidency of research, which Becker held. Robert C. Hickey remained as executive vice president and director of M. D. Anderson, with the charge of managing the day-to-day affairs of the office of the president. Fred Conrad was named vice president for patient care. Before coming to M. D. Anderson in 1978, Conrad had managed several air force hospitals. Conrad was charged with creating a new design for ambulatory care to allow the majority of patients to be treated without ever being admitted as an inpatient. Patient convenience was to be the driving force behind the design. That design also contained the seeds of what later became known as the organ site centers in the clinics with all the professional expertise readily available to outpatients. LeMaistre appointed Robert D. Moreton interim vice president for patient affairs and began a search for a permanent occupant. A new vice president for academic affairs would supervise all educational programs, which included more than fifteen hundred residents, fellows, and interns. The virologist James Bowen, who had been serving as associate vice president for research, would fill the post.[16]

⁓⟪⟫⟫⁓

Before he could undertake other issues, LeMaistre had a financial fire to extinguish, a scandal to clean up, a physical plant to expand, a nursing shortage to address, a horrific tragedy to manage, and a key vacancy to fill.

On May 14, 1976, the board of regents had approved Clark's plan to centralize meal preparation in the Texas Medical Center. He anticipated that eighteen thousand meals a day would be prepared at a central facility. Most meals would then be frozen, delivered by refrigerated truck, and reheated in each hospital. The menu would include three hundred different items, far more variety than standard preparation allowed. They required no more preparation than heating, and the aluminum trays were disposable. The board approved construction of a large plant capable of preparing, freezing, and storing the meals just when Clark retired.

It was the worst idea of Clark's career. When he approved it, gas prices in Houston were at fifty cents a gallon. Utility costs reflected the cheap energy prices. But in 1979, the Islamic revolution in Iran undercut Iranian oil production. Gasoline prices spiked to more than $1.10 per gallon. What once seemed a good proposal had morphed into an albatross. Delivering eighteen thousand meals a day by truck would consume an ocean of gasoline, and the central production facility, with its immense refrigeration capacity, would gulp power to maintain indoor refrigeration temperatures. LeMaistre had to either cut back on research and patient care or consign the meal plant to oblivion. He chose the latter course and renovated the building as a research facility.[17]

Clark thought he had left his successor with a physical plant sufficient to take M. D. Anderson well into the 1990s, but in 1980, 1981, and 1982, the number of clinic visits increased by 30,000 annually, and in 1983 the clinics recorded more than 420,000 individual visits. M. D. Anderson needed more clinic space. To fulfill the promise of ambulatory care, LeMaistre went to the UT board with plans for another 280,000 square feet, at an estimated cost of $39 million.[18]

LeMaistre also had to appear before a congressional committee. In June 1981, NCI accused M. D. Anderson of violating rules on experimentation with human subjects. The stakes were enormous. Through a pipeline connected to NCI and NIH, millions of dollars coursed to M. D. Anderson. Even temporary interruptions could cripple research.

The accusation arrived on the heels of a similar incident two years earlier. On November 23, 1979, Marvin Williams, a young man with an advanced leukemia, died from an accidental overdose of the experimental drug deoxycoformycin. Four weeks earlier, an M. D. Anderson oncologist had given Williams the drug, but when the young man's leukemia failed to respond, the treatment was discontinued. The oncologist then instructed a fellow to treat Williams with daunorubicin, a standard chemotherapy agent, but a misreading of the prescription resulted in Williams receiving continuous infusions of deoxycoformycin in lethal doses. "I'm sure the death was drug related," an M. D. Anderson oncologist told the NCI on April 23, 1980.

Then early in 1981, oncologists at M. D. Anderson sought permission from NCI to resume clinical studies of deoxycoformycin in combination with ara-A, another experimental drug. In a conference call on January 6, 1981, an advisory board at NCI rejected the proposal because they considered the planned dose to be excessive. FDA then weighed in as well, denying the request on the grounds that the drug combination had not yet been tested on animal models. M. D. Anderson oncologists proceeded anyway, prompting on April 1, 1981, an ominous response from Daniel Hoth, head of experimental therapeutics at NCI: "It has recently come to the attention of the National Cancer Institute that M. D. Anderson has instituted a trial of the combination of Deoxycoformycin and ara-A. There is no protocol on file with the NCI—This situation is unacceptable."

A state of war existed between M. D. Anderson on the one hand and the FDA and NCI on the other. During congressional hearings in 1979, Emil J Freireich had testified, "One of the real tragedies of development treatment is the conflict between safety and innovation and in the case of the Food and Drug Administration, we have now gone too far in the safety of drugs. We need some increased aggressiveness." As a result of Williams's death, as well as other incidents, the *Washington Post* concluded that "M. D. Anderson's research and testing programs have a reputation for being among the most aggressive in the nation."[19]

In that atmosphere of mutual recrimination, LeMaistre learned that clinical scientists in the Department of Developmental Therapeutics had grown enthusiastic about radiolabeled 5-methyltetrahydrohomofolate (MTHHF), a drug that had shown some efficacy against tumors resistant to methotrexate. MTHHF's toxicity in animal models was low, and the drug appeared to breach the blood-brain barrier. In May 1978, pharmacologist Ti Li Loo, chief of the DT section of pharmacology, had designed a protocol and secured the approval of the Surveillance Committee. With NCI funding, he requested and received small amounts of radioactive and nonradioactive MTHHF for preclinical studies. Early in 1981, erroneously assuming that FDA and NCI approvals had been secured, Loo proceeded, and a physician administered the drug to six patients. Loo submitted an abstract for presentation at the 1981 meetings of the American Association for Cancer Research, and a reader from NCI noticed the unapproved drug. After concluding an investigation, LeMaistre admitted that FDA, NIH, NCI, and institutional guidelines had been broken, although Loo had not done so knowingly. Part of the problem could be traced to an understaffed institutional review board with an inadequate system for compliance review.[20]

LeMaistre needed to act decisively, and following the recommendations of an ad hoc faculty hearing committee, he disciplined Loo for "failure to adhere to institutional rules for clinical investigations." Loo and an associate received formal reprimands and were relieved of any responsibility as primary investigator or coinvestigator on federal grants or contracts involving human subjects.[21]

Some clinical scientists shared Freireich's frustration, but others found the NCI and FDA investigations justified and the disciplinary action taken appropriate. After all, a patient had died in clinical trials because of carelessness, and to some federal officials, M. D. Anderson seemed petulant and refractory. In the words of Ralph Freedman, a gynecologist at M. D. Anderson, "We may have come as close as ever to being shut down."[22]

The NCI inquiry inspired thoughtful concerns. On November 6, 1981, LeMaistre appeared before Senator Paula Hawkins and her Subcommittee on Investigations and General Oversight. To smooth the way for LeMaistre, Senator Lloyd Bentsen appeared unexpectedly. Steve Stuyck, assistant to the president and head of public affairs at M. D. Anderson, accompanied LeMaistre to Capitol Hill. LeMaistre's smooth voice and calm demeanor masked intense anxiety revealed only by trembling hands, but he managed the moment, explaining the relevant issues and helping to clean up M. D. Anderson's smudged reputation. In the spring of 1982, LeMaistre prepared for a site visit from NCI and HEW's Office for Protection from Research Risks (OPRR). The OPRR team spent a week examining M. D. Anderson's Institutional Review Board. They recommended only modest changes, and LeMaistre assured them that the investigators had been disciplined and that the violation had been an isolated incident. In their final report, they concurred.[23]

The Ti Li Loo episode soon rose from the dead. Later in the year, accounts of the episode appeared in *The First Biennial Report of the President's Commission for the Study of Ethical Problems in Medicine and Biomedical and Behavioral Research*, which cited the Ti Li Loo episode as evidence of poor institutional oversight. LeMaistre, appalled and aggravated, considered the report a blatant affront, "biased" and "terribly distorted," without "objectivity," and loaded with statements taken "out of context." He wondered "why no effort was made to arrive at the truth of these matters, when guesses are so potentially damaging."[24]

To beef up surveillance, LeMaistre established an Office of Protocol Research and installed Gerald Bodey as its watchdog. Bodey was the perfect choice. He was a highly respected medical oncologist with a proven track record in clinical research using human subjects, and he had deeply held religious convictions, with an elevated moral consciousness about using experimental drugs on people. In the Office of Protocol Research, Bodey sat comfortably as the fulcrum balancing the competing demands of scientific progress and humane treatment of patients. Bodey proved to be proactive. In October 1982, he reported problems in the Bone Marrow Transplant Program. Karel Dicke, the physician in charge, had allegedly been vague in reports to NCI about remission and toxicity data. Dicke was removed from protocol report preparation, and Frederick Becker reminded LeMaistre that the "Bone Marrow Program has grown too autonomous in recent years and must be brought back into total conformation with the regulatory structure of our Center."[25]

Six months later, and still managing damage control, LeMaistre was confronted with the murder of an M. D. Anderson physician. Fred Conrad, the vice president for patient care, always arrived at work early. December 17, 1982, was no exception. "He was very smart and very personable," remembered his colleague Joseph T. Ainsworth. Fred Conrad, however, had an enemy. Soon after he sat down at a table near his desk that morning, an intruder stepped out of the dark hallway and shot him in the head. Conrad's body was discovered moments later. UT police and Houston police swarmed, and pandemonium spread through the hospital. When nurse Joyce Alt heard the news, she rushed to Conrad's office and saw blood splattered and pooled everywhere. Worried that other names might be on the killer's hit list, doctors, nurses, secretaries, and administrators scrambled to lock office doors and remove name tags. The police sealed the exits, hoping to trap and apprehend the murderer.

Speculation filled every office at M. D. Anderson. Some wondered if a disgruntled patient or the family member of a deceased patient had exacted revenge. Since Conrad managed the hospital formulary and selected the drugs used at M. D. Anderson, perhaps the killer was a pharmaceutical company representative or a deranged salesman. Others wondered whether the killer had been gunning for LeMaistre and happened on Conrad as a target of opportunity.

To veteran detectives, the crime scene smacked of deliberation, not rage—more likely the work of a professional than of the enraged or demented. The killer had negotiated the dark labyrinthine hallways and exited the building before the police dragnet fell. He left behind no fingerprints or footprints. The police had no suspects and no witnesses. Days stretched into weeks, weeks into months, and months into years, and every promising lead hit a dead end. The murder became a cold case, but it took years for the chill to leave M. D. Anderson.[26]

A nursing shortage also engulfed the hospital. And Renilda Hilkemeyer's retirement in 1978 left a real void. Hilkemeyer had helped to change nursing in America. She encouraged nurses to perform clinical research, engage in scholarly publication, participate actively in professional societies, and develop clinical applications from recent research. In 1973, at the behest of the American Cancer Society and NCI, more than twenty-five hundred nurses had gathered in Chicago for the first National Cancer Nursing Conference, and Hilkemeyer delivered the opening address. She had established one of the nation's most ambitious oncology nursing education programs. The curriculum included hospice care, management of postsurgical and postradiation recovery, rehabilitation, pediatric oncology, pain management, and clinical training in cancer detection.[27]

In her own clinical research, Hilkemeyer had embraced cancer rehabilitation long before most physicians. Surgeons and radiotherapists fashioned technologies that often cured cancer, but nurses dealt with the collateral damage. Patients receiving a colostomy for colon cancer, for example, might leave the hospital cured, but for the rest of their lives they had to deal with the stoma. Hilkemeyer developed techniques for long-term management of the stoma and the appliance. She did the same for urinary cancer survivors facing a lifetime of incontinence, catheters, and bladder stomas. And for all of these, she built education programs to teach cancer survivors how to care for themselves.[28]

To fill the void left by Hilkemeyer, LeMaistre hired Joyce M. Alt. In the year Hilkemeyer retired, M. D. Anderson had a high turnover rate for nurses and a shortage of nurses that delayed the staffing of the new Lutheran Pavilion. Conventional wisdom held that turnover would always be high at a cancer hospital, but Alt knew better. She had written a master's thesis on job satisfaction at M. D. Anderson. Most nurses had come to terms with the disease and left for more prosaic reasons. They had embraced oncology nursing as a calling. "Actually," Alt remembered, "the reason they stayed as long as they did is because they enjoyed dealing with patients and their families." Salaries, schedules, promotion problems, and the working environment explained the turnover. With LeMaistre's blessing, Alt addressed each problem.

Alt established a cooperative program with Houston Community College, which had its own licensed vocational nursing curricula and associate degrees. The students became employees of M. D. Anderson while completing clinical training and were transported back and forth to the college for laboratory courses. Credit-bearing academic courses met at the hospital, and upon graduation, many of the students signed on permanently at M. D. Anderson. She convinced Elmer R. Gilley, vice president for finance and administration, that salary increases would pay for themselves because a lower turnover rate would reduce retraining costs. Gilley was skeptical at first, but Alt was persistent and prevailed, making sure that nurse salaries at M. D. Anderson were competitive. "Gilley was very helpful to us in bringing that about," remembered Alt.[29]

Schedules were even more important than salary. Many nurses considered scheduling a major component of job satisfaction. Alt visited fire stations to learn about flexible schedules. Eventually, she worked out dozens of distinctive shifts, often accommodating individual needs. One of the most popular was forty hours of pay for three twelve-hour shifts on consecutive days. Alt had nurses from Louisiana, Dallas, San Antonio, and Austin who lived in Houston three days and returned home for the next four. In many instances, two nurses shared one job, each with full benefits, permitting them to juggle family and job responsibilities.

Alt also improved the working environment. She hired a full-time psychologist to conduct group therapy and individual counseling. Alt cautioned nurses not just to "spill out their hearts in my office but to talk to others." When nurses received shabby treatment from physicians, she urged them to complain to the doctor's supervisor, promising no repercussions. She convinced LeMaistre that too many acutely ill patients ended up on the floors and that intensive care facilities needed to be expanded. With Gerald Bodey's help, nurses secured more control over the medications they administered. Although Hilkemeyer had battled the problem, some doctors, in Alt's words, still "simply pulled research medications out of their pockets and told nurses to give them to patients." And Alt liberated nurses from the rule requiring that all nursing research, even projects not involving human subjects, have a physician collaborator.

She also determined that nurses faced not a glass ceiling but a steel dome on their careers. The only way they could improve their salaries was to abandon patient care for administration, even when their talents were more suited to the bedside. Alt developed a career ladder for nurses who wanted to make more money but remain in patient care. Nurses were challenged, through formal training, to acquire new competencies by certain dates, with commensurate salary increases when they succeeded. Even the most senior, seasoned nurses now had incentives for professional growth.[30]

Alt also added a new dimension to nursing oncology. Surgeons, medical oncologists, and radiotherapists managed patients with an eye toward the grade, the stage, and the biology of tumors, and Alt decided that nurses need to be equipped with some of that knowledge as well. The better they understood the pathophysiology of the disease, the better they would serve patients and keep physicians informed.

The new program enjoyed stunning success. Turnover rates fell 60 percent in the first year, and nurses acquired an unprecedented professional status reflected in their treatment at the hands of doctors and innovations in patient care. Nursing shortages at M. D. Anderson had been almost as common as metastasis. The new program relieved the acute crisis, allowed the hospital to fully staff its new facilities, and improved job satisfaction.

Nursing had never fully incorporated research into its own mission. Hilkemeyer had hinted at the importance of research, but Alt institutionalized it. Millie Lawson, for example, perfected the technique for inserting a long-dwelling catheter into the subclavian vein and suturing it in place without infection or clotting. She published the results and held workshops around the country. Alt patented an idea that resolved a problem facing oncology nurses. The age of combination chemotherapy and infection control had left bedsides tangled with multiple drip bags and IV lines delivering a variety of drugs. The odds of making a mistake, such as air bubbles in the lines or dose and schedule errors, increased. Alt eliminated the jerry-rigged mess with what came to be known as the multiport tubing system, which made eliminating air from the line easier and provided for a closed system as well as multiple consecutive and multiple simultaneous infusions. Patricia Tedder, a nursing associate of Alt's, worked with the Quest Medical Company to test the prototype.[31]

In 1980 and 1981, LeMaistre watched time catch up to several leading figures in M. D. Anderson history. In 1981, Gilbert Fletcher stepped down as chair of the Department of Radiotherapy. Fletcher's hair had lightened in color over the years, and so had his moods. Always brilliant but now less edgy, he left with a smile on his face. NCI had awarded M. D. Anderson $2.8 million, and in April 1981, engineers began installing the country's first medical cyclotron, a device capable of producing protons with its 42-million electron volts of energy. No radiotherapist had ever wielded such a powerful weapon.

The selection of Fletcher's replacement constituted a litmus test for LeMaistre to consolidate the achievements of M. D. Anderson's first generation and then lead the institution to a new level. Fletcher was widely recognized as the man who had married physics to clinical radiotherapy and designed not only new technologies but the protocols that had either cured many patients or reduced the long-term side effects of cancer treatment. LeMaistre needed a radiation oncologist up to the task of assuming Fletcher's mantle.[32]

Over the years, a few heirs apparent had emerged. Under Fletcher's tutelage in the 1960s, Herman Suit had made a name for himself in the treatment of osteosarcomas and cervical cancer. Employing brachytherapy techniques, he had fashioned methods for shielding the bladder and the rectum, which allowed for higher doses to the cervix without damaging rectal and bladder tissues. In 1970, however, Suit had left M. D. Anderson for Harvard and Massachusetts General Hospital. Robert Lindberg headed up the soft-tissue sarcoma clinic at M. D. Anderson. In the early 1960s, Lindberg had trained as one of Fletcher's fellows and staked out soft tissue sarcomas as a subspecialty. In 1964, Fletcher added him to the faculty.

The search committee, however, recommended Australian radiotherapist Lester J. Peters, a graduate of the University of Queensland who between 1975 and 1979 had worked as a faculty member under Fletcher. Early on, Peters had displayed to Fletcher the prowess that would in 2003 win the gold medal of the American Society for Therapeutic Radiology and Oncology. Like Fletcher, Peters nurtured a deep interest in head and neck cancer, and he enthusiastically embraced the use of combined surgery, radiotherapy, and chemotherapy in its treatment. And Peters foresaw on the horizon the ways in which molecular biology and radiation oncology might merge in a multidisciplinary assault on the disease. In Peters, LeMaistre found a worthy successor to Fletcher. Lindberg soon left M. D. Anderson for the University of Louisville Medical School.[33]

M. D. Anderson said goodbye to A. Clark Griffin, Cliff Howe, Richard Jesse, Leon Dmochowski, and Wataru Sutow. A pioneer in chemical carcinogenesis, Griffin died on December 13, 1982, of a pulmonary embolism associated with a ruptured esophagus. Cliff Howe, the longtime leader of the Department of Medicine, died on October 1, 1980. Even when a routine physical revealed a spot on his lung, he ignored it, perhaps, as one M. D. Anderson veteran remembered, "because he did not want to hear an 'I told you so' from [Lee] Clark." On August 10, 1981, head and neck surgeon Richard Jesse died of liver cancer caused by hepatitis acquired in surgery from a patient. Several weeks later, cigarettes caught up with Leon Dmochowski, who succumbed to a heart attack while vacationing in Mexico. Wataru Sutow died several months later. Sutow had helped develop successful chemotherapy and radiation regimens for Wilms' tumor, rhabdomyosarcoma, Hodgkin's disease, and osteosarcoma. Perhaps years spent studying the long-term health effects of the atomic bombs and hunting down survivors in the radiation-drenched environs of Hiroshima and Nagasaki had given Sutow the lung cancer that took his life. A nonsmoker, he died on December 20, 1981.[34]

———※———

As retirements and deaths took M. D. Anderson's founding generation, it fell to LeMaistre to transcend the work of the pioneers, to make sure that the hospital proved to be more than the institutional embodiment of R. Lee Clark's personality.

The UT Board of Regents had approved a plan by LeMaistre to strengthen the basic sciences at M. D. Anderson, mend the breach between Medicine and DT, and decentralize decision-making in what had become a complex organization. After years in higher education, LeMaistre understood the virtues of soliciting opinions and giving employees ownership in decisions. Academics and physicians could be extraordinarily conservative and slow to embrace change. If treated cavalierly or not consulted at all, they could undermine the best laid plans through passive resistance. LeMaistre moved methodically.

The discovery of DNA had sent the basic sciences into a dizzying period of change. The new understanding revealed that in all living organisms, from viruses to humans, fundamental chemical and physical laws governed the interactions of atoms and molecules, the most important of which revolved around genetics and protein synthesis. All genetic information in all organisms was stored in the DNA molecule. That knowledge sent the basic sciences into a virtual revolution. Anatomy moved beyond morphology and microscopic histology to the functioning of subcellular elements revealed in an electron microscope. Biochemistry passed from nutrition and metabolism to enzyme systems and biochemical and molecular genetics. Physiology no longer trafficked in mammalian organ function but in the discourse of fundamental cellular processes. Bacteriology morphed into microbial physiology and genetics. Instead of focusing on the impact of drugs on intact animals, pharmacology ventured into the effects of chemical agents on the various structures of cells. As medical research merged with the new biology, the term "biomedical research" came into vogue.[35]

Biochemistry symbolized the challenges. Darrell Ward, the leader of M. D. Anderson's biochemists, candidly admitted as much in a letter to LeMaistre. "I have on several occasions voiced concern over the continued down-the-hill slide of the present Division and Departments. Tight budgets, retirements, and resignations have left biochemistry a shell of its former self." For Frederick Becker, biochemistry was the Achilles heel of the basic sciences and the starting point for reform. In 1982, Ward stepped down. One year later Becker replaced him with William J. Lennarz, a biochemist whose 1973 paper on glycoprotein biosynthesis and the structure of cellular membranes was already considered a classic.[36]

To compensate for deficits in tumor biology, Becker recruited Garth Nicolson. Nicolson had a résumé that grew longer by the month as he added papers, conference presentations, and journal articles. At Nicolson's suggestion, the Department of Virology was downgraded to a section and transferred to the Department of Tumor Biology. Classical virology was dead.[37]

On the recommendation of a faculty search committee, Becker then set his sights on the NCI's Frederick Cancer Research Facility, where Isaiah J. Fidler and Margaret Kripke worked. Fidler, a University of Pennsylvania–trained veterinarian, directed the cancer metastasis laboratory. He visited M. D. Anderson, and the search committee endorsed his candidacy. LeMaistre offered him an endowed chair and chairmanship of the Department of Cell Biology. For unspecified reasons, Fidler turned him down.

The search committee also brought Kripke to Houston for interviews. An immunologist, Kripke had established the effects of ultraviolet light on the immune system. Cancers on the ears and necks of laboratory mice, she demonstrated, displayed strong antigenic properties. Tumor cells transplanted to another mouse stimulated its immune system and prompted rejection. When

Kripke exposed the recipient mouse to ultraviolet light before the transplantation, however, the animal accepted the implanted cells and tumors grew. From there, she pioneered the field of photoimmunology, explained the role of the immune system in sunlight-induced carcinogenesis, and revealed the long-term effects of ultraviolet radiation on the skin. Felix Haas, soon to step down as chair of the Department of Biology, had told LeMaistre and Becker in 1980 that the immunology program was fragmented and the research effort "poor to mediocre." When LeMaistre and Becker offered Kripke a full professorship and chairmanship of the Department of Immunology, she replied, "I'll have to discuss this with my husband, Josh Fidler." Until that moment, neither LeMaistre nor Becker had known that the two were married. LeMaistre reassured her that M. D. Anderson wanted both of them on the faculty. They agreed. Fidler was about to be elected president of the American Association for Cancer Research. Lennarz, Kripke, and Fidler later came to be known as M. D. Anderson's "class of 1983," which helped shape the future of the institution.[38]

A. Clark Griffin's death had left the Environmental Science Park, now known as the Science Park Research Division, without a leader just as the NCI turned to chemical carcinogenesis. Becker wanted to recruit a major scientist to the post, and he found the right person in Thomas J. Slaga, whose international reputation in environmental carcinogenesis strengthened the Science Park and whose interest in the emerging field of chemoprevention fit in well with LeMaistre's vision.[39]

In addition to recruiting scientific stars from the outside, LeMaistre and Becker conducted a job satisfaction survey among M. D. Anderson scientists. After decades of NCI targeted research, some scientists had followed the money trail and not necessarily their own intellectual instincts. The survey revealed that many were doing research in fields of little interest and for which they had not been formally trained. LeMaistre charged Becker with the task of helping them reorient their careers.

LeMaistre also created the External Advisory Committee for Research (EACR) to conduct ongoing evaluations of the basic and clinical sciences and report to LeMaistre through Becker. The committee included leading scientists and clinicians from such institutions as Harvard, Chicago, and Johns Hopkins. EACR members made multiple visits to Houston. During the first two years of the evaluations, twenty-eight of thirty-two leadership positions changed. Half of the new appointments came from M. D. Anderson and half from the outside.

The evaluations also led to important internal reorganizations. Prior to the external examinations, for example, urology was a surgical department. After the evaluations and the appointment of Andrew von Eschenbach as chair, urology was converted into a research-oriented department containing both surgeons and medical oncologists as faculty members. Its success planted the seeds for M. D. Anderson's subsequent program of organ site centers.

Under Becker's leadership, both the clinical and the basic sciences had been strengthened. Members of the External Advisory Committee had only praise for him. Judah Folkman of Harvard wrote Becker, "You should be proud of the enormous progress in the development of your programs since our last visit. . . . Research at M. D. Anderson now ranks with the top institutes in the country." David Patterson of the Eleanor Roosevelt Institute for Cancer Research said: "It seems to me that M. D. Anderson is now an extremely competitive research institution. I commend you on a superb job."[40]

Becker still felt that genetics needed strengthening and reorganizing. Geneticists were scattered across a variety of departments and sections, and Becker was convinced that natural centrifugal forces made intellectual cross-fertilization difficult. He recommended the creation of a Department of Genetics with sections in cellular genetics, biochemical genetics, human genetics, and molecular genetics. LeMaistre concurred, and Becker formed a search committee to identify a leading geneticist to assume the chairmanship, one who would revitalize the research effort that, in the words of the geneticist Alfred G. Knudson, "does not include internationally front rank scientists . . . [on the] cutting edge . . . [of] research on cancer or fundamental biology." Geneticist Michael J. Siciliano, who agreed to serve as interim chair, urged Becker to hire from the outside.[41]

Reorganization of the Departments of Medicine and Developmental Therapeutics (DT) constituted a third item placed on the board of regents' agenda. Ever since its founding in 1965, Developmental Therapeutics had operated as a fiefdom with the latitude that comes with success. Resources had flowed to DT because of its pioneering work in medical oncology. In the process, however, the subspecialties in Medicine languished. Late in 1979, the Department of Medicine at the UT Medical School at Houston decided to send no more residents for training to M. D. Anderson. "The current house staff rotations at the M. D. Anderson," wrote Thomas Andreoli, the UT chair of medicine, "are not appropriate for our current teaching program."[42]

Changes in medical oncology complicated the situation. DT's reputation had come from its innovations in the treatment of lymphomas, leukemias, and plasma cell disorders. New chemotherapy regimens had emerged for some metastatic solid tumors, but DT maintained a tight grip, and many medical oncologists laboring in Medicine felt stifled. The Ti Li Loo episode had confirmed in LeMaistre's mind the need to rein in Developmental Therapeutics.

Any limits on DT, of course, implied constraints on Emil J Freireich, a man who ferociously guarded his prerogatives. In 1980, Freireich began a year as president of the American Society of Clinical Oncology, and in 1983 he shared with Emil Frei the prestigious Charles F. Kettering Prize for cancer research. The dilemma shouted for resolution, and the noise alerted oncologists everywhere. LeMaistre solicited the opinions of many. Eugene Frenkel, head of medicine and hematology-oncology at Southwestern, acknowledged the difficulty of running a department of medicine at a categorical cancer hospital. "An appropriate development under the umbrella of Medicine," Frenkel replied, "could effectively weld together the loose aspects of teaching and research along much broader lines. The absence of a hematology department needs to be addressed.... What I'm proposing, then, is not a Department of Medicine in the classic sense, but one which develops as a critical, non-hostile supplement to the primary focus of the mission of the Hospital."[43]

LeMaistre decided not on a department of medicine but on a division, an administrative superstructure that would break DT into its components, liberate other medical oncologists from its control, and find a home for cancer prevention. All across the nation, internal medicine departments were subdividing. Late in 1983, he established a Division of Medicine that included a Department of Cancer Prevention, with Guy Newell in charge, and sections of preventive medicine and epidemiology. Freireich presided over the new Department of Hematology and its sections of leukemia, lymphoma, bone marrow transplantation, and pheresis. Evan Hersh took charge of the new Department of Clinical Immunology and Biological Therapy and its sections of clinical immunology and regional therapy. The Department of Internal Medicine clustered together the sections of nuclear medicine, cardiopulmonology, endocrinology, gastroenterology, neurology, psychiatry, infectious diseases, and rehabilitation. Included in the new Department of Medical Oncology were sections of genitourinary oncology, gastrointestinal oncology, medical breast, and melanoma/sarcoma.[44]

To identify a leader for the Division of Medicine, LeMaistre asked Jim Bowen to chair a search committee. LeMaistre was inundated with unsolicited letters of recommendation promoting Freireich. Some of the letters even disparaged other applicants, prompting LeMaistre to inquire of Bowen about a breach in the search committee's confidentiality. LeMaistre was clearly annoyed. Those recommending Freireich read like a who's who of modern oncology, including Emil Frei of Dana-Farber, Robert Gallo of NCI, Gianni Bonadonna of the National Tumor Institute of Italy, James F. Holland of the Mount Sinai Medical Center, and Howard Skipper of the Southern Research Institute.[45]

LeMaistre eventually settled on an outsider, a physician without a personal history at M. D. Anderson, a man with no debts to repay and no institutional memory to appease. He hired Irwin H. Krakoff, who would also chair the Department of Chemotherapy Research, with its sections of new drug studies and pharmacology. Krakoff, director of the Vermont Regional Cancer Center, had been head of the Laboratory of Chemotherapy Research at Sloan-Kettering Institute, as well as associate chair of the Department of Medicine at Memorial Hospital and chief of the Medical Oncology Service there. The new Division of Medicine began operations on October 1, 1983.[46]

Within a matter of months, sparks flew. Freireich seemed to have little respect for Krakoff's work as an oncologist or for the tiny Vermont Regional Cancer Center. LeMaistre, of course, had been primarily looking for a good administrator, and he considered Krakoff more than qualified. Freireich, who was not always the most cooperative subordinate, rubbed Krakoff like sandpaper on skin. On a Friday afternoon early in the spring of 1984, just before boarding a plane to Tokyo, Krakoff fired Freireich as chair of hematology. The decision blindsided LeMaistre, who found himself caught in a tsunami of protest from medical oncologists inside and outside of M. D. Anderson. James Bowen likened Krakoff's actions to "pulling the pin on a hidden hand grenade and leaving the conference room before a scheduled meeting begins." LeMaistre telephoned Krakoff in Tokyo and instructed him to return immediately and repair the damage. Krakoff flew back and tried to mend some fences, naming Freireich head of adult leukemia research. Krakoff would spend ten years at M. D. Anderson, and in 1993 he won the David A. Karnofsky Memorial Award.

The reassignment of another medical oncologist, however, had LeMaistre's endorsement. George Blumenschein first arrived at M. D. Anderson in 1973 to direct education programs. He soon moved into medical oncology and the treatment of breast cancer. With Jeffrey Gottlieb, Blumenschein developed the first FAC protocol—5-fluorouracil, Adriamycin, and cyclophosphamide, with slow infusion of the Adriamycin to reduce cardiac toxicity—as adjuvant therapy for women with stage 2 breast cancer. Along with several other medical oncologists at M. D. Anderson, Blumenschein successfully added BCG to the FAC protocol for women with metastatic disease. In 1975, he had succeeded Nylene Eckles as chief of medical breast oncology. To his credit, Blumenschein was a gifted oncologist who spent a great deal of time with patients, answering their questions and soothing their concerns, but such sensitivity often backfired in the clinic, throwing off the schedules and eliciting complaints from nurses and patients. At times the clinic stayed open late into the evening. Krakoff and LeMaistre repeatedly asked Blumenschein to report to the clinic at 8:00 a.m., but in the words of James Bowen, "Blumenschein appeared indifferent." He was often late to work. In April 1985, Krakoff relieved him of his duties as chief of medical breast oncology. Blumenschein moved to Arlington, Texas, and built a successful private practice. In LeMaistre's recollection, "The action did have a resounding effect throughout the institution because everyone realized that the rule that patients came first was going to be enforced. Regrettably, we lost one of our most talented oncologists in the process."[47]

⚬

In 1978, a few physicians in the United States and Sweden reported a strange disease affecting some young gay men. A virtual zoo of bacteria and fungi inhabited their bodies, including flora and fauna usually confined to other species. The patients suffered from candida mucosal (thrush) infections, which usually appear in newborns with immature immune systems. They also had heavy cytomegalovirus infections and the deadly PCP—pneumocystis carinii pneumonia (jiroveci pneumonia). On July 4, 1981, the CDC reported a strikingly high incidence of Kaposi's sarcoma (KS), a deadly skin cancer, in gay men. As more and more afflicted young men reported to physicians around the country, oncologists and virologists took a closer look. The incidence of Hodgkin's disease, non-Hodgkin's lymphomas, anal cancer, and central nervous system malignancies was higher than expected. In December 1981, Guy Newell warned Debra Danburg, the woman who represented the Montrose area of Houston in the state legislature, of a new disease that would hit Houston's gay community within the month. Almost on cue, M. D. Anderson had a patient with KS. Early in 1982, CDC linked it to an infectious agent in the blood, probably a previously undiscovered virus, and labs around the world went on the hunt.

Guy Newell harbored a primal anxiety about the disease. As a cancer-causing infectious disorder that destroyed a patient's immune system, appeared to spread through sexual contact, and was universally fatal, acquired immune deficiency syndrome (AIDS) had the potential to threaten the world. Because homosexual men seemed particularly at risk, AIDS might become an epidemiologist's nightmare, where science, culture, religion, and homophobia coagulated into a dangerous public health clot. With Houston a center for gay culture in the Southwest, M. D. Anderson soon saw a stream of young men with AIDS-related cancers.

In 1983, the KS/AIDS Foundation and the AIDS working group in Houston, with the assistance of Peter Mansell, issued through the City of Houston Health Department the "Acquired Immunodeficiency Syndrome Criteria and Definition." The disease was blood borne, had a long incubation period, and could be transmitted from mother to offspring, from male intravenous drug users to female partners, and from homosexual men to their partners.[48]

At M. D. Anderson, Newell insisted on being included in all discussions regarding AIDS. He earned the praise of the gay community. Debra Danburg wrote, "Dr. Newell made several critical points. . . . First, the epidemic required two levels of response; one humanitarian and the other scientific. This roughly translated into sound education of the community . . . and scientific research into the problem because of its uniqueness in holding the key to some of the basic knowledge about cancer in general. Secondly, he stressed that the problem should not be treated . . . as a political issue. . . . Anyone could be afflicted. Third, he begged that the response to the epidemic should be a single, unified one from the Houston community." Mansell became the recognized community medical leader about AIDS and instrumental in the formation of the Kaposi's Sarcoma Committee of Houston.

At first, M. D. Anderson was the only hospital in the Texas Medical Center willing to accept AIDS patients. Little was known about the disease except its lethal outcome. Hospital administrators everywhere worried about whether AIDS posed a threat to other patients and to staff. When M. D. Anderson opened its AIDS Clinic in 1983, Newell monitored it carefully. "Of 300 Houston men who volunteered to visit the Preventive Medicine Clinic, 80% have measurable disease," he reported. "During the past 19 months, we have treated more than 35 cases of Kaposi's sarcoma. Suspected causes include viruses (CMV, EBV, other herpes viruses, etc.), chemicals, nitrites, marijuana, cocaine, multiple sexually transmitted diseases, and perhaps elements of sexual practices. The public health and scientific implications of understanding the cause(s) of acquired immune deficiency syndrome are enormous."

Fear spread through the staff, who worried about their own susceptibility and the fate of other patients, especially those with chemotherapy-compromised immune systems. Might AIDS patients put others at risk for the associated infections? Social worker Sue Cooper of M. D. Anderson recalled that there would be "one room that had an AIDS patient and one person who was having chemotherapy" and one nurse moving between them. Under Hickey's direction, the hospital and clinics developed elaborate precaution measures that involved protective clothing, rubber gloves, and specific procedures for handling blood and urine. "I hated it," says Cooper, "because when I would touch them or hold their hand or do whatever needed to be done, . . . it was just important for a patient not to be touched with a cold rubber hand."

M. D. Anderson opened a clinic in the Anderson Mayfair, performed immunological tests, and gathered information. "In those days," Mansell remembers, "we had absolutely no idea of what was going on. I mean, we suspected very early on that it was an infectious process but not that it was a virus, or that there were any cofactors. The clinic then grew so rapidly that we moved here to the hospital." Cooper and Mansell fanned throughout the community, established close ties with gay organizations, and formed support groups for men afflicted with the virus. "We went and collected blood and filled out questionnaires in gay bars. . . . It was a very strange time indeed. And people started flooding in." For Sue Cooper, "It was new, and it was a very frightening thing. So I think that's what helped bring so many people in so quickly. Once they really did get sick, they really did get *very* sick." The patient burden soon forced the AIDS clinic out of the Anderson Mayfair and to Station 16 in the main hospital. There Mansell grappled with the miseries and mysteries of AIDS.

The decision to treat AIDS patients subjected M. D. Anderson to severe criticism about such patients' eligibility for admission to the hospital. LeMaistre explained repeatedly to the board of regents that M. D. Anderson was charged with treating cancer and allied diseases, so an AIDS patient with KS qualified for treatment. M. D. Anderson internists, however, fought losing battles against the infections, and AIDS patients suffered from a disease whose cruelties ran as deep as its hopelessness. "They were just so sick and there was so little we could do," Mansell recalled.[49]

10

Chasing the Devil,
1984–1988

*We have reached a watershed when about half of all serious cancers
can be cured. Lung, breast, and colon cancer account for two-thirds
of all cancer deaths. . . . Taking control of one's life and destiny
can lessen the chances of developing cancer.*
—CHARLES A. LEMAISTRE, 1985

An acrid odor fouled the air at Station 19, forever distinctive to visitors—
a mix of soap, antiseptic solution, anticancer drugs, vomit, and the body
odor of people under stress. Station 19 served as the ambulatory stop
in the early and mid-1980s for M. D. Anderson patients receiving chemotherapy. All day long they gathered at Station 19—limping amputees and people
with artificial limbs; bald patients wearing babushka, wig, scarf, turban, baseball cap, or knit hat; and others whose external appearance belied the disrup-
tion inside. With their arms and hands connected to multiple infusion lines and
drip bags, the patients rested on beds separated by only the thinnest cotton cur-
tains. The patients enjoyed no privacy. Some passers-by in the corridor craned
for a peek inside.

In the mid-1980s, Becker and LeMaistre continued their efforts to strengthen
the sciences. With cancer research fully reoriented to molecular biology, genetics
was a priority, and Becker wanted to hire a star.

The search dragged on, partly because, in the words of biochemist William J. Lennarz, "it is of extreme importance that an individual of international reputation be appointed. . . . Building the basic sciences without a department containing strong molecular genetics is akin to remaining scientifically locked in the 1970's. . . . The future understanding of oncogenesis is irrevocably dependent on research in molecular genetics." The best-known geneticist in M. D. Anderson history no longer worked there. In the late 1960s, Alfred G. Knudson Jr. had specialized in studying retinoblastoma, a lethal cancer of the eye in children. Knudson noticed that the disease had two forms, one characterized by multiple tumors in infants and the other by a single tumor in older children. Working in clinical epidemiology and statistics, Knudson teased out a hypothesis—that in the early-onset version, infants inherit a defective gene from one parent, which pushes them one stage into the disease. Full-blown retinoblastoma, however, requires a second step, or "hit," in which a mutation of DNA or a replication error in a single cell of the eye causes a defect in the normal gene inherited from the other parent. Together, the two defects program the cell to proliferate wildly. Children with the later onset, Knudson hypothesized, were born with two good genes, one of which acquires a defect and causes the disease later in childhood. Knudson and Louise Strong in 1971 reported his "two-hit" theory of cancer causation, which requires mutations in two paired genes to initiate the cancer process. The research raised the theoretical possibility of the existence of tumor suppressor genes. But not until 1987 did molecular geneticists clone the gene responsible for retinoblastoma and confirm their hypothesis, which also helped explain neuroblastoma and Wilms' tumor. By that time, however, Knudson worked at the Fox Chase Cancer Center in Philadelphia.

Becker went to the outside, even though several biologists wanted as chair their own geneticist, Michael J. Siciliano. He was actively designing molecular probes—using the fluorescence in situ hybridization method (FISH) technique—to expose chromosomal rearrangements in chronic leukemia and other cancer cells. T. C. Hsu reminded LeMaistre of the old Chinese proverb, "Monks from far away places always preach better." Biologist Roger W. Hewitt endorsed Siciliano and expressed concern about the trend to hire stars instead of developing people from within. Garth Nicolson, however, told LeMaistre, "This critical position must be filled by a molecular geneticist who will take the lead at our institution in molecular biological approaches to cancer genetics. This is the most critical weakness at our institution in the basic sciences." Lennarz was asked to chair the search committee.[1]

He faced a real dilemma. Few applicants really had star power, and those who did won multiple job offers. LeMaistre grew frustrated. In a 1985 letter to Becker, after the withdrawal of several candidates, he wrote, "In keeping with my instructions to recruit only the most able, I anticipate that the [next] three new candidates will have credentials and potential equal to those who have withdrawn. Please recall that I stated from the outset that I was willing to commit the generous genetics resources package only to a stellar figure in the field."

It took another year for Lennarz and Becker to score, but the victory was complete. In 1986, LeMaistre tendered an offer to Benoit de Crombrugghe, the head of the Gene Regulation Section of the Laboratory of Molecular Biology at NCI. De Crombrugghe's article on viral oncogenes had recently appeared in *Nature*, and he was a leading geneticist. His decision to accept the position stunned Becker. "This is truly remarkable," he told Lennarz. The biggest hole in the basic sciences at M. D. Anderson had been filled.[2]

In the mid-1980s, M. D. Anderson sported one of the best leukemia programs in the nation. Emil Freireich had few peers, and in 1985, LeMaistre hired Don Pinkel to head pediatric leukemia research. Thirty years earlier, Pinkel had engineered external beam radiotherapy protocols to destroy leukemia cells in the brain, systems for injecting chemotherapy drugs directly into spinal fluid, and specific chemotherapy regimens for young adults with ALL. So in 1985, M. D. Anderson had on staff Emil Freireich and Don Pinkel, two clinical scientists destined for a place in medical history.[3]

To Freireich and Pinkel, Becker and LeMaistre added Albert B. Deisseroth. Fascinated with genetics and tumor biology, Deisseroth's facile mind was tinkering with the ideas of cancer vaccines, monoclonal antibodies, and chemotherapy protocols unique to the specific genetic and molecular composition of an individual's cancer cells.

M. D. Anderson also had on its staff the Israeli hematologist Moshe Talpaz, under whom the drug interferon finally found its greatest success. Talpaz completed a postdoc at M. D. Anderson in 1979 and then stayed on to work under Jordan U. Gutterman, concentrating his research in immunotherapy, biological response modifiers, and cytokines. Talpaz and others knew that interferon was effective in treating chronic myelogenous leukemia (CML), but its side effects left patients with what resembled an intolerable case of flu during treatment. As soon as their oncologists stopped treatment, the flu symptoms disappeared, but the CML cells rebounded. Talpaz joined with hematologist Hagop M. Kantarjian in convincing many patients to stay with interferon therapy for extended periods, and their survival periods lengthened significantly. Because of Talpaz, interferon would soon became the frontline therapy for CML. LeMaistre was convinced that the leukemia program at M. D. Anderson was "second to none."[4]

In 1987, he said exactly that to Vice President George Bush and his wife, Barbara. The Bushes scheduled a visit for late 1987, and they expressed keen interest in leukemia research. LeMaistre asked Irwin Krakoff to organize the presentation. Krakoff's presence added a poignant nostalgia to the visit. He had treated Robin Bush at Memorial Sloan-Kettering. Krakoff had Jordan Gutterman talk about the promise of interferon and Deisseroth the day when genetic engineering might join radiation, chemotherapy, and bone marrow transplants as frontline therapies. They gave the good news and the bad news about leukemia research—that the cure rate for acute lymphocytic leukemia in children had reached 70 percent but that the treatment of adult leukemia had stalled. The Bushes were impressed and dedicated the royalties from his autobiography *Looking Forward* to pediatric leukemia research at M. D. Anderson. LeMaistre promised, "Please know that your generous contribution will be used wisely to honor both Robin's memory and the trust you have placed in M. D. Anderson."[5]

LeMaistre had a vacancy in surgery to fill. In 1984, Richard Martin stepped down. Osteoporosis and series of compression fractures had given his shoulders and upper spine the profile of a question mark. Martin was also living with a time bomb. Thirty years earlier, during the Korean War, he had acquired hepatitis C from a wounded soldier, and he knew only too well that carriers of hepatitis C are likely headed for liver cancer.

Martin's decision to retire unsettled many surgeons. He had been the best of bosses, even-tempered and willing to leave people alone as long as they maintained high standards of patient care. Robert Hickey became acting chair and headed the search committee. LeMaistre wanted an outsider with strong research credentials. After thirty-five years under Ed White and Richard Martin, the Department of Surgery had ossified a bit. The search committee considered many candidates, and a group of surgeons tried an end run by petitioning LeMaistre to promote Clifton F. Mountain, an M. D. Anderson thoracic surgeon. LeMaistre deferred to the committee, however, and they recommended Charles Balch, a surgical oncologist at the University of Alabama at Birmingham. Balch negotiated an attractive salary package that included, thanks to the Houston businessman Ben Love, membership on the board of directors of Texas Commerce

Bancshares. Balch suggested that M. D. Anderson change the model for paying surgeons to compensation based on patient revenues generated. The practice plan, the Physicians Referral Service (PRS), collected all professional income. A PRS Council composed of clinical and basic science departmental chairpersons recommended how the money should be employed by the president. It was used largely for salaries and fringe benefits for both clinical and basic science faculty. The institution was not about to change this highly successful pay structure. Balch eventually signed on for less than he was making in Birmingham, but close enough with the board membership.[6]

LeMaistre also needed a new director of the pharmacy. When Clark retired, James McKinley had stepped down as the only director of the pharmacy in the history of M. D. Anderson. In the professional pecking order, the pharmacist enjoyed little prestige. Physicians considered the job mechanical and repetitive— not much more than "count, pour, lick, and stick"—counting pills, pouring them in containers, and then licking and sticking on labels.

In 1965, the arrival of Frei and Freireich upended pharmacy. Chemotherapy made the job of hospital pharmacist infinitely more complex. "Because of clinical trials," McKinley remembered, "our accuracy had to be measured very carefully. We knew that our medical errors could kill people and ruin research projects." According to McKinley, "The change made pharmacists more responsible and powerful." In 1975, McKinley launched clinical pharmacy services at M. D. Anderson, the first program of its kind in the nation. He established a Drug Information Center, which supplied data to health professionals.

The rise of medical oncology gradually blurred the line between pharmacy and pharmacology; pharmacists worked closely with physicians in clinical settings, communicating more with patients and helping to design new drug regimens. "During my career," McKinley recalled, "I watched the job of hospital pharmacist change dramatically. We became more involved in the health care team and more central to M. D. Anderson's mission. And because of that, we enjoyed much more respect." After a search committee identified a short list of candidates, LeMaistre hired Roger W. Anderson, who had received one of the country's first master's degrees in hospital pharmacy.

Station 19 was an eyesore to LeMaistre, the medical oncologists, and the patients, and its congestion prompted the development of portable devices that would allow patients to receive their chemotherapy treatments at home. As early as 1983, Yehuda Z. Patt, a clinical immunologist at M. D. Anderson, developed a new pump for liver cancer patients. The device, implanted near the diseased organ, delivered the chemotherapy drugs in high doses over the course of a day. Larger doses could be administered directly to the liver because the drugs were not circulating systemically and therefore did not affect other organs. Implanted pumps allowed patients to come to M. D. Anderson at three-month intervals instead of weekly.

Efforts around the country in the 1980s to develop other portable pumps produced a variety of devices, and the new pharmacist chaired a program to evaluate existing pumps and to fashion more efficient external portable devices. In 1982, M. D. Anderson maintained 250 portable pumps for use among inpatients and outpatients. The existing infusion pumps were cumbersome and unreliable. Each weighed fourteen pounds and had to be carried around in a satchel. The pump had only one channel for administering a drug. Roger Anderson's team, which included pharmacists, nurses, and bioengineers, designed a lightweight model with four delivery channels, which enabled patients to take chemotherapy infusions at home.[7]

The expansion of medical oncology at M. D. Anderson had complicated the task of pharmacists. The explosive expansion in the number of drugs and near geometric increases in the possible combinations placed a premium on speed and accuracy, two demands that often operated at cross purposes. Anderson developed a highly regarded, research-oriented pharmacy with a robotic delivery system that helped reduce drug errors to zero. Roger Anderson would soon bring Lawrence A. Trissel to the Department of Pharmacy. Trissel's reference work *Handbook on Injectable Drugs* was destined to go through fourteen editions and to become a highly regarded reference work used by pharmacists throughout the world to assess drug compatibilities of all types, including cancer chemotherapy.[8]

—⟨⟨⟨⟨⟩⟩⟩⟩—

Charles LeMaistre assumed the reins of M. D. Anderson just as two powerful political movements in the United States converged. Concerns about women's rights and the protection of human subjects in medical research forged a vigorous patient rights movement consistent with other civil rights campaigns of the era.

The modern women's movement targeted medicine as a critical arena. Instead of viewing physicians as demigods, the movement inspired women to regard doctors as professionals who marketed services and themselves as consumers of those services who were entitled to respect and consultation. In 1971, the first edition of *Our Bodies, Ourselves,* compiled by the Boston Women's Health Book Collective, warned that "doctors are not gods, but human beings." Women should "act as independent consumers of professional services. Don't . . . be stampeded into sudden decisions. . . . It's your body." The women's health movement developed a Bill of Rights to govern the patient-doctor relationship. Patients were entitled to answers to any of their "questions about any examination or procedure . . . in advance of or at any time during the performance of it. [Doctors must stop] any examination or procedure at any moment, at your request." The best physicians would display a ready "willingness to accept and wait for a second medical opinion before performing any elective surgery which involves alteration or removal of any organ or body part." In 1973, the American Hospital Association issued its Patients' Bill of Rights, which enthroned respect from physicians as a natural right of patients.[9]

The new consumerism in medicine also expressed itself in fresh concerns about protecting the rights of human subjects in medical research. In 1982, LeMaistre asked James Bowen, vice president for academic affairs, and Jan van Eys, the physician who had replaced Grant Taylor as head of pediatrics, to develop a code of ethics. They were good choices. Bowen had lost his first wife to Hodgkin's disease, and van Eys had for years nurtured deep concern about the ethics of clinical trials among children unable really to give consent. They spent two years on the project.

Rather than draft a set of regulations, they crafted a body of moral principles "to bond patients and staff together in the difficult task of contending with cancer." It called on employees to treat patients with a reverence "that affirms the value and dignity of life" and "to help them come to terms with their altered lives," to promise to guard patient confidentiality, to keep patients fully informed about the risks and benefits of treatment, and to engage in no behavior that undermines a patient's therapeutic needs. In addition, each employee "bears an individual moral obligation to each patient."

Patient care and research complemented one another, but the code recognized the potential conflict between the two and acknowledged that "the immediacy of patient care tends to obscure the relevance of basic biological research. . . . We affirm that research, responsibly conceived and scientifically sound . . . encourages realistic prospects of eradicating cancer, thus promoting a favorable balance of risks and benefits." The code reaffirmed the commitment to treating patients aggressively, as long as such treatments held out better survival odds than less radical procedures. "We dedicate ourselves to provide our best care and to use our knowledge to attempt cure of the disease in each patient while pursuing understanding of the basic biologic nature and eradication of cancer." And when aggressive treatments failed, M. D. Anderson acknowledged that "the presence of cancer may justify but not demand, heroic measures. Curing disease, reducing suffering, and sustaining an acceptable quality of life, as defined by the patient with the help of health care professionals, are the central goals of this institution." In June 1984, LeMaistre accepted the code of ethics.[10]

Almost as important as the ethical treatment of patients was efficient treatment. During the last few years of Clark's tenure, clinic operations had badly deteriorated. A patient satisfaction survey in 1984 revealed that many people were having difficulty navigating M. D. Anderson's burgeoning bureaucracy, an inevitable consequence of spectacular growth. Robert Moreton had long focused on patient satisfaction as a special concern and over the years had worked to sooth patient frustration, massage complaints, and improve the patient experience. In 1986, M. D. Anderson opened the Office of Patient Advocacy and assigned its staff the role of ombudsmen for patients.[11]

Back in 1978, after LeMaistre had toured M. D. Anderson for a few days on his own, he had dedicated considerable effort to streamlining patient care operations. Lack of communication between different clinic sites was endemic. During a typical visit, a patient might have several clinic stops—one to draw blood, one for an x-ray, one for a radiotherapy or chemotherapy consult, one to pick up a prescription, and a final visit with the physician supervising the case. A delay at one station guaranteed delays at subsequent stops. If a physician arrived late for work in the morning, patients paid the price for the rest of the day. In some departments, faculty meetings were scheduled in the middle of the day, without regard to clinic schedules. Physicians popped in and out at random. Holding department meetings early in the morning or after hours helped. Securing and moving medical files had to be streamlined. Since the late 1970s, patients had often been asked to carry their own medical records from appointment to appointment, and it was not uncommon for the files to be left behind in clinic waiting rooms, restrooms, or the cafeteria. LeMaistre and his staff began to computerize patient charts, laboratory and pathology reports, and x-ray readings, and they built a preregistration system to ease the frustration of first-time patients. Streamlining clinic visits also planted the seeds for construction of a faculty office building across the street from the hospital, which would free up clinic space and relieve congestion.[12]

One decade later, the situation had greatly improved, even though the burden of patient care had doubled. More then 16,000 patients were admitted to the hospital that year, with more than 350,000 outpatient visits in the clinics. The average length of inpatient stay had fallen from fourteen days to eight days, even though the percentage of acutely ill patients being hospitalized had increased. More than 85 percent of all radiotherapy treatments were delivered on an outpatient basis, and the miniaturization of portable pumps, along with expansion in the physical plant and construction of the Ambulatory Treatment Center, had allowed most chemotherapy to be administered on an outpatient basis. By 1988, more than 9,000 patients had used the portable pumps, and 300 people were receiving chemotherapy every day. Same-day admissions and outpatient surgery had increased tenfold. In 1946, R. Lee Clark had envisioned what became known as ambulatory cancer care, but full realization of the dream had fallen to Fred Conrad and LeMaistre. The number of outpatient visits at M. D. Anderson was four times greater than at any other cancer center. Patient attitudes reflected the changes. The Gelb Consulting Group concluded that the "M. D. Anderson Hospital ratings are the highest Gelb has ever obtained in 20 years of conducting hospital surveys."[13]

During the early to mid-1980s, M. D. Anderson explored exciting new areas of treatment, with scientific assumptions that promised to unlock some of cancer's mysteries.

The cyclotron had arrived with great fanfare just when Gilbert Fletcher stepped down as chair. Initially lauded as the first hospital-based cyclotron in the country, the contraption soon became fodder for Rube Goldberg jokes, a machine so complex that it could not be used in a real world setting. The Cyclotron Corporation experienced delay after delay in its installation and testing and failed to resolve problems with hardware and software. The company blamed radiation hazards in the installation area, but LeMaistre denied the charge and threatened to enforce the deadline penalties in the contract. Hapless and strapped for money, the Cyclotron Corporation in 1982 filed for bankruptcy.[14]

The cyclotron went online in 1984 but quickly proved to be more boondoggle than boon. Setting up patients for treatment was labor intensive and expensive. Since the advantages of fast neutron therapy had not been clinically demonstrated, M. D. Anderson could not bill insurance companies. The problems with its operations mirrored those of its installation. The machine was unreliable, the worst in M. D. Anderson's stable of radiotherapy devices. Frequent breakdowns upset clinic operations. The only thing higher than its maintenance costs was the frustration level of patients and staff. Worst of all, as data accumulated it became clear that the cyclotron offered no clinical advantages to patients. In 1984, NCI pulled funding.[15]

A fourth fundamental modality in cancer therapy had appeared—harnessing the body's immune system and developing biological therapies. NCI concocted the term "biological response modifiers" (BRM) to classify cancer treatments designed to stimulate the immune system to fight malignant tumors. In the early 1980s, three particular BRMs captured the imagination of oncologists—monoclonal antibodies, BCG (bacillus Calmette-Guérin), and interferon.

Under the direction of Karel Dicke, M. D. Anderson maintained one of the largest bone marrow transplant programs in the world. Massive doses of radiation and chemotherapy killed cancer cells, but they also proved toxic to a patient's bone marrow. Bone marrow transplantation involved aspirating bone marrow from a patient and storing it before treatment. Once the highly aggressive treatments were complete, healthy bone marrow was reinjected. Dicke assumed that the treatments had attacked the cancer cells, and the healthy, undamaged bone marrow then restored the patient's immune system.

In leukemia patients, however, cancer cells often cloister deep in the bone marrow. After reinjection they start proliferating rapidly, and the patient relapses. Dicke began working on the concept of monoclonal antibodies. At Cambridge University in 1975, immunologists César Milstein and Georges Köhler had invented a laboratory technique for producing antibodies, the proteins generated by lymphocytes in response to the invasion of foreign bodies. Since millions of different invaders exist in nature, the immune system manufactures a specific molecule to detect the presence of the invader and destroy it. Milstein and Köhler called the technology the "hybridoma" method because they took mouse lymphocytes that had been immunized against human leukemia cells and fused them with cells from mice suffering from myeloma, a cancer of the bone marrow. The fusion produced a hybridoma that secreted its own unique antibodies. The nomenclature "monoclonal antibody" derived from the fact that there was one antibody for each hybridoma. Milstein and Köhler provided the means to manufacture to order supplies of pure antibodies called monoclonal antibodies. In doing so, they paved the way for the modern biotechnology industry and in 1984 for their own Nobel Prize. Dicke understood that antibodies could be targeted at the antigens present on the surface of human leukemic cells. In his research, he identified dozens of monoclonal antibodies that bind specifically to leukemic cells and not to other blood cells, and he anticipated the day when the technology could attack leukemia.

In 1988, M. D. Anderson performed its one thousandth bone marrow transplant. About 75 percent had been autologous transplants in which patients received their own bone marrow, and the rest had been allogeneic, in which the marrow of a compatible donor was transplanted. At first, bone marrow transplants had been confined to leukemia patients, but in the mid-1980s, the treatment was gradually expanded to some patients with lung cancer, breast cancer, multiple myeloma, Hodgkin's disease, and other lymphomas.[16]

Since the early 1970s, immunologist Evan Hersh had conducted NCI-financed research into BCG, a bacterial-based drug long employed in Europe to treat tuberculosis. Along with other investigators, Hersh had learned that BCG held out promise for treating bladder cancer and lung cancer, and he counted one near miracle as proof. In 1978, Richard Bloch, the founder of H&R Block, the income tax company, went to a physician in Kansas City complaining of shoulder pain. Tests revealed terminal lung cancer. "You have about ninety days to live," the physician told Bloch, "so get your affairs in order." A friend suggested that Bloch seek a second opinion at M. D. Anderson, and in a matter of days he was sitting in an auditorium there listening to physicians discuss his case. "'I'll do this,' he remembered one saying, and 'I'll do that,'" said another. They decided first to remove the tumor surgically. "I had surgery," Bloch later said. "It was performed by an Arab fellow. I can't remember his name, but I was terrified because I'm Jewish!"

The surgery went well and Bloch ended up with Evan Hersh, who recommended rigorous BCG therapy. "Hersh taught me so much about the disease and about myself," Bloch stated. "He told me to get to the hospital . . . without delay, because he wanted to know if I was willing to do whatever it took to get well. I was treated with BCG, an immunological therapy." The residual tumors melted away, and three months later, when he should have been dead, Bloch was back at work in Kansas City. He had follow-up treatments in the fall of 1978 and became an improbable cancer survivor, living another twenty-five years.[17]

The drug interferon, however, dwarfed BCG and monoclonal antibodies in the attention it attracted. In 1957, British and Swiss virologists had uncovered an agent secreted by chicken embryo cells infected with a virus. The substance protected the chicks from a variety of other infections, and the virologists named it "interferon." Interferon does not itself kill viruses; it alerts other cells that a virus is present and prompts some to produce antiviral proteins. Scientists soon learned, however, that interferon is species specific, and that mouse interferon had no effect on human cells. Nor could it be mass produced. The antigens had to be harvested from white blood cells almost cell by cell and at great expense. The cells could not be grown or sustained in a laboratory culture or be used more than once. It took a river of blood to produce one milligram of interferon, which was highly diluted and labeled "impure."[18]

Early in 1979, Congress provided $13.5 million to study biological response modifiers. NCI already had $3.5 million in active interferon research projects under way, one of them under the direction of Evan Hersh. The section of immunology in DT was scheduled soon to be upgraded to the Department of Clinical Immunology and Biological Therapy.

Hematologist Jordan Gutterman approached LeMaistre and indicated that it would take $250,000 to purchase what was then the world's supply of impure interferon from the Swedish Red Cross. The interferon was active one part in a thousand and therefore had to be given in very large doses to obtain a clinical effect. Some research indicated that it might work especially well in combination with chemotherapy. The greatest obstacle was going to be the expense of acquiring sufficient amounts of interferon for meaningful clinical trials. Mary Lasker and the American Cancer Society raised funds, and in 1978, with $4 million from Shell Oil, the ACS distributed interferon for clinical trials at ten institutions.[19]

LeMaistre grappled with the economics of interferon. "We must seek other ways of getting pure interferon to learn exactly what it does. The development costs to the pharmaceutical industry will be staggering and the risk of failing to recover the research investment high." Money would have to be raised to purchase interferon. Like Gutterman, LeMaistre walked a balance beam between the need to motivate possible donors and to prevent unrealistic expectations. "Interferon is no panacea," he wrote a potential donor. "Its real value is likely to be in its ability to regulate cell growth and permit the uncovering of the cell growth regulators, of importance in the prevention and treatment of cancer." LeMaistre made sure, however, not to understate the importance of interferon studies either. To the Houston attorney Leon Jaworski, who had won fame a few years earlier prosecuting President Richard Nixon for the Watergate crimes and now headed the M. D. Anderson Foundation, LeMaistre wrote, "There is no question that funds are urgently needed to purchase interferon so that the initial clinical trials may be completed. . . . I can certify without qualification, that obtaining the interferon necessary to complete the clinical trials remains as one of our top priorities."

Immunologists began wondering how to purify interferon. LeMaistre approached the pharmaceutical company Hoffman-LaRoche to develop a synthetic version of interferon. Dr. Emmanuel Grunberg, who served as director of research at Hoffman-LaRoche and was a close friend of LeMaistre, agreed. Thanks to the blood cell separation technique developed by Emil J Freireich and Jeanne Hester, M. D. Anderson shipped approximately 30 billion white cells a week to Nutley, New Jersey. There, using the *E. coli* technique for the production of purified interferon, Hoffman-LaRoche synthesized alpha interferon initially and then later gamma interferon, two of the three Type 1 interferons that were then known to be present in humans.

While the development of pure interferon was in progress, the Interferon Foundation, a philanthropic organization founded by Houstonians Roy Huffington and Leon Davis, provided $1.5 million to purchase interferon. The foundation ultimately donated $17 million to support research during the interim until Hoffman-LaRoche purified interferon, which lowered the cost and provided pure and fully active interferon.[20]

Not since the Salk vaccine for polio in the 1950s had a medical treatment garnered such media attention. Thousands of articles on interferon appeared in newspapers and magazines around the world. If oncology had a celebrity, it was Gutterman. He classified interferon with chemotherapy because it inhibited tumor cell proliferation, but it also seemed to activate macrophages, natural killer cells, and cytotoxic T lymphocytes. Under proper clinical conditions, it might modulate the immune response and achieve modest therapeutic benefit for some cancer patients."[21]

The world, though, yearned for more. Patients and their families inundated cancer centers with requests for interferon. Sometimes they crossed the line between request and demand. In the summer of 1984, for example, a middle-aged man with metastatic melanoma checked into M. D. Anderson. Because melanoma lesions pocked both lungs, surgery was not an option. Medical oncologists were prepared to treat him with a standard chemotherapy regimen, but they held out little hope. In a moment of frustration, he screamed to the charge nurse, "Give me some of that interferwhatever! It's my only chance. I have three kids, for God's sake."[22]

Actually, he had no chance. As a cancer treatment in 1984, interferon was a loose peg on which to hang one's life. Not unexpectedly to veteran oncologists, cancer demonstrated its typical resiliency. In two decades, recombinant genetic engineering would produce a substance known as alpha interferon that stimulated the immune system, and medical oncologists would find ways to employ other forms of interferon as treatments for several neoplastic diseases, but as a weapon to cure sick people, interferon in the 1980s was a pipe dream.

———

By 1985, LeMaistre was president-elect of the American Cancer Society, and he dedicated his tenure to cancer prevention. He argued that 85 percent of all cancers are caused by environment and lifestyle—tobacco use, poor diet and nutrition, obesity, and sun exposure. "We have reached a watershed when about half of all serious cancers can be cured," he said after his election as president. "Lung, breast, and colon cancer account for two-thirds of all cancer deaths, and each could be reduced by lifestyle changes. . . . Taking control of one's own life and destiny can lessen the chances of developing cancer." LeMaistre confessed that he had a weakness for Blue Bell nutty coconut ice cream, but he ate it judiciously, filled his plate with plenty of fruits and vegetables, and watched his weight. He urged other Americans to do the same.[23]

And the time had come for M. D. Anderson to clean up its own house. LeMaistre could not abide cigarette smoke. Back in 1975, when the city of Houston passed legislation restricting smoking indoors at public venues, Clark had implemented the policy at M. D. Anderson. He banned the sale of tobacco products in M. D. Anderson gift shops and forbade visitors to smoke in patient rooms. Patients could smoke in their own rooms, but only with the permission of roommates. Special zones in the hospital were set aside where people could smoke; they turned the air light blue. The smoke quickly drifted outside of the smoking zones and assaulted nonsmokers. Smoking's days at M. D. Anderson were numbered. LeMaistre belonged to state and national antismoking groups, regularly testified against the tobacco companies, and celebrated plaintiff victories. In 1985, he formed a committee to develop a new smoking policy. On January 1, 1989, M. D. Anderson became smoke free.[24]

———

Like Lee Clark before him, LeMaistre constantly monitored the physical plant, the escalating patient load, the quality of patient care, and the financial bottom line, trying to manage the growth of M. D. Anderson while preserving its culture.

In 1983, construction began on a new eight-story basic research building as well as additions to the Bates-Freeman Building, and dedication ceremonies were held for what became known as the R. E. "Bob" Smith Research Building on Knight Road. The building had been home to the ill-conceived food preparation facility. In 1984, contractors began work on the clinic building that would bear Lee Clark's name. In 1986, the Basic Research Building was dedicated.

On November 19, 1987, Secretary of the Treasury James Baker was scheduled to dedicate the R. Lee Clark Building. Baker's grandfather, Captain James Baker, had once owned the downtown Houston property that had been known as the Baker estate, the first site of M. D. Anderson. Unexpected developments at the Department of State, however, kept Baker in Washington, D.C., and Secretary of Commerce Robert Mosbacher substituted. The R. Lee Clark Clinic brought the physical plant to more than 1.5 million square feet. Clark attended the dedication. A debilitating stroke the year before had stolen his ability to speak. At the dedication, Clark walked laboriously to the podium, flashed a buoyant smile, and uttered to an audience in tears words that had taken a week of practice: "My good friends, I am honored today. Thank you for your fight against cancer and for hard work, loyalty, and friendship. Thank you. Thank you. Thank you. Thank you."[25]

—

Between 1984 and 1986, LeMaistre confronted a financial profile that tested his mettle. Indigent care, which had totaled just under $13 million, or 9.8 percent of total revenues, rose to $29.1 million, or 19.6 percent. The percentage of bad debts and indigent care in 1986 totaled $39.1 million, or 26.4 percent of total collections. More than 78 percent of the funds M. D. Anderson received from the state legislature annually now went to indigent care, compared to only 23 percent ten years earlier. The collapse in oil prices from forty dollars a barrel to ten dollars had slashed state tax revenues in the mid-1980s and raised the specter of financial exigency. The Republican governor William Clements directed all state agencies to cut their budgets.

An era of intense competition to sell patient care services had engulfed medicine. As consumers, patients enjoyed choices for cancer treatment. In addition to M. D. Anderson in Houston, they could report to the surgical and oncology services of Baylor, the UT Medical Branch, or the offices of local surgeons and oncologists who had entered the market only recently. Increasing competition characterized medicine throughout the nation, and medical schools, hospitals, and categorical institutions had to think increasingly in terms of market shares, capital expenses, and profit margins. As academic medicine competed in the marketplace, the bottom line became more important than ever before. Without a healthy spreadsheet, they could not fulfill the mission of research, patient care, and education. More and more, when questions of medical economics were discussed in America, the phrase "no margin, no mission" was heard.

As a first step toward more effective marketing, LeMaistre succeeded in restoring the name "M. D. Anderson" to the institution's title. Early in the 1970s, the board of regents had renamed M. D. Anderson "the University of Texas System Cancer Center," depriving the institution of a moniker recognized around the world. In 1988, the president's executive board recommended that LeMaistre ask the board of regents to rename the institution "the University of Texas M. D. Anderson Cancer Center." The change restored use of the renowned M. D. Anderson name while reflecting the institution's status as part of the University of Texas System. The board of regents agreed.[26]

LeMaistre asked Joseph Painter to develop a marketing plan to increase the number of out-of-state patients. M. D. Anderson was obliged to care for indigent Texans but had no such obligation for non-Texans. In 1988, thirty-five hundred out-of-state patients were treated. By increasing those numbers, the percentage of patients from whom costs could be recovered would automatically increase. Between 1983 and 1987, the number of patients from Texas had increased from 58.3 percent to 68.3 percent. Many were indigents.[27]

Louisiana was a natural market. The Interstate 10 corridor from New Orleans to Houston provided relatively easy access, and M. D. Anderson could accept only those patients with health insurance or money to pay for care. When Painter established committees to assist with the marketing, some physicians grumbled, considering the business end of medicine a distraction from research and patient care. At a meeting where such concerns surfaced, Charles Balch remarked, "No margin, no mission."

International patients were even better for the bottom line. Always well heeled, they put down hefty deposits and paid their bills promptly, without third-party involvement. Many international patients represented elites in their own countries, and their word of mouth about M. D. Anderson spread quickly. In 1988, LeMaistre signed an affiliation agreement between M. D. Anderson and the city of Madrid in Spain to develop a joint program of research, patient care, and scientific exchange.[28]

Private industry offered another option. Patents deriving from new discoveries in cancer diagnosis and treatment held out the possibility of rich returns, and LeMaistre forged financial arrangements with interested private businesses. In 1986, he established the Office of Technology Development and quickly entered into an agreement with the Macrophage Company to develop immunological treatments for cancer. Two years later, M. D. Anderson signed a similar agreement with Molecular Diagnostic Associates to develop new diagnostic tests for cancer. The agreements were initial forays into an effort that in the next two decades would intensify dramatically throughout academic medicine; critics began to decry the "commercialization of medicine."

Some bioethicists envisioned danger in financial ties to private companies. Regulations for the protection of human subjects could more easily be bypassed when using drugs from the pharmaceutical industry than drugs from NCI. Critics feared that by climbing into bed with private industries, conflict of interest problems would multiply. The lure of "commercial[izing] products that could cure some forms of cancer or serve as tools to diagnose" was powerful, admitted an M. D. Anderson physician. The Macrophage Company, for example, had received $600,000 in start-up money. "The investors are taking a calculated chance," admitted the physician Gabriel Lopez-Berestein. "They risk some seed money now, yet they could make a thousand percent return if we produce something to control cancer." With such returns possible, unbiased decisions about research and treatment would be more difficult.[29]

AIDS treatment illustrated the budget crisis. M. D. Anderson had seen a steady stream of young men with AIDS-related cancers. At first, the institution welcomed them as a cohort whose cancers had a viral origin and were part of an infectious process. M. D. Anderson received a $5.8 million contract from the National Institute of Allergy and Infectious Diseases to study AIDS, but the sheer number of patients soon taxed hospital resources. Intensive care beds filled with AIDS patients and spilled over into regular hospital rooms. Expenses mounted. They remained inpatients for months, clogging a hospital never designed for such a crisis.

In 1985, American Medical International (AMI) saw in AIDS a business opportunity. Because so little was known about the disease and because patients suffered such a variety of illnesses, hospital stays could be quite lengthy. AMI sought to establish in Houston a hospital dedicated to the treatment of AIDS. Because of the size of the gay community there and the infectious nature of the disease, AMI anticipated no difficulty in filling beds. In New York City, AMI had learned, 40 percent of AIDS patients, from the onset of symptoms to death, spent at least half of their days in the hospital at an average cost of $50,000 each.

Healthy profits could be extracted from the situation. AMI also saw an opportunity to elevate the company's profile to become "the first U.S. health care corporation to respond to the international AIDS epidemic."[30]

On the north side of Houston just off Interstate 45, AMI identified Citizens General Hospital as a suitable site. Citizens General had 150 beds with enough room to expand its facilities for emergencies and laboratory medicine. AMI suggested renaming Citizens General the Southwest Institute of Immunological Diseases and Infectious Disorders and providing M. D. Anderson with $1 million in research funding. In return, LeMaistre would loan physicians to the institute to treat patients and conduct clinical research. Irwin Krakoff welcomed the opportunity to transfer the AIDS patients. Oncologists could treat the AIDS-related cancers, but most of them did not have the expertise to manage the variety of other infections afflicting AIDS patients. For Krakoff, the AMI proposal would "provide some relief . . . while permitting us continued access to the material for clinical research, epidemiologic studies, [and] laboratory tests." He urged LeMaistre to pursue the relationship. "This project can be justified because of the high incidence of cancer in the patient population and in the population at risk. In our opinion, that would offset the possibly negative public aspects of this affiliation."

Eugene McKelvey, the associate vice president for research, voiced to Frederick Becker his concerns. "It is important to keep the AIDS research in perspective," he wrote. "We currently justify this research on the basis of what it has to offer our cancer patients. This type of proposal could very easily shift that balance so that cancer immunology is secondary to AIDS activities. We would then have the tail wagging the dog." McKelvey also sniffed problems in the AMI business model. "Frankly I doubt that AMI will be financially sound. It has been estimated that the treatment of one AIDS patient costs approximately $140,000 per year. Most AIDS patients do not have financial resources of their own."[31]

The proposal also carried ponderous political baggage. Community leaders on the north side of Houston sounded the alarm. In addition to Citizens General Hospital, AMI contracted with Parkway Hospital for services. Many residents feared north Houston would become a gathering place for gay men. Others worried about medical issues. Would the disease spread from the hospital to their homes? State Senator John Whitmire asked LeMaistre not to proceed with the plan, and in January Roger Bulger, president of the UT Health Science Center at Houston, told LeMaistre, "I think it's a good project, but it may not be worth the pain." The black physician James Haughton opposed the proposal for fear that it would "lay the groundwork for quarantining or isolating those afflicted with the deadly disease . . . and treat [them] differently from victims of other diseases." The most intense opposition erupted from blacks in the Acres Homes neighborhood, who asked Congressman Mickey Leland to block the proposal. "Believe me," said black businessman C. S. Roper, "resistance is a very mild word for what the public is going to hear from a very concerned group of citizens."[32]

Guy Newell urged LeMaistre to proceed, although the board of regents refused to sign any contracts with AMI and forbade company officials from advertising an affiliation with M. D. Anderson. Newell considered the AIDS initiative a logical dimension of cancer prevention. Young gay men in urban areas were dying in disturbing numbers. The Reagan administration seemed apathetic. Had AIDS been ravaging heterosexuals, Newell seethed, the nation would have mobilized a Manhattan Project against it. Instead, gay men were suffering horrible, lingering deaths and the epidemic was spreading. M. D. Anderson was morally bound to do something, and AMI provided that opportunity. Early in 1986, LeMaistre agreed and physicians Peter W. A. Mansell and Adan Rios were assigned to provide the expertise. LeMaistre asked José Trujillo to oversee laboratory medicine there.

Mansell and social worker Sue Cooper organized a model approach to AIDS, with a team that included physicians, nurses, nutritionists, physical therapists, and psychologists to treat patients at home until the end stage. In some ways, Mansell's success spelled AMI's demise. The company had predicated its success on having a hospital full of patients, with revenues of $140,000 to $200,000 per year each. Mansell treated them at a fraction of the cost. By the late summer of 1987, AMI reported losses of $7.12 million on its AIDS program, with a projected loss for 1988 of more than $6 million. They should have listened to McKelvey. Indigent care explained the losses. AMI pulled the plug.[33]

LeMaistre confronted a touchy issue. He asked Dan Oldani, the executive director of hospital and clinic operations, to provide a cost estimate should M. D. Anderson assume the responsibility. Oldani's report was brutally honest. Although Mansell's outpatient program worked well for the early stages of the disease, end-stage costs ranged from $25,000 to $30,000 per patient. "The cost of the clinics and sixteen beds in the hospital would . . . result in a major financial loss for us both short and long term. . . . We simply do not have either inpatient or clinic space to absorb this population in a separate discrete unit."

Taking care of AIDS patients also might stigmatize M. D. Anderson. "I believe we will have major problems in recruiting a broader private paying referral base of cancer patients if we assume this population," Oldani told Krakoff. LeMaistre decided that M. D. Anderson would treat only those patients who had cancer as part of their complex of AIDS-related diseases. Its categorical mission allowed such latitude.[34]

At the same time, M. D. Anderson developed new initiatives in pain management. In 1971, the same year that he promoted the war on cancer, President Richard Nixon had also declared a "war on drugs" to destroy "public enemy number one." During the 1980s, First Lady Nancy Reagan declared her own war on drugs. Hyperbole often replaced reasoned debate, and at the state level, the war on drugs collided with the war on cancer, particularly over the use of analgesic drugs to control intractable pain.

Late in the 1940s, Raymond Houde of Memorial Sloan-Kettering had pioneered the clinical pharmacology of opioid analgesics to relieve the suffering of cancer patients. Confusion reigned. The treatment guidelines for opioids had been established in postoperative settings, where patients were steadily improving and just as steadily in need of less medication. Cancer patients, on the other hand, steadily deteriorated with escalating pain. Because of those treatment guidelines and because many patients were too ill to speak or because of such factors as patient stoicism, anxiety, and individual differences in drug metabolism, most physicians had little sense of the best drugs and the most efficacious doses. "No one knew how much morphine to give a patient, since most information about narcotics was on drug addiction, not analgesia," remarked nurse Ada Rogers of Memorial Sloan-Kettering in 2007. "People did not realize that the street addict who takes drugs for a high and the patient who needs medication for pain are two different populations." Houde developed ways to measure subjective response to pain, and he characterized each of the major opioid analgesics. In 1981, Kathleen Foley, a protegée of Houde, established at Memorial the nation's first pain service for cancer patients.

At M. D. Anderson, physician C. Stratton Hill assumed Houde's mantle. A graduate of the University of Tennessee College of Medicine, Hill had later trained at Memorial. He came to M. D. Anderson in 1963, and in 1974 he was appointed head of the outpatient clinics. His own research at first revolved around thyroid disease, but he gradually shifted to brain peptides, from which he developed clinical interests in pain management. When LeMaistre came to M. D. Anderson, Hill moved to the Pain Clinic, which since the days of William Derrick had been housed in the Department of Anesthesiology. Several years later, the Pain Clinic became part of the new Neuro-Oncology Center under the direction of neurologist William Fields, and Hill had a raison d'être for the rest of his career. In March 1988, M. D. Anderson sponsored the conference "Drug Treatment of Cancer in a Drug-Oriented Society: Adequate or Inadequate?" The conference brought to Houston leading political and medical figures, including Raymond Houde and Kathleen Foley, who called for the "decriminalization of cancer pain treatment."[35]

Even when physicians were well informed about pain-killing drugs, many hesitated to administer them for fear of making addicts of their patients or being prosecuted by drug enforcement personnel, who were generally unenlightened about the difference between drug addicts and cancer patients. "Almost all pain is undertreated," Hill remarked. "The results of this can be devastating for patients and their families." He coined the term "opiophobic" to describe such physicians. Hill called for new legislation distinguishing "between people who need to take drugs for legitimate reasons and those who take them for recreational reasons."

After the 1988 conference, Hill crusaded to enlighten the state legislature and to weave a legal cocoon. Lieutenant Governor William Hobby, a personal friend of Hill, endorsed the Intractable Pain Act of 1989, which authorized physicians to use controlled substances for treating intractable pain, prohibited health care facilities from limiting use of such drugs, and prohibited the Texas State Board of Medical Examiners from disciplining physicians who employed them in treating pain. The legislation was the first of its kind, and Hill traveled widely endorsing its provisions. His film *My Word Against Theirs* to educate physicians won the Patient Care Award at the John Muir Medical Film Festival in 1990, and for his efforts in relieving the suffering of hundreds of thousands of cancer patients, Hill in 1995 received the Humanitarian Award from the American Cancer Society.[36]

—◦◦◦◦◦◦◦—

In 1988, as LeMaistre counted ten years at M. D. Anderson, Gary Hood counted his lucky stars. A gregarious professor of education who enjoyed acting in community theater, Hood had in 1984 been diagnosed with a melanoma on his foot. The tumor was removed surgically, and, to save his leg, surgeons followed up with a procedure that M. D. Anderson had perfected over the years—regional perfusion of his leg with massive doses of chemotherapy. The drugs attacked residual melanoma cells without frying major organs. Surgeons then removed the lymph nodes. Hood and his physicians waited for signs of life in his melanoma, and they waited and waited. The disease was gone. Hood was cured.

11

——◅▨▨▨⌇∩▨▨▨▷——

Victory, Defeat, and
an Elusive Enemy,
1988–1996

*He tells me that Waldenström's has a median survival time of nine
years after diagnosis, but that the disease is very complex. Each
patient has a unique molecular profile, which makes treatment
results unpredictable. There is no cure, but perhaps I have another
decade or more to live. I'll take that. It would be nice to collect
some Social Security.*

—JOAN COFFEY, 1995

At Sam Houston State University in Huntsville, Texas, in 1995, Joan Coffey's office resembled a salon. As befit the office of a French historian, France was everywhere, permeating the room and hanging in the air—pictures of castles from the French countryside, colorful homes along the Seine, milk cows in a pen, prints from the French Revolution, and a porcelain portrait of the Mona Lisa. The shelves were heavy with books, not ornamental books but artifacts of a true bibliophile. Joan Coffey was a meticulous scholar, in the old-fashioned meaning of the word, before deconstruction let historians hurtle beyond the evidence. Her husband, Ed, was there too, in a photo taken on December 15, 1995. He stands beside Joan whose hair is pulled back tightly and anchored with a band. She beams with a $1 million smile, happy but not healthy.

In 1995, Joan had lab tests as part of a pre-op workup for bunion surgery.

When the technician noticed a high sedimentation rate, he contacted a hematologist who identified an incurable, tongue-twisting case of Waldenström's macroglobulinemia, a non-Hodgkin's lymphoma of B lymphocytes today known as lymphoplasmacytic lymphoma. Since Joan was not yet symptomatic, the hematologist recommended wait-and-see, but patience stretched only so far, especially with lymphoma cells coursing through her veins. Joan went to M. D. Anderson and became a patient of medical oncologist Raymond Alexanian.

She soon had more to worry about than Waldenström's. Radiologists detected a small lump on one kidney. Joan's condition and the fact that cancer was no stranger to her family called for surgery. Pathologists diagnosed a malignant sarcoma. The surgeons performed a partial nephrectomy, lifting out the sarcoma with a portion of the kidney. During the operation, however, while inserting a central line, they nicked the artery, and Joan began bleeding into her pleural cavity. They closed without noticing, and radiologists missed the bleeder in postoperative x-rays. Joan soon ended up in intensive care complaining of chest pains. The surgeons had to repair the damage. M. D. Anderson officials apologized, and the Coffeys forgave the error.

Joan then entered the murky world of modern oncology, with its ironies and Hobson's choices. Raymond Alexanian was a kind, soft-spoken man with a résumé listing hundreds of articles and a lifetime of treating Waldenström's and multiple myeloma. He too recommended wait-and-see, and when Joan pressed for details, she remembers the response: "Waldenström's has a median survival period of nine years after diagnosis, but the disease is very complex. Each patient has a unique molecular profile, which makes treatment results unpredictable. There is no cure." He put Joan on interferon, hoping to postpone more toxic chemotherapy regimens. To a friend, Joan mused, "Perhaps I have another decade to live. I'll take that. It would be nice to collect Social Security." And then Joan Coffey waited for act 2.[1]

⸺⸙⸻

While Joan set out to fight cancer, Charles LeMaistre contended with a financial crisis. On the surface, all seemed well. The M. D. Anderson Outreach Corporation had established a formal relationship with the Orlando Regional Hospital System, and in 1989 the Orlando Cancer Center emerged from that collaboration.

The main campus appeared flush, in 1994 registering its three hundred thousandth patient. Several large philanthropic donations and the hoopla surrounding them masked trouble. (Large one-time gifts rarely solve fundamental problems.) In December 1989, M. D. Anderson had accepted a gift of $24 million from the estate of retail auto dealer C. P. Simpson and his wife, Anna Crouchet Simpson. In December 1993, Texas oilman Albert B. Alkek and his wife, Margaret, donated $30 million, which helped finance construction of the Albert B. and Margaret M. Alkek Hospital, a 550,000-square-foot patient care center scheduled to open in 1996. Then, early in 1995, oilman W. A. "Tex" Moncrief and his wife, Deborah, both friends of Robert D. Moreton, worked for and succeeded in transferring to M. D. Anderson the Moncrief Radiation Center in Fort Worth, a nonprofit cancer treatment facility valued at $44.2 million. The Moncriefs had lost a six-year-old daughter to leukemia. M. D. Anderson had also recently concluded the most successful capital campaign in its history, raising $151 million.[2]

Meanwhile, managed care represented what some called the "revolt of the payers." Ever since World War II, academic medicine had enjoyed reliable revenue streams from third-party payers. Health insurance had emerged as a benefit for workers during the war years when the federal government had imposed wage and price controls in order to stem inflation. Labor unions subsequently made health insurance a key element in contract negotiations. The number of Americans covered by health insurance steadily increased. The implementation of Medicare and Medicaid beginning in 1965 proved a boon to medical care providers because the federal government now funded medical care for the elderly and for some of the poor. Revenues from patient care soared.

By the late 1980s, however, revolution roiled medical economics. In 1960, health care spending as a percentage of pretax business profits in the United States had amounted to 5.3 percent. Medical costs far outpaced the consumer price index. In 1970, health care costs chewed up 7.4 percent of business pretax profits, a number that grew to 9.1 percent in 1980 and 12 percent in 1990. To rein in skyrocketing medical costs, many companies turned to systems in which patients surrendered the right to select family physicians and specialists. The preferred provider organizations (PPOs) and health maintenance organizations (HMOs) contracted for services. Patients had to use the physicians dictated by the HMOs and PPOs. Without entering into such contracts, M. D. Anderson risked being frozen out of the business model sweeping medicine. Between 1986 and 1995, the number of Americans insured under managed care went from 23 million to 60 million, and in Houston, 90 percent of the employees of major companies were enrolled in managed care. LeMaistre in 1993 had established an Office of Managed Care at M. D. Anderson, and in the next two years, the percentage of M. D. Anderson patients covered under some form of managed care jumped from 11 to 23 percent. M. D. Anderson was hardly alone. Managed care had an impact on medical centers, hospitals, and medical schools throughout the nation, threatening the solvency of many. Revenues fell while expenses mounted. During the 1970s, for example, the New England Medical Center annually hemorrhaged $7 to $10 million a year, and in 1989, Presbyterian Hospital in New York lost a colossal $50 million.[3]

To prepare M. D. Anderson for the threat, the institution had to lower costs, become more efficient, and cap the burgeoning cost of indigent care. David Bachrach, M. D. Anderson's chief financial officer, was given the task of developing a contract with the Rice University School of Business to accept four classes of twenty students, each at a cost of five thousand dollars. The students were to be drawn from the medical and nursing faculty, staff, and administrative personnel, with classes held on Friday afternoons and Saturdays. At first, the moans reverberated widely from staff members loathe to sit through a training course in business, but when it became known that LeMaistre intended to choose from that group those that he anticipated would become future leaders at M. D. Anderson, selection became an anointing. Participants maintained their full-time employment responsibilities. Eventually, LeMaistre had eighty business graduates to streamline clinical services throughout the institution.[4]

In spite of the philanthropy and training, early in the 1990s economic reality blindsided LeMaistre. Between 1985 and 1994, the cost of indigent care had ballooned by 470 percent and a $30 million shortfall loomed in the 1994–95 operating budget. The daily census, the average number of patients each night in the hospital, had nosedived from 430 in 1992 to 310 in 1996.[5]

LeMaistre asked Joseph T. Painter to help market M. D. Anderson services to corporations. When managed care companies did refer to M. D. Anderson, the patients were often at the end stage, with little to be done for them. "We continue to see," stated a financial summary presented to the president's executive board in 1990, "a modest number of 'out of network' patients referred from HMOs who desire our services. Typically, these patients are late stage disease or second opinion only patients with questionable opportunities for substantive intervention on our part."[6]

To address the fiscal crisis, LeMaistre formed a task force that included several of the Rice graduates. They first identified the need to reduce employment, if possible by attrition and early retirement at full benefits. Because so much of M. D. Anderson's budget went to employees, reductions in force proved inevitable. David C. Hohn, vice president for patient care, estimated that $30 million would have to be cut from the institution's $586 million budget in order to "right-size" M. D. Anderson. "The goals of these efforts," he noted, "are to ensure that our financial and human resources are directed where they are most needed and that we have the funds to support quality patient care and research." LeMaistre balanced a realistic view of the present while casting a cautious eye to the future. "Downsizing" [can be expected]," he said. "Much has already been done. In addition, we will abolish many vacant positions and reallocate dollars. We will explore early retirements and reduction in the work force by attrition.... This will not be a painless process, but it will be done with objectivity, sensitivity, and compassion." It was quite a change for an institution that had experienced only growth and felt itself to be immune to the vicissitudes of the medical marketplace.[7]

Instead of basing the cuts on seniority, department heads were asked to identify positions and workers no longer deemed essential to the mission. A three-year plan cut $90 million from the budget and took the number of employees from 8,500 to 7,595. "The layoffs were brutal," remembers one social worker. "We went from forty-eight FTEs [full-time equivalent employees] to twenty-six." Lois Woodard, an African American nurse with thirty-two years experience, lost her job. "I got caught in the downsizing of 1996," she recalls, "but it was easy for me to retire because of all the years I had accumulated in the retirement system. Many of the others ... felt betrayed by an institution they loved. It was like being banished by one's own family."[8]

Tension thickened and rumors raced down the corridors and in and out of offices and hospital rooms. Many faculty members put their résumés on the job market; nurses began lining up job possibilities in the Texas Medical Center; and blue-collar workers waited for pink slips. An insecurity previously unknown at M. D. Anderson gripped the staff and dampened the congenial optimism so central to the culture.

To hold down costs, the task force urged LeMaistre to secure exemptions from state purchasing regulations, which required all state agencies to seek competitive bids from suppliers and then purchase from the lowest bidder. The equipment needs of most state agencies were quite routine, and the system worked efficiently, but M. D. Anderson often needed highly specialized equipment, supplies, and pharmaceuticals available from only one company and had to buy at monopoly prices. Anderson needed to be able to circumvent purchasing regulations and negotiate prices directly with suppliers.

Fiscal stability also demanded limits on indigent care. M. D. Anderson had long supplied high-quality, free care to indigent Texans, but by the late 1980s, the expense rivaled the institution's total annual appropriation from the state legislature. Some major medical centers around the nation found a solution in transferring indigent patients to county hospitals, hospital districts, and Veterans Administration Hospitals. LeMaistre needed a similar option. The task force recommended securing from the legislature the right to contract with county hospitals the care of indigent patients with uncomplicated cancer cases.[9]

Cutting costs, however, would not by itself ease the crisis. New income had to be generated. Otherwise, the squeeze would only tighten. The task force quickly identified managed care as a culprit, especially since by law M. D. Anderson could only accept patients referred from other physicians. Managed care companies, however, had a financial interest in keeping their clients within the system. Patients could not self-refer to M. D. Anderson. The task force recommended that M. D. Anderson secure legislative approval for self-referral.

LeMaistre needed help, and he assembled a core team that included Harry D. Holmes, a former professor of urban development and director of government affairs at M. D. Anderson, David C. Hohn, David J. Bachrach, Charles M. Balch, and others. They helped design a strategy to deal with the crisis and worked closely with William H. Cunningham, the chancellor of the UT System, in developing new legislation. A former professor of marketing and dean of the UT Graduate School of Business, Cunningham understood the economics of M. D. Anderson's situation. "Indigent care was going to break us," Holmes remembers. "We were having difficulty recruiting new patients because the law required that they be physician-referred, and the state purchasing process cost us money because our needs were so specialized." The UT System office drafted the bill. To liberate M. D. Anderson from the clutches of HMOs, the bill allowed for self-referral. The measure also authorized M. D. Anderson to negotiate prices directly with suppliers. And to relieve the pressures of indigent care, the bill permitted M. D. Anderson to contract with county governments for care of the poor and the uninsured. The needs of Harris County were by far the most onerous. Finally, with Cunningham's assistance, the measure included language giving M. D. Anderson wide latitude to negotiate partnerships with private industry.[10]

In 1994, M. D. Anderson had friends in the highest places. Over the years, some of the richest and most powerful ranchers, bankers, oil men, and politicians in Texas had populated the board of visitors, and none was more influential than former president George H. W. Bush. Back in July 1990, when the Economic Summit of the Industrialized Nations convened in Houston, Barbara Bush hosted a tour of the Texas Medical Center for the spouses of ambassadors, diplomats, and heads of state from France, Italy, the United Kingdom, Germany, Canada, Japan, and the European Community. She selected M. D. Anderson for her own stop on the tour.[11]

George W. Bush, the son of the former president, was in the midst of a successful campaign for governor in 1994, and the family loaned its name and political clout to M. D. Anderson. So did Robert A. Mosbacher, the wealthy Houstonian, friend of the Bushes, and former secretary of commerce, and James Baker, the former secretary of the treasury and secretary of state. On May 4, 1994, George and Barbara Bush hosted at their new home in Houston a meeting of the Harris County legislative delegation. Thirty-one of the thirty-three senators and representatives attended, and Barbara served coffee and her special orange muffins. In subsequent weeks, M. D. Anderson staff met with every member of the state legislature or their staff. Few pieces of legislation in Texas history had been more thoroughly vetted. Don Henderson sponsored the bill in the Senate, and Tom Craddick in the House. Lieutenant Governor Robert "Bob" Bullock, a cancer survivor, great admirer of M. D. Anderson, and arguably the most powerful politician in Texas, endorsed the measure, which the House and the Senate passed unanimously. Governor George W. Bush signed it on March 28, 1995.[12]

The results spoke to the importance of the bill. M. D. Anderson soon signed its first agreement on indigent care with the Harris County Hospital District. A new oncology service, staffed by M. D. Anderson oncologists, was set up at LBJ Hospital, with UT Health Science Center faculty providing a triage mechanism to handle overflow. Within one year, M. D. Anderson's costs for charity care dropped by more than $22 million, and savings on purchases exceeded $8 million.[13]

The omnibus bill provided M. D. Anderson with the tools it needed to confront the changes in medical economics, and the other health science centers in Texas soon clamored for similar legislation. They needed the same privileges M. D. Anderson now enjoyed.

—◆—

Late in the 1980s and early in the 1990s, as LeMaistre grappled with the fiscal crisis, he attended funerals and memorial services for some of M. D. Anderson's founding mothers and fathers. Felix Haas, longtime head of the Department of Biology, passed away in 1989. Arthur F. Kleifgen, "the General," died in 1990, as did Joe B. Drane, M. D. Anderson's pioneer in dental oncology. Robert Moreton died in December 1992 and Gilbert Fletcher of heart failure associated with leukemia. Some faculty members wondered if, like Marie Curie, he had simply been overexposed to radioactive materials. Ed White succumbed three weeks after Fletcher, a victim of the cigarettes Lee Clark had badgered him about for so long. Charles M. McBride, the surgeon who worked at the side of Richard Martin, died in 1996 at M. D. Anderson of pancreatic cancer. José Trujillo, the head of laboratory medicine, died in December 1992 at M. D. Anderson of prostate cancer.[14]

Ethel Fleming, the African American nurse who pioneered the desegregation of the institution, was diagnosed with pancreatic cancer in May 1991. Fleming had succored enough pancreatic cancer patients to know just how short the tether of life now was. At first, she wanted to avoid having former colleagues observe her decline and death. Instead, she stayed home under the care of her daughter, Martha Jones, and underwent palliative treatment at Station 19. The disease, however, soon overwhelmed them. In February 1992, Ethel told Martha, "Take me home." Martha gently replied, "You are home, Momma." Ethel insisted, "No, take me home." Martha then checked her mother into M. D. Anderson. Fleming spent her final days under the tender care of nurses she had loved.[15]

On May 3, 1994, R. Lee Clark came home too. Cancer caught up with the boss. The grapevine had never operated so efficiently; news of Clark's illness worked its way at warp speed into every office and clinic station. M. D. Anderson employees hoped to see that he supped from the same banquet of multidisciplinary care, aggressive treatment, and reverent care that he had offered to so many others. Clark had entered the hospital suffering from colon cancer. The staff enjoyed little time to serve him. He died at M. D. Anderson of complications from surgery to remove the tumor. Since his retirement, Clark had served as president emeritus. Surrendering power had not been easy. For thirty-two years, the disease had served as catalyst to his ambition. Clark had built the frames, poured the cement, and erected the modern edifice of cancer treatment, and in doing so with grace and compassion had come to be regarded as a secular saint in Texas. If a man had ever personified the benevolent dictator, it was Lee Clark, who had ruled M. D. Anderson, in the words of James Bowen, with the style of a "lead pipe wrapped in silk."

The transition had not been easy for many staff either, who out of habit continued for a time to show up at Clark's new office for advice and decisions. He always had good advice, but decisions he had none. LeMaistre could not have been more understanding and deferential, even when Clark occasionally forgot his new place, but the arrangement, with Clark hovering around, had its awkward moments.[16]

Until his stroke in 1987, Clark had maintained an active interest in NCI's comprehensive cancer center program, which he had pioneered. He consulted regularly with institutions seeking comprehensive cancer center status. During the decade between his retirement in 1978 and the stroke in 1987, NCI had awarded comprehensive cancer center status to the H. Lee Moffitt Cancer Center and Research Institute at the University of South Florida in Tampa; the Herbert Irving Comprehensive Cancer Center at Columbia University; the Barbara Ann Karmanos Cancer Institute at Wayne State University in Detroit; and the Rebecca and John Moores Cancer Center at the University of California, San Diego, bringing the total to seventeen.

The deaths of Fleming and Clark broke the last two human links in a chain connecting M. D. Anderson to its beginnings. A few years after the establishment of the hospital in 1941, little Ethel Fleming had made a playground of the Baker estate, sneaking in through gaps in the hedges and frolicking on the grounds with her friends until being chased away each summer day at dusk. And five years later R. Lee Clark had commenced his remarkable career there.

Guy Newell died on November 2, 1994. Since 1978, he had presided over cancer prevention at M. D. Anderson, with AIDS and AIDS-related cancers getting his personal attention. Newell had reason to obsess about AIDS. Early in the 1990s, he battled fatigue and wondered why exhaustion set in at the beginning of the day. He seemed to catch every bug in Houston, missed work more and more often, and died after a long illness.[17]

New stars replaced the old in M. D. Anderson's firmament. In 1992, Joseph Painter was named president of the American Medical Association. In 1990, Charles Balch became president-elect of the Society of Surgical Oncology and Leland W. K. Chung president-elect of the Society for Basic Urologic Research. J. Taylor Wharton, the gynecologist who had succeeded Felix Rutledge as chair of the department, became president of the American Radium Society. Isaiah J. Fidler ascended to the presidency of the International Society of Differentiation. Helmuth H. Goepfert became president of the American Society for Head and Neck Surgery and Oscar M. Guillamondegui of the Society of Head and Neck Surgeons. In 1991, Gerald D. Dodd was named president-elect of the American Cancer Society, and President George H. W. Bush appointed Frederick Becker to the National Cancer Advisory Board (NCAB). One year later, the American Association for Cancer Research selected Margaret Kripke as its president. Immunologist Jordan Gutterman headed up the Lasker Award Program for the Albert and Mary Lasker Foundation. In 1993, Thomas S. Harle was elected president of the Radiological Society of North America and Lester J. Peters of the American Society for Therapeutic Radiology and Oncology. Louise C. Strong in 1995 was elected president of the American Association for Cancer Research. She also succeeded Becker as a member of the NCAB. Ralph Freedman, an M. D. Anderson gynecologist, would be appointed to the NCAB in 2000 when Strong's term ended. Luis Delclos became president of the American Endocurietherapy Society. S. Eva Singletary, an M. D. Anderson physician with an international reputation for innovations in breast cancer surgery, was appointed to the President's Cancer Panel Special Commission on Breast Cancer.[18]

The old hotel and its residents went the way of the old staff. Many of the residents were old and infirm. Robert D. Moreton wrote to the daughter of one, "I have known your nice mother, Mary Robinson, throughout the years. . . . In watching this nice, little lady . . . I have been worried about the possibility of her falling and injuring herself. . . . We have no accommodations for the care of bedridden persons." A number of other guests suffered similar maladies. "The Anderson Mayfair is not staffed to care for the elder sick," LeMaistre decided, "and every effort must be made to see that the needs of such residents are met elsewhere." In September 1979, he decided to convert all unleased apartments to transient use. The rumor mill, of course, had him throwing the older residents out on the street. The Anderson Mayfair had grown as feeble as its elderly residents. Pipes burst all too frequently, flooding rooms and requiring expensive redecoration. The ventilation was poor. Robert Moreton, who lived in one of the apartments, noted that "cooking odors from one apartment drift into another. Occasionally, this upsets the guests, particularly if they are experiencing nausea from their treatment." Late one night Jean Copeland went downstairs to mail a letter. The elevators stopped between the thirteenth and fourteenth floors, and she spent the night there, frantically, and to no avail, shouting and repeatedly pressing the alarm button. In a letter to LeMaistre, Moreton noted, "I would point out to you that there are patients of the Mayfair who, I am sure, would not fare nearly as well incarcerated in an elevator for that duration of time without contact with any human being [and] without any hope of immediate help." Contractors pegged complete renovation at $10 million. The Anderson Mayfair was scheduled to be demolished by a series of ten-second explosions.[19]

The Rotary Club of Houston offered to help raise the necessary $17 million to construct a new hotel. Club members tapped friends throughout Texas and the Southwest, and the Rotary Club of Houston Foundation raised $8 million, including a $6 million gift from the Houston Endowment. To provide the $17 million needed for the project, the UT Board of Regents authorized revenue bonds to raise another $9 million. On June 12, 1991, ground was broken at 1600 Holcombe Boulevard, directly across the street from M. D. Anderson, for the Jesse H. Jones Rotary House International, an eleven-story structure with 198 rooms.[20]

In 1991, M. D. Anderson marked its golden jubilee with a yearlong series of celebratory events. During that time, only three men had presided: Ernst W. Bertner, R. Lee Clark, and Charles A. LeMaistre. Putting the stamp of individual identity on a large institution demands focus, repetition, and credibility in leaders, and it requires time. When musical chairs becomes the administrative game, with leaders turning over every few years, institutions sputter in performance and stutter in message. Between 1941 and 1991, Ernst Bertner, Lee Clark, and Charles LeMaistre had shaped a culture and a sense of mission that had enthroned aggressive treatment, multidisciplinary medicine, and reverence for patients. Each man had been given more to consistency than to caprice. Each understood that a soft answer "turneth away wrath."

———

Of the three men, LeMaistre was by far the most committed to cancer prevention as a subdiscipline of oncology. When M. D. Anderson celebrated its golden anniversary, cancer prevention had been added to research, patient care, and education as core missions, and LeMaistre had discussed the issue frequently enough to begin to embed it in the institutional culture. To a 1988 meeting of the U.S. Public Health Service Professional Health Association, LeMaistre argued that "most cancers are avoidable and most cancers are preventable." Therapeutic advances, he argued, had been dramatic, "but curative medicine—despite its remarkable (legitimate) achievements . . . will not single-handedly lead us to significant control of cancer. . . . A cure rate of [a] perfect 100 percent . . . is an inspiring and worthy goal . . . but would not *eliminate* the disease—or the emotional, physical, and financial toll that it produces." By the late 1980s and early 1990s, cancer prevention had evolved beyond just the surveillance and early diagnosis phase to include genetic counseling, behavior modification, and chemoprevention.[21]

In cancer politics in the 1990s, few issues generated as much heat as chemical carcinogens. Environmentalists made the issue a cause célèbre. Breast cancer was especially volatile. As it turned out, however, the concern first raised by Rachel Carson in 1961 in *Silent Spring*—that the increases in cancer morbidity and cancer mortality could unequivocally be attributed to the saturation of the American environment with man-made chemicals—did not stand up to scientific scrutiny, despite the shrill rhetoric of advocacy groups. The organization 1 in 9, for example, had its origins in a dispute with NCI and CDC. In 1990, Senator Alphonse D'Amato of New York learned that Long Island breast cancer rates surpassed the national average by 15 percent. CDC officials informed that breast cancer was common on Long Island not because of environmental factors but because of the concentration there of Jewish women, who had higher rates of breast cancer than average. Outraged at what they considered to be smug complacency, several women formed the group "1 in 9," alluding to the well-publicized if somewhat misleading statistic that one in nine American women would eventually contract breast cancer.

As the political battle lines formed in the 1990s, many breast cancer advocates, like New Dealers of old, took aim at corporate America and its agents in the "cancer establishment." Devra Lee Davis, founding coordinator of the Breast Cancer Prevention Collaborative Research Group, insisted that "it makes sense to say the environment may be playing a role in human breast cancer. The weight of evidence, the hundreds of articles on animals, and the growing literature on humans all point in the same direction." According to Joe Thornton, a specialist on environmental carcinogens, "The worldwide increase in breast cancer rates has occurred during the same period in which the global environment has become contaminated with industrial synthetic chemicals."[22]

Gabriel N. Hortobagyi, chairman of medical breast oncology at M. D. Anderson, was not quite so sanguine. Those with careers in clinical cancer medicine learn to hedge their bets when dealing with a jigsaw puzzle of a disease with thousands of pieces. Cancer humbles those who fight it. "Most of us believe that every breast cancer is the result of some genetic injury," Hortobagyi continued, "[but] there will probably never be, at least in the foreseeable future, a telltale 'smoking gun.'" Researchers at the University of Utah in 1993 sequenced BRCA1, a mutated gene responsible for six thousand cases of breast cancer a year in the United States. One in every two hundred women carries the gene, and 80 to 90 percent of those women will get breast cancer. Thirty to 40 percent of women carrying the gene will contract ovarian cancer as well. Women of East European Jewish descent are far more likely than average to carry the gene, which helps explain the high incidence of breast cancer on Long Island, not greedy businessmen and their coconspirators in government. The subsequent discovery of the BRCA2 gene for breast cancer fixed another piece in the puzzle, but the effect of environmental pollutants remained elusive.[23]

Thoroughly immersed in the whirlpool of cancer politics, LeMaistre had fourteen years earlier pushed M. D. Anderson into cancer prevention, just when *Ms.* magazine attacked the "cancer" establishment—the coalition of NCI, American Cancer Society, major pharmaceutical companies, and comprehensive cancer centers, accusing this "mostly male establishment of ignoring prevention and focusing on 'cancer management' and search for a cure. What we have is a golden circle of power and money, where many of the key players are connected . . . [and] have much to gain or lose." An article in *Mother Jones* noted that General Electric and DuPont manufactured hundreds of millions of dollars of mammography equipment and film. Prevention programs did not get their due, some argued, because success would cut billions in revenues from major corporations. "What else could explain the fact that only 3 percent of the annual budget of the NCI was dedicated to the study of environmental carcinogens?" The critics had confused coincidence and conspiracy.[24]

Cancer politics had come full circle. In the 1950s and 1960s right-wing critics of American medicine, such as the John Birch Society, had accused physicians and hospitals of concealing potential cancer cures in order to jack up fees. A cure for cancer would put the cancer establishment out of business. Now, in the 1980s and 1990s, left-wing ideologues disparaged cancer practitioners for exactly the same reason, accusing them of ignoring cancer prevention in order to line their pockets. The critics overlooked the interplay between political reality and human nature. Those most concerned about the development of a cure are rarely those who have not suffered the ravages of cancer, whose concerns are at best an abstraction. Cancer survivors, those with the most to lose from the disease and the most to gain from a solution, fuel the engine of cure. Their feelings are visceral and primitive.

Controversies over hormone replacement therapy (HRT) stirred up shouting matches too. The link between diethylstilbestrol and vaginal cancer and breast cancer was unequivocal, but connections between birth control pills and breast cancer were sewn from thinner thread. Rose Kushner, a breast cancer survivor and fire-breathing advocate for women, crusaded for warning labels on birth control dispensers. Evidence continued to mount about a correlation between birth control pills and breast cancer. In 1986, Swedish and Danish epidemiologists reported an increase in breast cancer among women who took birth control pills for more than seven years. A study in 1988 revealed a 600 percent increase in breast cancer among women who had their first period before age thirteen, took the pill before their twentieth birthday, and remained on it for eight years. Many epidemiologists countered that the increase in the breast cancer incidence among users of the pill had less to do with the hormones than with the fact that women who postponed childbirth to a later date, pill or no pill, had increased risks. For Rose Kushner, however, denials were naive at best and criminal at worst. Between 1960 and 1970, more than 10 million American women had adopted the pill, and Kushner could not help but wonder.[25]

HRT tied another knot in the string of estrogen conundrums. A 1976 issue of the *New England Journal of Medicine* reported a 39 percent higher incidence of breast cancer among women receiving HRT, and CDC concluded that "approximately 4,708 new cases and 1,468 breast cancer deaths would occur each year because of estrogen use."[26]

Tobacco stirred little debate. Its use explained as many as 200,000 cancer deaths a year in the United States. LeMaistre had spent his professional life crusading against its use and could not have agreed more with King James I of England, who in 1604 described smoking as a "custom loathsome to the eye, hateful to the nose, harmful to the brain, dangerous to the lungs, and the black stinking fume thereof, nearest resembling the horrible Stygian smoke of the pit that is bottomless." If Americans never put in their mouths another cigarette, cigar, or pinch of tobacco, as many as 135,000 people annually would never develop cancer.[27]

Margaret Spitz, who chaired the Department of Epidemiology, tackled the question of why some heavy tobacco users are susceptible to tobacco carcinogenesis and why others seem immune. With a medical degree from the University of Witwatersrand in South Africa and later a master's degree from the University of Texas School of Public Health, Spitz had employed in vitro mutagen-induced chromosome sensitivity assays and determined that a sensitive phenotype constituted an independent risk factor for developing cancers

of the lung and upper aerodigestive tract, as well as for subsequently developing other primary tumors. Her research had important implications for screening and chemoprevention.[28]

Cancer prevention also called for reductions in exposure to ultraviolet light.
Were Americans to limit their exposure, many of the country's annual 7,400
deaths from skin cancer would disappear from epidemiological charts. Throughout M. D. Anderson clinics in the 1980s and 1990s, physicians and nurses harped
on the issue, especially with patients whose work or recreation took them out-
doors regularly.[29]

Sexual practices sometimes cause cancer. In 1985, cancers of the cervix, vagina,
vulva, and uterus killed more than 10,000 American women. Epidemiologists
viewed cervical cancer as a sexually transmitted disease, with a strong correlation to promiscuity and multiple sexual partners. Early in the 1980s molecular
biologists identified the human papilloma virus (HPV) as a cause of cervical
cancer, and clinicians implicated it in some squamous cell carcinomas of the
vulva, vagina, penis, and anus as well. Life in a monogamous relationship with
a partner free of HPV was an obvious way to avoid those diseases, and safe sexual practices could prevent AIDS-related cancers.

Guy Newell had included diet and nutrition on the agenda of cancer prevention. Nutrition, not chemical pollution, he believed, was the real environmental culprit. British epidemiologist Richard Doll estimated that 35 percent
of cancers could be attributed to poor diet and nutrition. Esophageal cancer
seemed connected to heavy alcohol consumption. Teasing epidemiological evidence about cancer incidence in certain ethnic groups and countries piqued
Newell's interest. Between 1950 and 1978, for example, the incidence of colorectal
cancer more than doubled in France, probably because of the increased consumption of pork, beef, and refined flour as well as reductions in bulk-supplying
vegetables and grains. In the United States, Latter-day Saints and Seventh-Day
Adventists displayed a lower incidence of colon cancer, perhaps because both
religions urged members to eat meat sparingly or not at all. The Japanese have
high rates of stomach cancer, probably due to a diet heavy in rice consumption, such salted foods as seaweed, shellfish, and soybean sauce, and the practice of burning fish when cooking. In the twentieth century, as the Japanese
adopted a fat-rich Western diet, the incidence of stomach cancer declined while
the rates of colorectal, breast, and prostate cancer soared.[30]

Newell had also considered obesity a carcinogen. Between 1959 and 1972, the

American Cancer Society determined that individuals more than 40 percent overweight were at increased risk of colon and rectal cancer in men and gallbladder, breast, ovarian, cervical, endometrial, and uterine cancer in women. Different rates of cancer incidence and cancer mortality between blacks, whites, and Hispanics in the United States also hinted at diet as a factor.[31]

For Newell, cancer prevention had to include nutrition. In the mid-1970s, he had planted nutrition in the infrastructure of NCI, and during the 1980s his efforts blossomed. Late in the decade NCI declared nutrition a key element in cancer prevention. Diets rich in fresh fruits, vegetables, and grains supplied vitamins A and C, iron, calcium, and bulk, all of which helped prevent some cancers. Consumers of such foods tended to eat less fat, reduce their consumption of salty foods, consume fewer calories, and avoid obesity. Were Americans to enrich their diets and cook food properly, significant reductions could be realized in the 60,000 annual deaths from colorectal cancer, the 38,400 deaths from breast cancer, the 25,500 deaths from prostate cancer, and the 14,300 deaths from stomach cancer.[32]

Fully exploiting early diagnosis technologies proffered fewer cancer deaths. Regular Pap smears promised to prevent most cases of cervical cancer. For men over the age of fifty, an annual digital examination and prostate specific antigen (PSA) test in some cases can identify prostate cancer before a patient becomes symptomatic. An annual occult blood test detecting blood in the stool can raise warning flags about the presence of colorectal cancer. Colonoscopies allow for surgical removal of precancerous polyps. And regular mammograms can reveal early-stage breast cancer. Newell and LeMaistre strongly advocated the screening of asymptomatic women with mammography for the early detection of breast cancer. Among women whose tumors were still in situ or minimal in size, ten-year survival rates exceeded 90 percent.[33]

Genetic counseling offered another strategy. Plans were well under way for the launch in 1996 of the breast and ovarian cancer genetic predisposition pilot program and genetic testing for susceptibility to breast and ovarian cancer. Women with breast and ovarian cancer in their family tree, for example, might consider genetic testing to see if they carry the BRCA1 and BRCA2 genes. Such knowledge encourages greater vigilance about breast lesions and the possibility of prophylactic mastectomy and hysterectomy. Genetic counseling might reveal the presence of Li-Fraumeni syndrome—mutations in the CHEK2 and p53 genes associated with a higher risk in children and young adults for osteosarcomas, soft tissue sarcomas, leukemia, brain tumors, and adrenocortical carcinoma. Li-Fraumeni families would be well served to implement all elements of cancer prevention. Multiple cases of colorectal cancer in a family might indicate Lynch syndrome, an inherited proclivity for colon cancer.[34]

Cancer prevention required lifestyle changes in patients and those at risk, which put a premium on behavior modification research. Kicking a tobacco addiction or consuming less alcohol is a complex psychosocial and physiological problem. LeMaistre had elevated cancer prevention from departmental to divisional status, and within the new Division of Cancer Prevention, he created the Department of Behavioral Science. In 1993, LeMaistre recruited from UCLA Ellen R. Gritz, an internationally known psychologist and specialist in smoking issues. Under her direction, the Department of Behavioral Science conducted research into lifestyle changes, especially smoking cessation and reductions in exposure to sunlight.[35]

M. D. Anderson assumed the national lead in chemoprevention—the use of drugs to prevent, delay, or reverse tumors. In 1983, NCI had initiated a research program for the study of natural and synthetic inhibitors of carcinogenesis. At M. D. Anderson, medical oncologist Waun Ki Hong, an emigrant from Korea, took advantage of the program. Using 13-cis-retinoic acid, a derivative of vitamin A, Hong had reversed leukoplakia, a precancerous condition primarily in the mouths of tobacco users by inhibiting the Cox-2 enzyme. In 1986, his landmark article appeared in the *New England Journal of Medicine*. Because the Cox-2 enzyme expresses itself in a variety of head and neck, lung, prostate, and gynecological tumors, Hong's paper inspired investigators around the world and put M. D. Anderson in the middle of the new discipline. In 1996, NCI awarded Hong a research grant of $6 million, one of many research grants and accolades he was to receive for his pioneering contributions to cancer chemoprevention.[36]

M. D. Anderson was also heavily involved in testing tamoxifen. A product of the British pharmaceutical industry, tamoxifen in the early 1990s had demonstrated a capacity for binding to estrogen receptors in breast cancer cells and for preventing the binding of estrogen to cells that crave the hormone. In a British study, 75,000 women taking a two-year regimen of tamoxifen enjoyed better ten-year survival results. Bernard Fisher at the University of Pittsburgh in 1992 initiated a prophylactic tamoxifen study, enrolling 16,000 high-risk women who had started menstruating before age thirteen, had close relatives with the disease, or who had never had children. M. D. Anderson medical oncologists participated in the study, which definitively determined that high-risk women taking tamoxifen prophylactically enjoyed significantly longer periods of disease-free survival.[37]

Cancer prevention in the 1980s and 1990s was good medicine and good politics. The media throbbed with criticism of the war on cancer, resurrecting the unfulfilled predictions of 1971. After the oncogenes responsible for "turning on" some cancer cells were discovered, Lewis Thomas, the head of Memorial Sloan-Kettering, promised the disappearance of cancer "before the century is over," a prediction akin to a Jeanne Dixon tidbit in the *National Inquirer*. Cancer prevention seemed less likely to attract criticism and more likely to promote control.[38]

In some ways, the cancer advocacy movement could be its own worst enemy. On the one hand, advocates painted a portrait of dramatic scientific progress to sustain optimism, while on the other hand, to prime the pump of government appropriations and private donors, they hyped the unrelenting march of the disease. In 1985, for example, the American Cancer Society warned that 462,000 Americans would die of the disease, one person every 68 seconds, and that nearly 1 million Americans that year would be diagnosed with cancer. Critics found ample data to skewer the war on cancer. Empty promises attracted new skepticism. Cancer had killed 48,000 Americans in 1900, 158,000 in 1940, 330,000 in 1970, and 472,000 in 1986. Without adjusting for age, cancer mortality rates had mounted from 64 per 100,000 people in 1900 to 120 in 1940, to 163 in 1970, and to 193 in 1985. In 1990, approximately 26 per 100,000 women in the United States died of breast cancer, a fraction unchanged since 1930. After sixty years of research and a $25 billion investment, long-term survival rates for the major killers—lung, breast, and colon cancer—had changed very little. In 1986, John Bailar, an epidemiologist and former editor of the *Journal of the National Cancer Institute*, argued that the war on cancer was being lost. "The ugly fact," he announced, "remains that overall cancer mortality is rising. . . . This cannot be explained away as a statistical artifact obscured by the clear evidence of progress here and there or submerged by rosy rhetoric about research results still in the pipeline." Many prominent scientists echoed Bailar's dismal assessment. President Donald Kennedy of Stanford University likened the war on cancer to a "medical Vietnam," and James Watson dismissed the so-called "war . . . as a bunch of shit." At the fall 1993 meetings of President Bill Clinton's Cancer Panel, after listening to a laundry list of rosy pronouncements, Bailar reacted again: "I . . . again conclude, as I did seven years ago, that our decades of war against cancer have been a qualified failure."[39]

Although criticizing cancer prevention seemed to some like assailing the unassailable, the discipline had its detractors, primarily concerning the economics of screening, the utility of certain diagnostic tests, and the value of chemoprevention.

R. Lee Clark had learned in the late 1940s that the finances of widespread screening often did not add up. Physicians could be tied up day after day doing exams on otherwise healthy people in hope of discovering cancer in its early stages, with relatively little to show for it. Three decades later, John Cairns, a physician and molecular biologist, performed a cost-benefit analysis on screening. He analyzed a New York City program that since 1969 had annually screened 31,000 women, ages forty to sixty-four, and compared them to a control group of 31,000 untested women. Nine years into the study, 91 of those screened had died of breast cancer, compared to 128 deaths in the control group. Cairns calculated that each life saved cost $50,000 (at $25 per test). He then suggested that if similar screening programs were implemented for ten other forms of the disease on all Americans over forty-five, the cost would exceed $15 billion a year, four times the total in 1977 to treat all forms of cancer in the United States. More than 350,000 Americans died of cancer that year.[40]

Mammograms and PSA tests had dark sides. Early in the 1990s, controversy bubbled about the use of mammograms by younger women without known risk factors. The American College of Surgeons called for an end to routine screening, as did the American College of Family Practice. The National Women's Health Network and the Center for Medical Consumers steered premenopausal women away from mammograms. NCI and ACS, on the other hand, held out for annual exams. M. D. Anderson endorsed the ACS guidelines that women perform monthly breast self-examination beginning at the age of twenty; that women between twenty and forty receive a physical breast examination every three years; and that women forty and older have a mammogram once a year.[41]

In 1993, NCI and ACS responded, urging women in their forties to get a mammogram "every one or two years." Late in 1993, NCI dropped even that guideline, advising women to discuss the issue with their physicians. Cindy Pearson of the National Women's Health Network was furious. "The NCI is the repository of public trust," she said. "As a consumer, I want the government to say what they think the state of the science is." Only the American Cancer Society continued to support routine screening for younger women.[42]

To some critics, enthusiastic urologists had badly overestimated the value of PSA tests, which could leave patients and physicians with a false sense of security because the test all too frequently failed to detect tumors. At the same time, it produced too many false positives. No evidence existed, others claimed, that it either reduced the morbidity of prostate cancer or lengthened survival time. It might also encourage some patients to seek aggressive treatment when wisdom dictated caution.[43]

Chemoprevention invited scrutiny. Waun Ki Hong employed 13-cis-retinoic acid inhibitor drugs to reverse the progression of leukoplakia, but when medical economists put the data into a calculator, questions of cost effectiveness surfaced. In only 2 percent of cases does leukoplakia progress to squamous cell carcinoma, and in only 1 percent of lichen planus. Might not the money be better aimed at reducing tobacco and alcohol use? Other scientists worried about long-term side effects. Might patients be putting themselves at risk for other diseases?[44]

The prophylactic use of tamoxifen drew fire. Although few questioned its value in treating tumors with estrogen receptors, employing it on asymptomatic women seemed reckless to some. Tamoxifen's daunting side effects—hot flashes, vaginal discharge, irregular menses, skin rashes, phlebitis, thrombocytopenia, blood clots, and an increased risk of liver, uterine, and endometrial cancer—justified caution. Women had to perform their own risk-benefit analysis to determine whether a reduced risk of developing breast cancer was worth enhancing the odds for liver and endometrial cancer. Although research indicated a reduced incidence of cancer in the other breast of patients taking tamoxifen, the National Breast Cancer Coalition throughout the 1990s preached, "We do not know whether the benefits outweigh the risks in women who take these drugs to reduce their chances of developing breast cancer. Scientists must design clinical trials that will provide this information."[45]

Critics might throw darts at small targets within the larger field of cancer prevention, but the discipline was here to stay. Perhaps the impact of tamoxifen said it best. In 1991, mortality for breast cancer peaked and then began a slow, steady decline of 2 percent annually until 1995, when the death rate fell even faster. Richard Peto, an epidemiologist at Oxford University, remarked, "This is the first time that improvements in the treatment of any type of cancer have ever produced such a rapid fall in national death rates. They really are remarkable trends." Between 1989 and 1996, the breast cancer death rate among women in the United States between the ages of twenty-nine and sixty decreased from approximately 40 in every 100,000 women to less than 34, a decline of historic proportions. Broader statistics were equally encouraging. Overall age-adjusted cancer mortality had been increasing in the United States for as long as the statistics had been collected, but in 1991 the tide of history reversed; mortality declined at a rate of 4.2 deaths per 100,000 people.[46]

For LeMaistre, the growing emphasis on cancer prevention explained some of the decline, and when the time came to replace Guy Newell, a search committee identified Bernard Levin. A graduate of the University of Witwatersrand Medical School in South Africa, Levin had arrived at the University of Chicago for a residency in internal medicine; he remained there to complete fellowships in gastroenterology and biochemistry/pathology. Gifted and eclectic in his medical expertise, Levin had in 1971 joined the University of Chicago, published widely in the area of gastroenterology, and in 1984 arrived at M. D. Anderson to develop a multidisciplinary program in gastrointestinal cancer, a field where progress against tumors of the esophagus, stomach, pancreas, small intestine, and colon had been agonizingly slow. Levin soon distinguished himself. At the time, with twelve faculty members working on twenty-three projects, M. D. Anderson had one of the most active programs in the nation, a statistic revealing the miserable state of the discipline. Levin remembered those modest beginnings when "many of us working in cancer prevention have begun our primary papers with something resembling . . . 'It is estimated that x cancer will account for y deaths . . . this year. These grim statistics persist despite advances in cancer therapy, arguing forcibly for a new approach, such as cancer prevention, to help in controlling this devastating disease.'" Levin would become a leading figure in the evolution of cancer prevention into a recognized subdiscipline of oncology.[47]

In 1995, Charles A. LeMaistre anticipated another stage in his life. He had celebrated his seventieth birthday in 1994 and took satisfaction in having brought cancer prevention to the forefront of people's awareness. The financial future seemed under control. And his wife, Joyce, to alleviate respiratory difficulties, needed to leave Houston. They picked San Antonio because of its climate and because it was home to several children and all six grandchildren. LeMaistre announced his decision to step down on August 31, 1996.

As LeMaistre approached the end of his career, the war on cancer scored a long-awaited breakthrough, although ironically, cancer survivors were not the beneficiaries. What antibiotics did to infectious diseases in the 1940s and the Salk and Sabin vaccines to polio in the 1950s, protease inhibitors did to AIDS in the 1990s. The formerly incurable and inexorable disease suddenly, and suddenly is the only word to describe it, met its match.

In 1983, Robert Gallo of NCI and Luc Montagnier of the Pasteur Institute identified HIV as a retrovirus. Molecular biologists knew that in retroviruses, the enzyme reverse transcriptase polymerizes RNA into DNA. Because of NCI-financed retroviral research, molecular biologists learned that the ten different proteins making up reverse transcriptase are not individual units but polyproteins, and that before individual proteins become fully functional within a cell, they must first be sliced out of the polyproteins, a process conducted by protease, one of the proteins in the polyprotein chain. Molecular biologists soon deciphered the genetic structure of the retrovirus, identified its proteins, and learned that inhibiting protease interfered with the virus's replication process. By 1996, they had designed a pharmacological cocktail composed of the protease inhibitors and the drug AZT, a nucleoside analogue. Death rates from AIDS plummeted.[48]

Cancer, however, was not about to go the way of infections, polio, and AIDS. The war on cancer had discovered no magic bullet, and few could see any breakthrough on the medical horizon. No disease rivaled cancer in complexity, its mysteries coiled tightly into a bewildering knot of proteins, enzymes, genes, mutations, DNA, RNA, viruses, and tissues—the stuff of life itself—and "life," as the chaos theoretician Ian Malcolm says in *Jurassic Park*, "always finds a way." In 1996, the year Charles A. LeMaistre emptied his office and headed for San Antonio, 554,740 Americans died of cancer. The disease remained the bogeyman of modern medicine, and millions of cancer survivors waited for help. Joan Coffey was one of them.

12

John Mendelsohn and the New Frontiers in Oncology, 1996–2000

I . . . learned what research is all about—really hard work, long hours, many hypotheses and many experiments, many times over, without the outcomes you expect. And complete intellectual honesty trying to figure out whether an experiment was truly negative or perhaps failed because you didn't do it correctly.

—JOHN MENDELSOHN, 2004

Charles and Joyce LeMaistre returned to M. D. Anderson much sooner than anticipated and not to visit former colleagues or to groom a potential donor. They came for treatment. Retirement in San Antonio had suited them. Although a gracious hostess with a southern touch, Joyce LeMaistre had always preferred the background to the limelight, basking in the presence of her four children and six grandchildren or the ambience of the backyard greenhouse, where she raised hybrid orchids with the green thumb of a master gardener.

Like her husband, she took cancer prevention to heart. Joyce regularly had Pap smears and mammograms. In 1998, during a routine Pap smear at M. D. Anderson, pathologists noticed highly atypical cells. J. Taylor Wharton, the gynecologist who had replaced Felix Rutledge as chair of gynecology, ordered a second Pap smear, with the results just as inconclusive. Ruth Katz, the institution's leading gynecological cytologist, reported that the cells were more than atypical but not enough to call them premalignant. She expressed her concern. Repeated Pap smears continued to show atypicality but were not deemed premalignant. Two more pathologists took another look, again without consensus. Years at M. D. Anderson had left the LeMaistres acutely cognizant of cancer's dangers, so they assumed the worst and acted aggressively. In discussing the matter, Joyce said, "We must settle this without questions and therefore I want Dr. Wharton to begin with the cervix and then take out each of the female organs in succession until he finds the cause. I want him to remove each of the organs, even if the frozen sections are negative so the pathologists can examine them in detail." Wharton did so, and thirty-five sections were negative. He then proceeded to an abdominal wash, flushing the general region with saline and then siphoning it out for the pathologists. Again nothing. Four days after Joyce's surgery, Elvio Silva, a gynecological pathologist, asked LeMaistre to view the path slides with him. He found two areas of cancer cells, one 1 mm and the other 1 mm by 7 mm. The cells were unquestionably malignant and believed to be all that were present; therefore, no treatment was recommended.

Oncology runs in the LeMaistre family, however, and doing nothing was out of the question. Two of the LeMaistre children—Fred LeMaistre and Anne LeMaistre—were physicians, both trained in oncology, and they too recommended systemic treatment. Medical oncologists mixed the chemotherapy, and the LeMaistres stood watch to see if modern oncology had a gift for them.

Although their paths never crossed, Joan Coffey, the French historian with Waldenström's, returned to M. D. Anderson at the same time the LeMaistres were there. Wait-and-see was over for Joan. In 1997, tumor markers indicated that the cancer was rebounding, and her blood counts soon went haywire. She was experiencing fatigue, enlarged lymph nodes, a swollen spleen, and a tendency to bruise. The time was at hand, according to hematologist Raymond Alexanian, to attack the lymphoma. In April, May, and June, she endured a chemotherapy protocol of 2cda/Cytoxan that sent the Waldenström's into remission, her hair into the toilet, and her immune system into near collapse. One month after the last dose, Alexanian treated her for idiopathic thrombocytopenic purpura, a disorder in which the number of blood platelets plummets. One crisis averted, Joan proceeded in September 1998 to a bout with bacterial meningitis. Antibiotics cleared it up. With her blood counts at least stable, Alexanian resumed the chemotherapy, this time a concoction of 2cda, Cytoxan, and Rituxan.

In January 1999, to reverse the loss of blood cells, surgeon Paul Mansfield removed Joan's spleen. Two months later, Alexanian began to plan for the last stand. Anticipating that Joan would eventually need a stem cell transplant, he recommended a new round of Cytoxan, vincristine, Adriamycin, and dexamethasone to suppress the disease before harvesting stem cells. Unless he could push Joan back into remission, a transplant was out of the question. In April and June, physicians executed several bone marrow aspirations without success. One month later, Joan struggled with pneumonia and then returned to work, learning that the University of Notre Dame Press had agreed to publish her biography of French industrialist Léon Harmel. Shingles cut short the celebration. Joan's blood counts then took a wrong turn and the Waldenström's rebounded. More 2cda/Cytoxan/Rituxan. In December 1999, while the Coffeys were on a long-anticipated trip to Colorado, the pneumonia returned, sending her to the emergency room of Rocky Mountain Cancer Center. Through it all she continued to teach and write with the stiff resolve of a medieval saint. Joan Coffey knew about the natural course of Waldenström's and its inevitable, slippery slope. The slide was under way.[1]

⟨⟨⟨ ⟩⟩⟩

The tall, slender young man standing outside the offices of the biology department at Harvard in 1953 radiated a confidence sprouting from the security of a happy childhood in Cincinnati, where he had been born on August 31, 1936. John Mendelsohn hearkened back to German- and French-speaking pre–Civil War immigrants, his father's people hailing from Alsace-Lorraine and his mother's from central and southern Germany. In the tradition of many Jewish immigrants, Mendelsohn's father, Joe Mendelsohn, built a prosperous business wholesaling men's accessories—belts, suspenders, wallets, and garters—to department stores, dry goods stores, and haberdasheries. He enjoyed a solid reputation among the employees. Mendelsohn's mother, the former Sarah Feibel, worked as a homemaker and a volunteer with the PTA, the Girl Scouts, and the temple sisterhood.

John Mendelsohn's uncle, the Rabbi Victor E. Reichert, presided over a Reform Jewish synagogue and wielded great influence. Known for erudite sermons leavened by faith, logic, and vivid anecdotes, he preached lessons that the young Mendelsohn absorbed—"live life so that you can look forward with the greatest possible hope and look back with the fewest possible regrets" and "do not unto others as you would not have them do unto you." Equally adept at science, mathematics, and the humanities, Mendelsohn graduated from Walnut

Hills High School, a selective college preparatory public school in Cincinnati. Still a teenager, his hazel eyes divulged a winsome personality. An affable smile usually graced his face, as if he were always greeting old friends.

As a freshman at Harvard, Mendelsohn contemplated majoring in physics or chemistry, but during his sophomore year, he settled on biochemistry. Eager to secure some research experience, he stepped into the biology office and inquired about a job. The receptionist nodded affirmatively. "We have a new professor here," she said. "He's very young, but he's supposed to be very good. His name is James Watson." Mendelsohn then ambled into history. Three years earlier, with Francis Crick at Cambridge University, Watson had illuminated the configuration of deoxyribonucleic acid (DNA), the fundamental molecule of all living things. Genes rest on the double strands of DNA, and during cell division, the double helix unzips into two complementary genetic components and gives life to a new cell. It was the stuff of Nobel Prizes, which Watson and Crick would receive in 1964.[2]

Watson offered students a soaring intellect graced by an unselfish heart; he refused to include his own name as first author on papers issuing from his own laboratory. Mendelsohn labored there in the presence of brilliant graduate students and scientists, including Alfred Tissieres, a leading expert on ribosomes, the particles in a cell that synthesize protein, and biochemist John Edsall, whose lab owned the university's only analytical centrifuge. Mendelsohn investigated the production and assembly of ribosomes in *E. coli,* a ubiquitous bacterium inhabiting the human colon. Edsall gave Mendelsohn a free hand with the centrifuge, and with it Mendelsohn helped decipher the information sequence chain from DNA to RNA to protein. He established that when bacteria in a resting phase move on to a growth phase, they produce abundant ribosomes, and when starved in the medium, the cells will enter a stationary growth phase, metabolize the ribosomes, and release the products into the medium. The research produced his first publication, a paper in *Biochimica et Biophysica Acta.*

In Watson's laboratory, Mendelsohn "learned what research is all about— really hard work, long hours, many hypotheses and many experiments, many times over, often without the outcomes you expect. And complete intellectual honesty, for example, trying to figure out whether an experiment was truly negative or perhaps failed because you didn't do it correctly. . . . In research, investigators often secure answers to questions they do not ask, and the purpose of research is to learn, not to justify hypotheses." DNA washed away the solid walls dividing biology, biochemistry, genetics, and medicine. The birth of molecular biology was at hand and complicated Mendelsohn's career plans—physician or biologist, medicine or science? At the end of Mendelsohn's junior year, the clairvoyant Watson helped sort it out. "He outlined what he thought would happen," Mendelsohn recalled, "in the field of molecular biology and genetics in the next decade—accurately, as it turns out. He urged me to get my Ph.D. because the field was going to blossom." Medicine, however, had a hold on Mendelsohn. "But after much thought, I told him that I really wanted to be a physician. . . . I liked people and the thought of applying science to the problem of human disease was very appealing."

Mendelsohn choose the career path of the physician-scientist, the doctor with a laboratory attached to the clinic, the bench chained to the bedside. After a stimulating year as a Fulbright Scholar studying molecular biology at the University of Glasgow in Scotland, Mendelsohn attended Harvard Medical School, graduating in 1963. The next seven years took him to Boston for a residency in internal medicine at Brigham and Women's Hospital, to Bethesda for an NIH research fellowship, and to the School of Medicine at Washington University in St. Louis for a fellowship in hematology and oncology.

The arc of his professional career streaked skyward in 1970 when Mendelsohn accepted an assistant professorship at the new University of California, San Diego (UCSD), School of Medicine. In the laboratory there, he mimicked the model of the research scientist he had learned in Watson's lab, at Glasgow, at the NIH, and at Harvard Medical School. With mentoring from outstanding physician scientists at the new medical school, Mendelsohn ran a well-organized lab fueled by a thoughtful research strategy. He racked up paper after paper in major journals and in only seven years rose from assistant to full professor. And his research transcended the didactic labors of many scientists whose hypotheses often lean more to the next grant cycle or to the dictates of a journal than to the logic of the evidence. His initial research focused on how external molecules stimulate the proliferation of lymphocytes, and he turned later to the growth factors that activate proliferation of epithelial and mesenchymal cells. His research discoveries generated a steady stream of grants, publications, and awards.[3]

When Mendelsohn went to work in 1956, biologists thought that DNA possessed the appropriate chemical configuration to convey the genetic code, but they did not yet know the code or how it could be transmitted into protein. But in 1970, as Mendelsohn settled into his lab at UCSD, messenger RNA was well described, and the virologist Howard Temin reported his discovery of reverse transcriptase. Classical virology came down faster than a circus tent after the last elephant has lumbered out, and molecular biologists began to fill in the blanks about DNA, RNA, cell division, and apoptosis, or cell death.

After several years at UCSD, Mendelsohn forged a fruitful collaboration with the cell biologist Gordon H. Sato. By 1981, evidence had accumulated that prompting cell growth and division depends on the role of growth factors in activating receptors on cell surfaces. The growth factors could be produced in an autocrine fashion, which the cell itself secretes. Activation of a tyrosine kinase in the intracellular portion of the receptor molecule generates a biochemical signal. When growth factors bind to their specific receptors, the tyrosine kinase becomes active. Mendelsohn and Sato hypothesized that interrupting the tyrosine kinase-mediated signal might inhibit cell proliferation and, if targeted at tumor cells, constitute a new weapon in the arsenal of cancer therapies. Because receptors for epidermal growth factor (EGF) and transforming growth factor alpha (TGFα) and EGF itself express themselves in a variety of human cancers, the two scientists decided to test the hypothesis by attempting to inhibit EGF receptors. They produced several monoclonal antibodies, including one labeled 225, that bind to EGF receptors and then block the binding of EGF or TGFα, preventing cell proliferation in cultures. Their initial findings appeared in a 1983 issue of *Proceedings of the National Academy of Sciences USA* and added a new dimension to modern oncology.[4] Publications on EGF receptor function and its mechanism of inhibition by antibody 225 continued to flow from Mendelsohn's laboratory for fifteen years.

The 1983 paper stoked the curiosity of cell biologists throughout the world. One year later, Mendelsohn and his colleagues reported that the monoclonal antibodies could indeed prevent activation of growth-signaling pathways mediated by the EGF receptor's tyrosine kinase activity. Later in 1984, they demonstrated that anti-EGF receptor monoclonal antibodies had the capacity to inhibit the growth of human cancer cells grafted on athymic, or "nude" mice. Athymic mice are especially useful in cancer research because they have been engineered not to reject the transplanted cells. These observations had enormous implications for oncology.[5]

Residents and fellows at UCSD found in him a mentor who paid attention and understood their circumstances, another lesson he had carried away from Watson's lab and his teachers at Harvard Medical School. "I also learned from Dr. Watson," Mendelsohn remembered, "[to understand] the different needs of a young investigator-in-training trying to become a fine laboratory researcher." From the well of Alfred Tissieres's research experience, Mendelsohn drew lessons on patience, critical thinking, and consistent "use [of] the scientific method, which anyone who has ever run a lab has to learn." At UCSD, he earned a respect bordering on devotion from students. Mendelsohn could push without being pushy, criticize without being condescending, and forgive without being forgetful.

On top of aptitudes for research and teaching, he displayed the gifts of a budding administrator, delegating authority, setting goals and monitoring progress, and communicating effectively with superiors and subordinates. In 1976, UCSD wanted to create a cancer program to garner a fraction of the funds flowing to NCI-recognized cancer centers, and the medical school asked Mendelsohn to chair the committee seeking formal recognition. Mendelsohn kept his laboratory operating in high gear, and in 1978 UCSD received acknowledgment as an official NCI Cancer Center. Mendelsohn became its first director; during the next nine years, he constructed it from the ground up, using as a template approaches similar to Lee Clark's multidisciplinary model. His experiences at UCSD benefited him greatly when he later moved to M. D. Anderson. The fruits of his own research had also inspired other scientists who embarked on a search for new ways to inhibit growth factor receptors and tyrosine kinases.[6]

Employing Mendelsohn's logic, other investigators turned their attention to HER2, a growth factor receptor closely related to the receptor for EGF. The earlier development of the drug tamoxifen had demonstrated the importance of receptors, located in this case inside of cancer cells and normal cells. Molecules of the drug tamoxifen fasten to estrogen receptors and prevent the binding of estrogen, which in turn deprives the malignant cells of an important growth factor. In 1985, scientists at the University of Pennsylvania and MIT demonstrated how a monoclonal antibody against HER2 inhibited the proliferation of murine breast cancer cells expressing the receptor, not only in culture but also in athymic mice. Soon the biotechnology company Genentech produced an anti-HER2 monoclonal antibody that they called Herceptin and introduced it into clinical trials. A few years later, a number of pharmaceutical firms designed low molecular weight, soluble inhibitors of the EGF receptor tyrosine kinase that acted intracellularly by blocking the binding site for ATP (adenosine triphosphate, a high energy phosphate molecule supplying energy for cellular function). Late in the 1980s, the biotechnology company Hybritech took the anti-EGF monoclonal antibody 225 into clinical trials.

By then, Mendelsohn chaired the Department of Medicine at Memorial

Sloan-Kettering Cancer Center, where he uncovered the mechanism by which the blocking of EGF receptors inhibits cell proliferation. The complete progression of cells through the cycles of DNA synthesis and cell division depends on the sequential activation of a series of cyclin-dependent kinases. The blockade of EGF receptors by monoclonal antibody 225 induces the inhibitor p27Kip1, which deactivates the kinases and keeps the cell from synthesizing new DNA. A similar mechanism was later found to explain the antiproliferation activity of Herceptin and the low molecular weight, soluble inhibitors of EGF receptors.

After other investigators in the laboratory of Michael Sela observed significant responses in a xenograft animal model treated with a combination of another anti-EGF receptor monoclonal antibody and the chemotherapy drug cisplatin, Mendelsohn tested 225 with the chemotherapy drugs doxorubicin, cisplatin, and paclitaxel. In nude mice with human tumor xenografts, maximum doses of chemotherapy alone could not produce complete regression, nor could the administration of 225 alone. The combined use of 225 with a chemotherapy drug or with radiotherapy, on the other hand, demonstrated a powerful synergistic effect in eradicating tumors. In 1995, Mendelsohn joined with clinical investigators in his department and Genentech in a phase 2 trial, which provided the first clear proof-of-principle for the efficacy of antireceptor antibodies. Of the women with advanced breast cancer, 10 percent experienced objective clinical responses when treated with Herceptin.[7]

The University of Texas System Board of Regents appointed a search committee to identify LeMaistre's successor. Charles Mullins, a cardiologist and executive vice chancellor for health affairs in the University of Texas System, chaired the search committee. The board of regents asked Mullins to recommend four finalists. Mendelsohn's was a dark-horse candidacy; his interest came in response to a letter from the search committee soliciting an application.

Four finalists surfaced: Edward M. Copeland, Charles Balch, Andrew von Eschenbach, and John Mendelsohn. Three were surgeons and one a medical oncologist; all were highly accomplished, but Mendelsohn was a true translational scientist.

Edward M. Copeland of the University of Florida was a strong candidate. A nephew of Murray M. Copeland, he had come to M. D. Anderson as a surgical fellow in 1963 after completing a residency at the University of Pennsylvania. During the next nineteen years, as a surgical resident and then surgical faculty member at M. D. Anderson, Copeland earned a national reputation for his work in total parenteral nutrition, a surgical technique for providing all of the nutritional needs of advanced cancer patients through intravenous feeding. "Malnourished patients," Copeland explained, "could not have received and/or survived the indicated antineoplastic treatment." In 1982, Copeland had left M. D. Anderson to lead the surgeons at the University of Florida, but he left behind strong contacts in Houston.[8]

Charles Balch brought to the candidacy a pioneering role in the development of surgical oncology, a personal reputation above reproach, and an insider's knowledge of the institution. After coming to M. D. Anderson in 1985, Balch had served as head of the division of surgery, executive vice president for health affairs, vice president of the hospital and clinics, and chair of the Department of Surgical Oncology, where he had undertaken the task of completing the transition from general surgery to surgical oncology, a challenge not unlike that of changing the starting lineup and playbook of an NBA team, with its highly talented, multimillionaire players sporting egos as big as their basketball shoes. In butting heads with several surgeons, Balch had bruised some egos. He had successfully recruited new players, such as Raphael Pollock, who would eventually succeed him as head of the division, but Balch's heavy hand, in the mind of some, had pushed too hard.[9]

Andrew von Eschenbach enjoyed some of Balch's strengths with none of the handicaps. A devout Roman Catholic raised in South Philadelphia, von Eschenbach is "a gentle man and a gentleman," in the words of James Bowen, the former vice president for academic affairs at M. D. Anderson. "He cares about people, always manages to find time for them, and carries his religion wherever he goes, not as a self-righteous evangelist but as a man of faith." Von Eschenbach also had lines on his résumé that, in the growing community of cancer survivors, bestowed special credibility. He was a two-time cancer survivor—melanoma and the rarely fatal basal cell carcinoma, and prostate cancer was in his future. Nobody speaks to cancer patients more eloquently than cancer survivors.[10]

In the end, the search committee looked to the future of oncology, not to the past. Surgeons, radiation oncologists, and medical oncologists had dominated the history of oncology, but to some the future rested in the hands and the laboratories of those designing targeted therapy, not just the older biological response modifiers like interferon or nonspecific vaccines like BCG, but new pharmacological and biological products engineered to be molecules specific to tumor cells or targeting the immunological status of the host.

As one of the founders of targeted therapy, Mendelsohn enjoyed special assets. At the campus visit and interview, where the faculty and the regents eyeball the candidate and the spouse, the Mendelsohns took M. D. Anderson by storm. Anne Mendelsohn was an accomplished woman in her own right. The two had met in 1961 while attending the engagement party of another couple in Harvard Yard and were married one year later. She had majored in chemistry and mathematics at Mt. Holyoke and worked as a chemist for Polaroid, where in a lab next to that of Edwin Land, she helped to develop the emulsions that coated the instant photographs. During the campus visit, she mingled comfortably at the lectures, receptions, and cocktail parties. She added an intangible dimension to Mendelsohn's candidacy, something that no résumé could reveal. They also learned that the Mendelsohns had teamed together to raise funds for the new UCSD cancer center. In academic medicine, where philanthropic donations, political savvy, and style can spell the difference between mediocrity and greatness, the regents saw in her a real asset.

Like his wife, Mendelsohn left few doubts. The basic and the clinical scientists fell in line because of his laboratory accomplishments and clinical experience. Mendelsohn owned a CV thick with papers in the premier journals of molecular biology and oncology, and his design of an effective monoclonal antibody added a fourth leg to the treatment tripod of surgery, radiotherapy, and chemotherapy. In the new age of targeted therapies, John Mendelsohn had been present at the creation. The clinicians, for their part, recognized that successful tenure at UCSD and at Memorial Sloan-Kettering qualified him for the calling. "He was," recalled James Bowen, "the billboard for translational research, for transferring discoveries in the laboratory to clinical treatments for cancer patients." His candidacy took flight, and when the board of regents announced his selection, the naysayers fell mute.[11]

Not securing the top post hardly interrupted the careers of Balch or von Eschenbach. Balch left M. D. Anderson to become CEO of the City of Hope in Los Angeles, and in 2000 he was named executive vice president of the American Society of Clinical Oncology, with a clinical appointment at Johns Hopkins. In 2000, von Eschenbach became president-elect of the American Cancer Society, an office he would relinquish when President George W. Bush named him director of NCI. In 2007, von Eschenbach would assume the leadership of a faltering FDA, then mired in a new round of political advocacy and disputes about its purpose.[12]

Between 1995 and 2000, fifty-one of Mendelsohn's articles found their way into major scientific journals. Other investigators derived from 225 the murine: human chimeric monoclonal antibody and labeled it C225. In 1993, the biotechnology firm ImClone Systems licensed C225 and three years later began clinical trials. Experiments from Mendelsohn's and other laboratories, and early clinical trials, revealed that C225, when combined with radiotherapy or chemotherapy against a variety of cancers, was safe as well as efficacious in some patients. Mendelsohn's research with Herceptin on animal models also had helped to generate the data and the momentum for clinical trials. In the laboratory, the team enhanced the effectiveness of Herceptin on xenografts of human breast cancer tumors in nude mice by adding doxorubicin or paclitaxel to the concoction, which led to the registration clinical trial with these agents. Mendelsohn and others subsequently revealed several mechanisms critical to the antitumor activity of monoclonal antibody 225.[13]

In 1996, as LeMaistre handed M. D. Anderson over to John Mendelsohn, modern oncology had just put to rest two longstanding debates about breast cancer—whether the radical mastectomy could be abandoned in favor of more conservative surgery and radiotherapy and whether the disease was systemic, requiring the systemic treatment of chemotherapy.

In April 1995, the *New England Journal of Medicine* proclaimed that an intellectual revolution had forever altered breast cancer treatment. "It has been almost exactly 100 years," the editor noted, "since William Stewart Halsted published his seminal report on the use of radical mastectomy. . . . [His] ideas and his operation dominated our thinking about breast cancer until approximately twenty-five years ago, when a major paradigmatic shift began." In 1979, NCI undertook a randomized, long-term study comparing lumpectomy plus axillary dissection and radiotherapy with modified radical mastectomies in 247 women. After ten years, more than 77 percent of the women receiving lumpectomy plus radiation were still alive, compared to 75 percent of the mastectomy group. More aggressive forms of local control could not improve survival rates because breast cancer was not a local disease. The paradigm shift was complete. Halsted moved from the surgical texts into the history books.[14]

Emil Freireich had for years insisted that solid tumors of all sorts were systemic from the outset and that only systemic treatments could fully address them. Preliminary studies in the 1970s and 1980s, including those completed at M. D. Anderson by such medical oncologists as George Blumenschein, had clearly demonstrated that various chemotherapy regimens could postpone recurrences of breast cancer for many women, but a few critics pointed out that disease-free survival and long-term survival were not the same. In the 1990s, however, convincing proof appeared that the disease was systemic at its beginning. The practice of confining chemotherapy to women with advanced, metastatic disease gave way to treating women with early stage breast cancer as well. After one hundred years of surgery and radiotherapy, the quest for cure was back in the apothecary.[15]

The same edition of the *New England Journal of Medicine* reported the results of Gianni Bonadonna's extended clinical trial at the National Tumor Institute in Italy. The Italian experiment divided 386 women into two groups. All of the women had positive lymph nodes and had undergone radical mastectomies, but half also received follow-up treatments of combined Cytoxan, methotrexate, and fluorouracil. Twenty years later, 34 percent of the chemotherapy patients were still alive, compared with only 25 percent of the others. Breast cancer was a systemic disease, and chemotherapy worked.[16]

After a century of increasingly aggressive surgical protocols, the scientific winds had changed direction. Damage control and quality of life, not just long-term cures, now dominated the thinking of surgeons and radiotherapists. Instead of removing and radiating more tissue in order to prolong life, surgeons and radiotherapists collaborated to remove less—to do less damage—while maintaining or improving existing long-term survival rates.

—⟨⟩—

When Mendelsohn took over, the immediate financial crisis had eased, but the institution was still reacting to the dire predictions from consultants and the staff cuts that had taken place. However, he had come to M. D. Anderson to build it into the world's best cancer center. The new vision statement he authored read, "We shall be the premier cancer center in the world, based on the excellence of our people, our research-driven patient care and our science." Mendelsohn spent his first six months creating an agenda for growth and planning to "raise the bar" in each of these areas. A new theme was introduced and trademarked: Making Cancer History. To accomplish this agenda he needed an aggressive business plan. Knowing that his experience and expertise were primarily in research and patient care, Mendelsohn set out, with wise advice from M. D. Anderson supporters on the board of visitors, to recruit a business executive with broad expertise in healthcare delivery and in negotiating deals.

After a national search, he replaced David J. Bachrach as executive vice president with Leon J. Leach, a former financial leader with Cornerstone Physicians Corporation and Prudential Insurance Company.[17]

Leach brought to the job an entrepreneurial spirit, openness, and business savvy previously unknown at M. D. Anderson. James D. Cox, head of the Division of Radiation Oncology, remembers that caution had long characterized the M. D. Anderson business plan. Financial data about internal operations had always been tightly controlled, but under Leach and Mendelsohn, information was more regularly and more widely disseminated, and division heads and department chairs were more readily able to compare their bottom lines and productivity with those of other entities. Leach seemed more dynamic in his approach, more willing to take risks, and Mendelsohn was highly supportive of the new financial culture.[18]

Mendelsohn had a huge business to manage, and he took to heart the ideas of Michael E. Porter, a professor of business at Harvard, whose book *Competitive Advantage* soon occupied a permanent spot on Mendelsohn's nightstand. "He taught me that you can be Kmart or Saks Fifth Avenue, but you can't be both," Mendelsohn recalled. "I decided we wanted to compete on quality." He spent his first three months meeting with each department and he asked one question: "What can I do to make it easier for you to do your work and accomplish your mission?" Over and over he received the message, paraphrasing Winston Churchill, "Give us the tools and we will finish the job." He was convinced that M. D. Anderson stood poised to grow in size and accomplishment, and he knew that it would take advantage of the recently legislated ability of patients to self-refer to the institution. Mendelsohn scrapped the advice of the consultants to downsize, and he decided to push forward. Instead of paring the 1997–98 budget by $70 million, which administrative leaders now suggested, Mendelsohn added $20 million, strategically hiring new physicians, nurses, and support personnel. M. D. Anderson then launched a marketing campaign. Although its charges for an overnight stay moderately exceeded those of some other hospitals in Houston, the daily census, which had fallen from around 420 a day in 1992 to 290 in 1996, surged, hitting 360 a day in 2000. One Care, a South Texas HMO, in 1998 had placed its 166,000 members under M. D. Anderson care and soon experienced a 15 percent drop in its cancer costs "because," in the words of the *Wall Street Journal*, "the hospital's high quality care has resulted in fewer complications that cost money to treat later."[19]

Between 1996 and 2000, Mendelsohn saw the completion of several major construction projects, including the new clinical services building named after Charles A. LeMaistre; the Margaret and Ben Love Clinic; the long-awaited Albert B. and Margaret M. Alkek Hospital; the Robert D. Moreton Imaging Center; the freestanding M. D. Anderson satellite Radiation Treatment Center in Bellaire; and the Faculty Center, a multistory office building located directly across the street on Holcombe and connected to the main campus by an elevated, air-conditioned walkway. On May 11, 2000, Mendelsohn secured from the board of regents authorization to construct a new ambulatory clinic building with 600,000 square feet and a price tag of $299 million, and later in the year they gave the go-ahead for a new South Campus Research Building. On November 21, 1997, the new Charles A. LeMaistre Clinic was dedicated, and M. D. Anderson maintained its position in the forefront of ambulatory care.[20]

Mendelsohn also began to put his own stamp on M. D. Anderson. On his rounds meeting with departments after joining M. D. Anderson, he had learned that there were silos that discouraged open communication between administrative units. An early goal was to eliminate barriers and enhance collaboration. For example, there were two financial officers, one for the hospital and another for the institution as a whole. The recruitment of Leach, who quickly consolidated those areas, solved that problem. Faculty in basic research reported to one vice president and faculty in clinical research to another. Mendelsohn's goals for M. D. Anderson included encouraging increased collaborations between laboratory scientists and clinical investigators. So he created a new position, a chief academic officer, with responsibility for all faculty, as well as their research and educational programs. Responsibility for all patient care activities was placed in the hands of a chief operating officer, whose title eventually became physician in chief. The new executive leadership team consisted of Mendelsohn and three new leaders overseeing business, academic, and patient care activities who became executive vice presidents.

For a vice president for academic affairs, he turned to immunologist Margaret Kripke. Some women had gained prominence at M. D. Anderson, such as radiation oncologists Lillian Fuller and Eleanor Montague; geneticist Louise Strong, the first woman faculty member to hold an endowed chair; and surgeon Eva Singletary. But no woman had ever exercised such power as Kripke now held. Political pressure to appoint more women had been mounting for years. In 1985, the UT System asked member institutions to "encourage greater participation in administrative roles for both minorities and women." LeMaistre responded with a committee to enhance the opportunities for women to "participate in senior positions and activities of the cancer center."

Many faculty women celebrated Kripke's selection, but others worried that the appointment would prove more symbolic than real. In 1996–97, remembered Elizabeth Travis, "we had another survey on women in leadership positions. The gender ratios at the various faculty ranks and in tenured appointments had changed very little. Salary equity was still an enormous problem. Instead of interpreting the data for Mendelsohn, we just showed him the raw data, and he said, 'There are a lot of "M's" there,' meaning a lot of men. After that, Mendelsohn really raised the bar."[21]

M. D. Anderson was hardly alone in the need to diversify. Back in 1969, only 929 women nationwide had entered medical school and therefore the educational pipeline leading to faculty positions. Title 9 of the Higher Education Amendments Act of 1972 banned sex discrimination in programs receiving federal funds, and the number of women medical students rose dramatically, reaching 6,851 in 1993, 27.2 percent of the total number of matriculants. Women planning careers as medical school faculty, however, faced daunting obstacles. Because completing medical school, internships, residencies, and fellowships could consume more than a decade for some specialities, many women had to postpone or forego starting families, a choice many refused to make. Because of prevailing social norms, most women medical school graduates specialized in pediatrics, gynecology, family practice, or internal medicine. They also faced a host of internal barriers to gender equity, not the least of which was the reluctance of male-dominated faculties to restructure medical education in such a way as to handle the needs of women raising children, including day care centers, maternal leave, split residencies, and longer probationary periods. Those women finishing medical school, internships, residencies, and fellowships then found themselves engulfed in a system that required seventy-hour workweeks in order to secure tenure and promotion. And as subtext to the structural problems, women encountered what Janet Bickel has called "microinequities"—from gender bias and stereotypes on the part of many male colleagues to outright sexual harassment. Women were also likely to receive lower salaries and fewer promotions than male colleagues of similar training and accomplishment. Not surprisingly, in 1995, less than 10 percent of medical school full professors nationwide were women and only 4 percent of department chairs, a figure that had not changed since 1980.[22]

The issue of minority faculty members was even more difficult. Between 1973 and 1987, approximately 8 percent of new medical students were minorities, even though minority groups constituted 14.2 percent of the U.S. population. In medical school, they had found few role models among faculty members around which to fashion themselves, and many educational institutions in the South had only recently desegregated. During medical school, internships, residencies, and fellowships, minority students experienced the effects of outright racism and the more subtle consequences of institutional racism. In 1994, only 2.4 percent of faculty members were black, and relatively few minority physicians were in the job market. Competition for them was fierce. Every medical school in the country wanted to diversify, and only the most proactive would find and keep them. As the historian Kenneth Ludmerer wrote, "Equality of opportunity in medicine . . . ultimately required not only a sound educational pipeline but a profession and society that welcomed all qualified individuals and recognized that it was in the national interest to make good use of all the country's human capital."[23]

The wheels of change turn very slowly, but at least the issue of diversity was now on the radar screen. In 1990, the Brooklyn-born, Harvard-trained cardiologist Harry R. Gibbs became the sixth African American faculty member at M. D. Anderson. "I arrived in Houston," Gibbs recalls, "on 'Go Texan Day,' and I wondered if I had made a mistake. All of these people were walking around in cowboy hats and cowboy boots. It was a long way from Brooklyn. In fact, it was mind-boggling." When Gibbs and his wife came down to look for a home, a white realtor was to meet his wife in the lobby of the hotel. She did not connect with Ms. Gibbs because she was not looking for a black woman. While Ms. Gibbs sat there, the realtor talked to every white woman she saw and kept asking at the desk. Gibbs found at M. D. Anderson in the early to mid-1990s an institution with much yet to do in building a diversified faculty. For several years, he worked with the black cardiologist Joseph Swafford, and they enjoyed an inside joke. Born of African American and German ancestry, Gibbs had a light complexion, and at six feet nine inches, he was an imposing figure. Swafford had a darker complexion and was short in stature. The two men could not have looked more different. Gibbs remembered how "people kept confusing us, even some in senior management, as if all blacks look the same. More than once Swafford and I laughed over it." In 1999, Mendelsohn named Gibbs the first vice president for institutional diversity and provided resources to create a new Office of Diversity.[24]

The new president also had to address the emergence of integrative medicine as a discipline. The field of cancer medicine had always attracted more than its share of hucksters and con men who knew that their remedies did not work but matched each other lie for lie and sleazy promise for sleazy promise, selling their products to vulnerable patients bereft of hope and medical options. Phony tonics such as Krebiozen and Laetrile always lost traction when it became obvious that using them to fight cancer was like shooting ducks with blanks.

Lee Clark had enjoyed unmasking cancer frauds. But the all-too-common mutual contempt on the part of orthodox physicians and nontraditional practitioners divided American medicine, and during the twentieth century, the crack opened into an abyss. Ever since its founding in 1847, the American Medical Association had ruthlessly suppressed all competing models of medicine. To describe them, the AMA, with pens dripping in acid, bandied about pejoratives—quackery, nonscientific, unproven, irregular, unorthodox for the disciplines and charlatans, pretenders, and con artists for their practitioners. With rare exceptions, the AMA extended them no respect. M. D. Anderson had only occasionally flirted with nontraditional therapies, such as its tests of acupuncture in the early 1970s.

Late in the twentieth century, however, the AMA observed a groundswell of Americans not so dismissive. Biomedicine conceives of itself as a science, but health care is a social and political construct; millions of Americans avail themselves of the thousands of nontraditional practitioners and believe that they are experiencing outcomes good enough to return again and again. In the 1990s they became a potent political force. Congress established the Office of Alternative Medicine (OAM) within the NIH in 1991 to evaluate promising unconventional treatments. The National Institutes of Health Revitalization Act of 1993 placed the OAM within the office of the director of the NIH. In 1998, Congress reorganized the OAM into the National Center for Complementary and Alternative Medicine (NCCAM) and elevated its status to an NIH Center. The NCI at the same time established its Office for Cancer Complementary and Alternative Medicine.[25]

Lorenzo Cohen, an M. D. Anderson faculty member trained in medical psychology and interested in psychoneural immunology, represented the face of integrative medicine at M. D. Anderson. Soon after his arrival in 1997, Cohen began to promote integrative medicine. He determined that more than 50 percent of M. D. Anderson patients used integrative treatments but that only half shared that information with their oncologists, as if the subject were taboo. Because of the new NCCAM and research funds available for research in integrative medicine, Cohen wanted M. D. Anderson to stake out a claim. Mendelsohn convened a meeting of key faculty members and inquired whether M. D. Anderson should embrace integrative medicine. When they reacted positively, he formed a committee of such prominent faculty members as surgeon Raphael Pollock and medical oncologist Waun Ki Hong. To Mendelsohn's pleasant surprise, they recommended the move into integrative care. The senior medical staff agreed.[26]

Although many patients came to M. D. Anderson already interested in integrative medicine, others acquired it after exhausting standard treatments. For most patients, the diagnosis of cancer is like taking a sucker punch to the gut. Because the standard treatments often take place over the course of weeks and months, patients can acquire a sense of empowerment, enduring the discomfort and even disabilities of side effects but tolerating them in the name of survival. The hospital and clinics—with helpful staff and the M. D. Anderson culture of hope—become emotional umbilical cords, reassuring patients that they are doing everything possible to extend their lives. And when the day comes that the last radiation treatment has been delivered or the last chemotherapy infusion administered, many patients leave the hospital with a strange ambivalence, relieved that the ordeal of treatment is over but at the same time, in the oddest ways, missing them too. Many needed more.

In September 1998, the "Place . . . of wellness" was opened at M. D. Anderson, the first such program at a comprehensive cancer center, where patients could investigate and integrate traditional and nontraditional cancer treatments. Integrative medicine became a section in the Department of Palliative Care and Rehabilitation Medicine. Lorenzo Cohen emerged as the point man for integrative oncology. In 2003, he became a founding member of the Society for Integrative Oncology. He endowed the program with a compelling motto: "There is not alternative medicine and traditional medicine. There is good medicine and bad medicine."[27]

—◆◆◆—

Late in the 1990s, the plight of two professional athletes, Lance Armstrong and Kim Perrot, put M. D. Anderson in the spotlight. A world-class cyclist from Austin, Texas, Armstrong was diagnosed in October 1996 with embryonal testicular cancer and metastases to the lungs and brain. After surgery to remove the diseased testicle, he began combination chemotherapy to treat the metastases. An Austin oncologist put him on bleomycin, the drug M. D. Anderson's

Melvin Samuels had pioneered for testicular cancer, and etoposide and cisplatin, which had been added years later. A few days into the chemotherapy, a physician asked Armstrong to get a second opinion from either M. D. Anderson or the University of Indiana.

Armstrong's visit to M. D. Anderson was a disaster. He experienced nothing but fear, confusion, and anger. The reaction of cancer patients to their disease depends greatly on their own emotional, financial, and personal circumstances. What one patient sees as insensitivity, another interprets as honesty and transparency. "Houston is a gigantic metroplex of a city, with traffic jams choking freeways," he wrote several years later. "Just driving through it was nerve-wracking." Things only got worse.

When Armstrong finally met with an oncologist, M. D. Anderson's culture of aggressive treatment sucked his breath away. The oncologist voiced grave anxiety and recommended combination chemotherapy that included bleomycin in doses even "Megadose" Melvin Samuels might have found excessive. "You will crawl out of here," Armstrong remembers hearing. "I'm going to kill you. Every day I'm going to kill you, and then I'm going to bring you back to life. We're going to hit you with chemo, and then hit you again, and hit you again." In such doses, the oncologist continued, Armstrong "would have to give up any hope of having children and of cycling competitively any more." No longer would he be able to complete the Tour de France's numbing uphill climbs through the Pyrenees and the Alps, where some of the best-conditioned athletes in the world end up gasping for breath like end-stage lung cancer patients. When Armstrong asked why the treatments had to be so harsh, he was told, "You're [a] worst case. But I feel this is your only shot, at this hospital. . . . Your chances aren't great. But they're a lot better if you come here than if you go anywhere else." He insisted that Armstrong start treatment immediately. "This is the only place to get this sort of treatment, and if you don't do it, I can't promise you what will happen."

Petrified with fear, and with his wife and mother in tears, Armstrong left thinking, "'I won't walk, I won't have children, I won't ride.' Ordinarily, I was the sort who went for overkill, aggressive training, aggressive racing. But for once I thought, 'Maybe this is too much. Maybe this is more than I need.'" Armstrong punched some numbers into his cell phone and placed a call to Lawrence Einhorn, the Indiana University oncologist who had devised the curative treatment of embryonal testicular cancer. Einhorn had trained at M. D. Anderson. In August 1974, his oncology group at Indiana had begun testing the therapeutic effects of adding platinum to Samuels's two-drug formula of bleomycin and vinblastine. The results were stunning. Thirty-three of forty-seven patients achieved complete remission and fourteen partial remissions. Thirty-one patients treated with what became known as the BVP protocol were still alive four years later. Craig Nichols, one of Einhorn's colleagues, took Armstrong's call and agreed to see him.

Nichols recommended surgery to remove the brain metastases and a chemotherapy regimen that did not include bleomycin and its risk of pulmonary fibrosis, certain to kill the career of a professional cyclist. Instead, they suggested vinblastine, etoposide, ifosfamide, and cisplatin, a regimen Armstrong would find extremely caustic but that would spare his lungs. Armstrong warmed to the prescription and then survived with flying colors. In 1998, he returned to competitive cycling. He won the Tour de France in 1999, 2000, 2001, 2002, 2003, 2004, and 2005, becoming the only man ever to win seven consecutive times. The public could not get enough of him. Setting a world's record in the most grueling event in sport, after surviving the most grueling treatment in medicine, made for legends. By the fall of 2004, after signing a multimillion-dollar endorsement contract, Armstrong appeared on television sets everywhere. He was not pitching Nike shoes like Michael Jordan or Buicks like Tiger Woods but chemotherapy drugs for Bristol-Myers Squibb.[28]

Mendelsohn had read about Armstrong's unfortunate encounter at M. D. Anderson in the cyclist's book *It's Not about the Bike*. He was shocked and has pointed out that Armstrong's experience was and is the exception at M. D. Anderson. Patient after patient gives kudos for the humane and supportive environment they encounter. Mendelsohn often said, "We treat the cancer and we care for the patient." Since 2000, Armstrong has worked closely with Margaret Kripke, Mendelsohn, and others at M. D. Anderson, spreading the word about the need for supporting cancer research and making cancer care accessible to all who need it. In 2005, as part of the Bristol-Myers Squibb Tour of Hope, Armstrong again spent some time at M. D. Anderson and came away with a completely altered perspective: "This is truly one of the greatest institutions in the world.... [T]his is not my first encounter with M. D. Anderson. But I'm proud to say... I think this place has changed... in talking with Dr. Mendelsohn and seeing his vision and seeing his passion for the illness, and just hearing him talk about clinical trials. He says we have the largest clinical trials program for cancer on the planet.... Nobody [else] can say that."[29]

Another marquee athlete also came to M. D. Anderson. Kim Perrot, a franchise player in the Women's National Basketball Association (WNBA), led the Houston Comets to WNBA championships in 1997 and 1998, earning a reputation for scrappy determination. An invisible opponent, however, lurked inside her. During the 1998 championship series against the Phoenix Mercury, Perrot suffered headaches and weakness in her left arm, which worsened during the off-season. A lung cancer had metastasized to her brain. For a superbly conditioned athlete and nonsmoker, the diagnosis left her thunderstruck. "I have the will to win," she told a television interviewer. "I won't be defeated. I just feel confident this is just a challenge, just a trial for me." She tried really hard to win, but brain cancer is an unforgiving opponent. "I've never been really sick or injured and now I'm faced with life or death," she told Houston television station KTRK. "All I can do is put it in the hands of the Lord." In February 1999, after neurosurgeons lifted out a golf ball of a tumor, she spoke frequently to cancer patients, usually refusing wigs and hats and letting her clipped hair and battle scar speak for themselves. M. D. Anderson oncologists, however, had no cure. She died on August 19, 1999. With a $1.25 million gift, the Comets memorialized Perrot at M. D. Anderson, which in 2002 opened Kim's Place, a retreat and respite for teenage cancer patients and survivors.[30]

―――

By the turn of the millennium, as Mendelsohn completed four years at M. D. Anderson, talk of curing cancer by 2000 had faded. The disease had defied medical logic for so long that people had endowed it with metaphorical power, as if it were a conscious entity. Cancer is "malignant" and "insidious," many described, and "eats" its way into "victims," "devouring" them until, skin and bones, they expire. In 1893, British poet Rudyard Kipling, in his poem "Children of the Zodiac," warned, "Cancer the Crab lies so still that you might think he was asleep if you did not see the ceaseless play and winnowing motion of the feathery branches around his mouth. . . . It is like the eating of a smothering fire into rotten timber in that it is noiseless and without haste." More than a century later, the social critic and breast cancer survivor Susan Sontag wrote, "Any important disease whose causality is murky and for which treatment is ineffectual, tends to be awash in significance. . . . The disease itself becomes a metaphor." Until the end of the twentieth century, a lexicon of defeat, doom, and desperation cloaked oncology.

Some of the gloom, however, had started to lift. The 2001 edition of the *Oxford English Dictionary* (OED) changed the definition of cancer for the first time in 111 years. In 1889, the OED had defined cancer as "a malignant growth or tumour in different parts of the body that tends to spread indefinitely and to reproduce itself and also to return after removal; it eats away or corrodes the part in which it is situated and generally ends in death." The new definition labeled cancer as "what happens when a group of cells grow in a disorderly and uncontrollable way and invade neighbouring tissues. They may or may not later spread into distant parts of the body. The cancer process is shared by over 200 diseases." Gone were the metaphors of decay and hideous consumption. Gone was the near certainty of death.

While attempts to find cures for cancer were being pursued, hope for controlling the disease, for converting cancer from an acute to a chronic illness, also gained momentum. In 1999, Stephen B. Baylin, associate director of research at Johns Hopkins University, remarked, "One aims for cure, certainly, but we should do well to simply delay the progression of cancer. You could live with cancer, as long as you knew it wouldn't spread. I think that offers the greatest hope for the near future, to slow the progress of cancer."[31]

For Joan Coffey, Joyce LeMaistre, and millions of cancer survivors, however, Baylin's "near future" was frustratingly nebulous, like an Atlantic Ocean fog bank through which they were to sail.

13

New Offensives,

2001–2007

The days of the heroic genius working alone in an
isolated laboratory and discovering a cure for cancer are over.
Cancer is just too complicated.
—JOHN MENDELSOHN, 2007

Back in the summer of 1985, Page Lawson, the director of Volunteer Services at M. D. Anderson, had noticed a distressed middle-aged woman in the gift shop. Lawson asked if she needed help. Recently operated on for colon cancer, the woman's colostomy seal had failed, soiling her clothing and fouling the air. Humiliated and desperate, she needed to find a restroom. Lawson showed her and waited outside for fifteen minutes. When the woman did not reappear, Lawson found her in a stall sobbing, the floor littered with dirty paper towels. Lawson reassured the woman that the accident was a trifle. She invited her to sit down on the commode, and promised to return shortly. Lawson called the laundry and asked for clean towels and two sets of surgical scrubs. She summoned a volunteer colon cancer survivor. Together they went to the restroom. They dressed the woman in the first set of scrubs and escorted her to a locker room in the hospital where she showered. They helped her to reset the stoma appliance, and the woman put on the second set of scrubs. Lawson then walked her to the parking lot.

Lawson became director of Volunteer Services in 1973, inheriting 125 volunteers and a culture of hospitality. She had a keen eye for opportunities. In 1974, for example, a group of volunteers working with pediatric patients held a Christmas card design contest for the children, and three designs were selected for reproduction and sale in the gift shop. "The reaction we got was amazing," she recalled. "We got the printed cards late, just a couple of weeks before Christmas, but they [nine thousand copies] were nearly all sold before we got them to the Gift Shop." The Christmas cards, or what became known as the Children's Art Project, would within a decade be selling more than 2 million copies a year and become a signature of M. D. Anderson. By 1979, the year she won the Joan Waite Hanlon Memorial Award for professionalism in volunteer administration, Lawson had 700 volunteers working in thirty-four different programs. In Steve Stuyck's recollection, "By the time she [Lawson] retired, she and her volunteers had built a wide variety of new programs that are still going strong: the Volunteer Endowment for Patient Support, the Children's Art Project, and the volunteer concierge desk. She had a million ideas, virtually all of them good, and a great way of rallying others to the cause."[1]

———

Lawson also catered to the power of survivorship. Over the years, she had become impressed with Reach to Recovery, a body of breast cancer survivors that counseled mastectomy patients. On many occasions, she had seen a depressed patient perk up in the presence of a mastectomy survivor. "Nobody else can give a breast cancer patient as much support," she often confided. Lawson applied the same logic to other patients.

When John Mendelsohn arrived at M. D. Anderson in 1996, the culture of cancer survivorship had bloomed. Between 1971 and 2001, the number of cancer survivors in the United States would increase from 3 million to nearly 10 million people. They were not really patients anymore, and they increasingly rejected the label "victim." Richard Bloch, the founder of H&R Block and a survivor of lung cancer at M. D. Anderson, had joined with his wife, Annette, and christened the survivorship movement. In 1981, they sponsored their first cancer survivor rally in Kansas City, with the theme "There is life after cancer." Cancer Survivors Day rallies and celebrations spread to other communities. The Blochs founded the National Cancer Survivors Day Foundation to promote the event and constructed cancer survivor parks in dozens of communities. In 1982, President Ronald Reagan appointed Bloch to the National Cancer Advisory Board.[2]

The lives of cancer patients may appear to return to normal, but cancer is not like a case of strep throat banished by antibiotics. Each of oncology's treatment protocols exacts a price, and survivors pay the bills, which can include long-term deficits in fertility, sexual function, cognition, and mobility, to say nothing of new, treatment-induced cancers. Many face discrimination in the workplace and the inability to secure life and health insurance.

Impressed with the success of gay rights activists in placing HIV/AIDS on the national political agenda, small groups of cancer survivors organized. In 1986, under the leadership of Ellen L. Stovall, an attorney and two-time cancer survivor, a group of physicians, social workers, and cancer survivors gathered in Albuquerque, New Mexico, and established the National Coalition for Cancer Survivorship (NCCS). The NCCS advocated quality cancer care for every American, an end to discrimination against cancer survivors, and support for such organ-specific advocacy groups as the National Breast Cancer Coalition, the National Ovarian Cancer Coalition, the National Prostate Cancer Coalition, the Leukemia and Lymphoma Society, and the Colon Cancer Alliance.

In Washington, D.C., as cancer survivorship advocacy gained steam, the NCCS exercised more muscle on Capitol Hill. Richard Bloch attended the NCCS's first assembly in 1987 and several successive national meetings. Stovall helped write the Americans with Disabilities Act of 1990 and the 1993 Rocke-feller-Levin Medicare Cancer Coverage Improvement Act, making sure to include language requiring Medicare to cover routine patient expenses associated with clinical trials. In 1994, the NCI acknowledged that "over the last decade, a strong infrastructure has . . . emerged to advocate for research, health care and social reform on behalf of cancer survivors. This is clearly seen in the establishment of the National Coalition for Cancer Survivorship and its increasingly sophisticated efforts to push for changes that improve the opportunity for cancer patients to successfully reestablish a normal life during and after treatment." In 1996, Stovall, as executive director of NCCS, was named to the National Cancer Advisory Board, and NCI established its Office of Cancer Survivorship. Two years later, on the National Mall, the NCCS assembled 250,000 cancer survivors to celebrate life, demand more funding for research and rehabilitation, and promote broader access to cancer care.[3]

The data on survivorship could be deceiving. Much of the increase resulted from more effective screening tests and earlier diagnosis, which skewed the number of survivors without necessarily implying better treatments. Pap smears, occult blood and PSA tests, and mammography had spiked the number of Americans diagnosed with cervical, prostate, and breast cancer. In 2003, for example, more than 30 million mammograms were performed in the United

States, revealing 211,300 cases of invasive breast cancer and 55,700 cases of the less severe carcinoma in situ, for a total of 267,000 women. Because breast cancer killed 39,800 women in 2003, the number of breast cancer survivors jumped by 227,200 people. Similar jumps for similar reasons characterized many other forms of cancer.[4]

At Station 80 of the Clark Clinic in the 1980s, sarcoma, melanoma, and breast cancer survivors waited to see a surgeon. For the resourceful, finding a compatriot was not that difficult. A sarcoma survivor might look for an amputee. A breast cancer survivor might notice a woman whose bosom seemed out of whack, one breast much larger than the other or a sag in her dress that a breast had once filled. Downstairs in the basement, where Gilbert Fletcher's machines buzzed all day long, men and women waited their turns, many displaying the telltale, purple pink Magic Marker radiation targets on their faces, necks, throats, and collar bones. At Station 19, patients could always find others in similar circumstances hooked to IVs.

In the summer of 1985, Joseph Painter, vice president for extramural programs, suggested to seven patients the need for a program to help others recently diagnosed. Painter's initiative became the Anderson Network, one of the first hospital cancer survivorship programs in the nation. "The Anderson Network," he later recalled, "was a response to the fact that people really went into shock when they were first diagnosed. They didn't know what they were facing and they needed some way of seeing that there was hope, that people do survive." At first, the Anderson Network was a telephone service matching new patients with survivors. In 1988, the Anderson Network opened the first Hospitality Room on the second floor of the Clark Clinic, providing patients and families a cozy, living room setting. Two hospitality rooms by 2004 were hosting nearly sixty thousand visitors a year. And in 1989, the Anderson Network sponsored its first annual Living Fully with Cancer conference, where long-term survivors meet annually. Within a decade, the Anderson Network linked by telephone more than ten thousand cancer survivors and also published *Network*, a quarterly newsletter whose circulation exceeded eighty thousand.[5]

———

In 2003, for three M. D. Anderson patients, the truce ended in their personal wars on cancer: Susan Cording, the teenager cured of Hodgkin's disease early in the 1970s; Joyce LeMaistre, the wife of the former president of M. D. Anderson; and Joan Coffey, the history professor with Waldenström's lymphoma.

Susan Cording had suffered multiple ailments stemming from massive doses of radiation thirty years earlier. The intervening years had been difficult. Radiation and surgery had cured the lymphoma, but she endured debilitating side effects. Tumorcidal doses to the lymph nodes in her upper torso and neck suppressed bone development, and she grew to adulthood with a shrunken frame. The radiation to the torso also stunted the development of her heart, damaged her kidneys, and diminished the capacity of her lungs, and the gifted athleticism that graced her childhood never found full expression. The radiation treatments stole her fertility. And through it all, as a cancer survivor, Susan made the most of life, becoming a successful teacher and assistant principal. She married, adopted a son, and built a loving family of her own, of the type she had enjoyed as a child.

Susan Cording died on July 10, 2005, in an Austin, Texas, hospital, suffering from thyroid cancer and congestive heart failure, two long-term side effects of the Hodgkin's cure. When former president John Adams notified Thomas Jefferson about the death from breast cancer in 1813 of his daughter Abigail, he wrote, "Your Friend, my only Daughter, expired, Yesterday Morning in the arms of Her Husband, her Son, her Daughter, her Father and Mother, her Husbands two sisters and two of her Nieces . . . She has been a monument to Suffering and to Patience." Jefferson poignantly replied, "There is no degree of affliction produced by the loss of those dear to us. . . . While experience has not taught me to estimate . . . time and silence are the only medicine, and these but assuage, they can never suppress, the deep drawn sigh which recollection for ever brings up, until recollection and life are extinguished together." The two presidents could have been talking about Susan Cording and her family.[6]

Joan Coffey began to touch bottom in February 2003, when hematologist Sergio Girault put her through a high-dose regimen of the antibody therapy Rituxan. One month later she was treated with Campath, and in May 2003, in hopes of controlling the disease in preparation for a stem cell transplant, Girault put Coffey on melphalan. Coping with the treatments and the disease all but swamped her life. Still, Joan managed to keep working on the biography of Léon Harmel, which the University of Notre Dame Press agreed to publish, and submitted to them a complete manuscript.

Finishing the book became a race against Waldenström's. In June 2003, a neurosurgeon installed in Joan's skull an Ommaya port, and Charles A. Conrad, the son of the internist who was murdered at M. D. Anderson in 1982, administered chemotherapy directly to the brain and spinal cord. Joan's crippled immune system collapsed, and her struggle with the lymphoma became a grim struggle against opportunistic infections. In July 2003, she developed fungal pneumonia and spent the rest of her life as an inpatient, receiving blood transfusions and antibiotics until further efforts were clearly futile. As the end approached, the editor at Notre Dame worked overtime to get the final manuscript to the printer. When he learned of Joan's plight, the printer worked nonstop to produce the first bound copy of the book wrapped in its glossy cover. A courier delivered the book to her home, and her husband, Ed, brought it to M. D. Anderson. Joan handled it and smiled before falling into a coma. She died on August 13, 2003. Ed placed the book in Joan's coffin.[7]

For Joyce LeMaistre, medical oncologists mixed a series of chemotherapy drugs, but her disease resisted stubbornly. Over the course of more than two years, she received the drugs through a catheter, and Mickey created in their home a small hospital, with home health care workers, an M. D. Anderson-designed diet, and regular deliveries of medical supplies and pharmaceutical products. Joyce did well until a metastatic intestinal obstruction forced her to intravenous feeding. Mickey made sure that she was free of pain. In the fall of 2003, Joyce LeMaistre gradually lost strength. On Thanksgiving Day, she invited her children and grandchildren to gather at home. Over Mickey's protests about the need to conserve energy, she spent eight hours talking with, playing with, and loving her family.

Joyce LeMaistre's condition steadily deteriorated, although the quality of her life remained intact. The strategies of pain control worked, allowing her to remain conscious and pain free almost to the end. She began sleeping for longer periods, and on December 5, 2003, her breathing grew shallow and irregular. Joyce LeMaistre died at home later in the day. "We could not have scripted a better death, especially from ovarian cancer," Mickey remembers. "She was the cheerleader for the family. She had known for two years that ovarian cancer would take her life, but the realization never really fazed her."[8]

But overall, the chances for surviving cancer were improving. Mendelsohn points out that the five-year survival rate has nearly doubled during his lifetime. These survivors, however, face risks of recurrence as well as higher odds of developing other cancers, and they are subject to risks of late-occurring consequences of the therapy they received. Despite the progress, in 2001 and 2002, cancer took the lives of surgeon Richard Martin, radiologist Robert Egan, and volunteer specialist Page Lawson.

On October 14, 2001, Richard Martin became a belated casualty of the Korean War. On a MASH unit's operating room table, he had saved the life of a wounded GI and in doing so had acquired hepatitis. Over the next fifty years, the infection irritated the cells of his liver and led to liver failure secondary to chronic active hepatitis. He died before liver cancer had a chance to take his life. During a lifetime of treating patients and training surgeons, Martin had earned the respect and devotion of thousands. The M. D. Anderson commitment to high-quality patient care was fully realized in Richard Martin, a general surgeon devoted to a quest for less and less radical surgery.[9]

Robert Egan had spent decades x-raying everything imaginable, including thousands of amputated breasts removed by mastectomy and thousands of x-rays of women with benign and malignant diseases of the breast. Indefatigable, or perhaps obsessed as some said, Egan and his colleague Gerald Dodd developed the diagnostic tool of mammography and saved or extended the lives of hundreds of thousands of women. Egan had left M. D. Anderson for Emory University years earlier, but his colleagues remembered him well and wondered if a lifetime spent in laboratories hot with x-rays had given Egan the same deadly leukemia that had felled former colleague Gilbert Fletcher a decade earlier.

Page Lawson embodied M. D. Anderson values. During her career, which ended with her retirement in 1991, she had transformed Volunteer Services. At the time of her death from lung cancer on July 24, 2002, Lawson no doubt recalled Lee Clark's urging her to give up smoking. Like the nurse Ethel Fleming, Lawson spent her last days surrounded by men and women devoted to her.[10]

Her legacy thrived. Lawson had directed Volunteer Services for seventeen years, and her successors—Tyrrell Flawn, Mary Nell Jeffers Lovett, and Susan French—had each raised the bar. With sixteen hundred onsite volunteers, fifteen hundred in the Anderson Network, and eighty separate programs, Volunteer Services at M. D. Anderson had become the largest in the world. Volunteers manned the surgical and intensive care waiting rooms, registration desks, concierge desks, gift shops, beauty shops, and the fourth-floor business center; they served as parallel patient advocates with full-time patient advocates, bilingual interpreters, and laboratory assistants; and they assisted chaplains as denominational volunteers, with strict prohibitions against proselytizing in any way. Each volunteer undergoes an annual performance evaluation that can result in being discharged from the program. "We really have an upbeat attitude," said Mary Nell Jeffers Lovett. "We must deliver compassion, be nonthreaten-ing, and give all volunteers a sense of ownership and pride in M. D. Anderson."[11]

In 1996, the year John Mendelsohn arrived at M. D. Anderson, ImClone began testing C225, the monoclonal antibody that had emerged from his laboratory, and over the next eight years, data from multiple clinical trials with multiple forms of cancer revealed the potential of targeted therapy. The phase 1 and phase 2 clinical trials employing C225 with chemotherapy and radiotherapy demon-strated safety and efficacy in some cancer patients. A phase 2 trial of C225 and radiotherapy in advanced head and neck cancers showed a dramatic 100 per-cent response. Another phase 2 trial, this time with C225 and the drug cisplatin, had a 67 percent response among patients with advanced head and neck tu-mors. Equally encouraging, the combination had worked on half the patients whose tumors had grown after cisplatin treatment alone.

Reports at the 2001 meeting of the American Society of Clinical Oncology indicated a 22.5 percent response when patients with colon cancer growing while on irinotecan were treated with irinotecan plus C225 and a response rate of 23 percent when head and neck patients progressing while on cisplatin were treated with cisplatin plus C225. More striking, patients with advanced pancreatic carcinoma treated with gemcitabine plus C225 had a response of 12.5 percent with a one-year survival of 32.5 percent, compared with the pivotal trial of gem-citabine alone, with its 18 percent one-year survival. Since one-third of solid tumors express high levels of EGF receptors and may depend upon this sig-naling path, C225's potential seemed huge.

Events in 2002 through 2004 confirmed the import of Mendelsohn's research. ImClone reported a response rate of 10.5 percent for patients with colorectal cancer progressing while on irinotecan therapy. At the request of the FDA, Im-Clone's partner, Merck KGaA, conducted a randomized clinical trial and re-ported on the results at the ASCO meeting of 2003. In patients with colorectal cancer progressing while on irinotecan, the addition of C225 to irinotecan pro-duced a response rate of 22.9 percent. When irinotecan was withdrawn from the regimen and C225 given alone, the response was lower. Both groups demon-strated the efficacy of C225 and the advantages of combination therapy. For his development of the monoclonal antibody 225 and pioneering the new era of targeted therapy, Mendelsohn received the David A. Karnofsky Memorial Award of 2002 from the American Society of Clinical Oncology. Two years later, the FDA approved C225 (Erbitux) for advanced, drug-refractory colorectal cancer. In 2006, it was approved for head and neck cancers, alone and in combination with radiation therapy.[12]

For several years in the early 2000s, Erbitux garnered some unwelcome attention. In June 2002, the *Washington Post* reported that between 1997 and 2001, M. D. Anderson oncologists had enrolled 195 patients in clinical trials involving Erbitux without mentioning that Mendelsohn had a financial stake in the drug's success. M. D. Anderson let it be known that they had also enrolled more than 300 patients in trials involving the drug Iressa, Erbitux's major competitor. And although Mendelsohn had neither a personal involvement in the treatment of the patients nor a role in selecting Erbitux, the potential for conflict of interest seemed real. When oncologist Roman Perez-Solar began to test Erbitux at M. D. Anderson, he was reported to have said, "I knew right away this was dangerous territory. You need a promotion. You need a salary increase. You need another lab. It distorts the normal conduct of things because you go all the way to please the boss."[13]

The controversy spoke directly to the conflict-of-interest debate. Erbitux was one of the most promising drugs in oncology's new pharmacopeia, with trials being conducted around the world. Soon after his arrival, Mendelsohn had beefed up the institution's conflict-of-interest policies. New regulations required full disclosure of potential financial conflicts of interest involving new treatments and the physicians studying and prescribing them. Physicians were no longer permitted to treat patients with experimental drugs in which they had a financial interest. The irony is that prior to the public airing of the issue, Mendelsohn had decided to add the policy of including statements in all clinical trials to inform patients when a senior executive or M. D. Anderson itself would receive income if the experimental drug proved successful.[14]

Until this time, according to gynecologist Ralph Freedman, "there was little understanding of conflict-of-interest issues at M. D. Anderson. Federal and state guidelines were just appearing to address conflict-of-interest charges." The controversy accelerated an internal policy review already under way. "I don't want to take any chances that a patient will feel they've been deceived by M. D. Anderson," Mendelsohn remarked to a reporter from the *Washington Post*. The conflict-of-interest committee and the institutional review board required that institutional leaders' conflicts be discussed in consent documents.[15]

For some medical ethicists, however, full disclosure was only a Band-Aid. Few patients would reject a recommended treatment for metastatic cancer just because their physician stood to profit from it. Marcia Angell, an instructor at the Harvard Medical School and former editor of the *New England Journal of Medicine*, worried about full disclosure. "There is a lot of talk," she said, "about 'managing' conflicts of interest [by simply informing patients] and not prohibiting them. I think disclosure is better than nothing, absolutely. But there should be a flat-out ban."[16]

Erbitux was also at the heart of the ImClone scandal. In 1991, Sam Waksal, the CEO of ImClone, had taken the company public at $14.00 a share. By 2000, with Erbitux racking up success after success in clinical trials and expectations abounding for FDA approval, ImClone stock climbed steadily, boosted as well by the infatuation of American investors with technology companies. In December 2001, Bristol-Myers Squibb offered up to $2 billion to ImClone for a 40 percent share of the profits from future sales of Erbitux. On the securities markets, the share price of ImClone soared beyond $60.00.

By the end of 2001, enthusiasm for Erbitux at the FDA had cooled. Some officials worried about the methodologies used in the studies. Further clinical studies were planned, and Mendelsohn was confident of the drug's efficacy. Waksal, however, could not wait. On December 20, 2001, ImClone and Bristol-Myers Squibb learned that the FDA had decided to postpone giving Erbitux its seal of approval, citing the need for more extended clinical trials. The decision was certain to pop the ImClone bubble. Waksal sold 79,797 shares of his ImClone stock for roughly $4.9 million. The markets soon learned that the FDA had balked on Erbitux. ImClone stock crashed, and Bristol-Myers Squibb watched $2 billion shrink in a few days to $1.3 billion. Waksal earned an 87-month prison sentence and an $800,000 fine. At the 2003 meetings of the American Society of Clinical Oncology, the pharmaceutical firm Merck KGaA confirmed that Erbitux in combination with chemotherapy drugs shrank colon cancer tumors and slowed the progress of the disease in terminally ill patients. ImClone stock surged.[17]

Early in 2000, four years into his tenure, Mendelsohn convened the faculty in a series of brainstorming meetings to assess the status of research at M. D. Anderson and to outline future strategies. Throughout the history of M. D. Anderson, the nature of its research mission has been a subject of intense, ongoing debate, with controversy swirling around three issues: the role of categorical institutions, the relative merits of basic and clinical research, and the place of investigator-initiated research versus targeted approaches. The debates intensified during and after passage of the National Cancer Act of 1971 and still rage today inside and outside M. D. Anderson.

Many biomedical scientists since 1941 have questioned the role of categorical institutions and whether focusing on a single disease really constitutes the best strategy for conquering it. Clark had always insisted that categorical institutions were best suited for a multidisciplinary assault on cancer because "cross-fertilization" between departments did not characterize medical schools. Medical school faculty countered that disease processes play out at the molecular level, and research strategies focusing on one malady risk overlooking basic biological principles.

The advocates of basic research insisted that clinical research too often puts the cart before the horse, testing potential therapies on patients before fundamental biological principles were fully understood and in the process investing enormous amounts of money for relatively little gain. Clinical researchers, however, had anxious patients in their offices, not just lab rats in cages and molecular models on computer screens. Most had seen how clinical research improved the capacity of physicians to cure or to control the disease. They heard the pleas of human beings, not just the squeaks of rodents, and yearned to bring relief to patients. Research bringing scientific discoveries to the clinic had given M. D. Anderson its reputation.

For decades, NCI, while supporting investigator-initiated research, had distributed a substantial portion of its millions primarily in a targeted fashion according to strategies developed in Bethesda, Maryland, and then had strictly monitored the work of grant recipients, leaving little latitude for investigators to make course corrections as data accumulated and conclusions suggested a different approach. The NIH, on the other hand, had pursued a more or less laissez-faire strategy in which investigators initiated their own research projects and then competed for funding, with peer referees external to the grant agency deciding on the quality of the projects. The best proposals would be funded. NIH then extended considerable latitude to investigators to follow the trail of their evidence. At the heart of the controversy lay the critical question— is the best science likely to be generated by targeted groups or from investigator-initiated projects, both of which must successfully compete in the scientific marketplace?

The career of Ralph Arlinghaus illustrates the odyssey of an investigator pursuing his own scientific instincts. After receiving a bachelor's degree in pharmacy at the University of Cincinnati in 1957, Arlinghaus remained there for a master's degree in pharmaceutical chemistry and then for a Ph.D. in bio-chemistry. Research fellowships took him to the University of Cincinnati College of Medicine and subsequently to the University of Kentucky. From there, Arlinghaus joined the staff of the Plum Island Animal Disease Laboratory. He arrived at M. D. Anderson in 1969 and redirected his research to the Rauscher leukemia virus, a murine retrovirus discovered at NCI in 1957 by Frank J. Rauscher Jr. The virus soon became widely used among scientists as an in vivo leukemia model because of the efficiency with which it acts on rodents.

At first, Arlinghaus hypothesized that the Rauscher virus caused leukemia by encoding a protein that acted once the viral genome had been inserted into the DNA of an infected cell. The discovery of reverse transcriptase in 1970 then toppled classical virology and, after five years of hard work on the virus and multiple papers published in such prestigious journals as *Cell, Virology,* and the *Journal of Virology,* Arlinghaus had to abandon the hypothesis and regroup, coming away from the experience without revealing the leukemic mechanism but with a much better understanding of how the virus manufactured proteins. As classical virology imploded under the weight of reverse transcriptase, Arlinghaus veered in another direction, investing his lab's resources in the analysis of protein synthesis, the role of reverse transcriptase, and a variety of retroviruses.

By that time, geneticists had revealed that 95 percent of patients with chronic myelogenous leukemia (CML) express the Philadelphia chromosome, which has a translocation defect. A portion of the BCR gene from chromosome 22 trades places with the ABL gene on chromosome 9. The hybrid gene stakes out a homestead on chromosome 22. The fusion gene has an activated tyrosine kinase and also activates several proteins and enzymes that regulate the cell cycle and accelerate mitosis. At the same time, the protein inhibits DNA repair and destabilizes the genome. The onset of CML could be traced not to a virus but to a genetic fusion.

In 1983, Arlinghaus left M. D. Anderson for Johnson and Johnson, where he hoped to study and help develop treatments for HIV and other disorders. Arlinghaus arrived in California just as a financial maelstrom sideswiped Johnson and Johnson. The year before, a lunatic in Chicago had sprinkled cyanide into bottles of Tylenol, a Johnson and Johnson product, and seven people had died after ingesting it. Sales and revenues evaporated, and Johnson and Johnson stock fell from sixty dollars to thirty dollars a share. Arlinghaus began looking for a job and John Batsakis, who had replaced Frederick Becker as chair of pathology at M. D. Anderson, rehired him.

Back in Houston, Arlinghaus reoriented his research from mice to humans, focusing on CML, a disease that hematologists Hagop Kantarjian and Moshe Talpaz studied in their clinical laboratories. Arlinghaus discovered that BCR, the partner of the BCR-ABL fusion protein, plays a major role in regulating this BCR-ABL tyrosine kinase. Talpaz helped develop the highly successful targeted therapy of Gleevec to treat CML. Success in treating patients with CML was measured by long-term survival without detectable levels of residual cells, a somewhat crude and unreliable tool.

As Gleevec became a frontline therapy for CML, oncologists needed a diagnostic tool requiring only small amounts of blood, not multiple bone marrow aspirations. Taking a cue from hematologist John Goldman of Imperial College, London, Arlinghaus and others at M. D. Anderson developed a comparatively simple diagnostic test employing tiny amounts of blood to detect residual leukemia cells via the presence of reverse transcriptase-polymerase chain reaction for BCR-ABL transcripts. With that knowledge, oncologists can assess the efficacy of treatment and the progress of the disease.[18]

In strengthening the scientific enterprise at M. D. Anderson, Mendelsohn saw merit in both investigator-initiated and targeted approaches. To the faculty, he posed the question in terms translational scientists would understand: "What research will best translate into earlier diagnosis and better treatments for our patients and what facilities will be needed to realize those objectives?" From such prominent M. D. Anderson faculty as Garth Powis, Isaiah J. Fidler, Juri G. Gelovani, Yong-Jun Liu, Gordon B. Mills, James D. Cox, and many others, a strategy emerged emphasizing metastasis, molecular imaging, targeted therapies, cancer immunology, molecular markers, and proton therapy, all six part of a South Campus initiative to establish an Institute for the Early Detection and Treatment of Cancer. The institute would push the research enterprise at M. D. Anderson in the direction of new approaches to oncology—toward basic research into the molecular biology of cancer cells and deriving from that research new targeted therapies to diagnose and treat cancer at the molecular level. Mendelsohn and the faculty crafted new approaches to carrying out their strategy. They envisioned a cluster of six Centers of Excellence, each with its own facility. Each center would be anchored by a basic science department with expertise in the targeted area, chaired by an outstanding and visionary leader who would direct the center. And each center would include faculty from other departments whose research explored that targeted area. The concept caught on quickly, and plans for recruitment, construction of new facilities, and fund-raising were put in place. Mendelsohn estimated a cost of $500 million to bring the concept to fruition.[19]

He undertook the task with typical energy and aplomb. During the next five years, the institute acquired a name, and its contours gradually took form. Red McCombs, a founder of the Clear Channel Communications empire, had served since 1986 on the board of visitors, and Mendelsohn had discussed with him a number of ideas, including the South Campus initiative. Late in 2005, McCombs and his wife, Charline, donated $30 million to the project. In return, the board of regents established the Red and Charline McCombs Institute for the Early Detection and Treatment of Cancer. The institute would encompass the six research centers of excellence.[20]

Early in 2006, Lowry and Peggy Mays contributed $20 million. A partner of McCombs and cofounder of Clear Channel Communications, Lowry Mays had served since 1993 on the board of visitors. In appreciation for the gift, the board of regents named the recently opened Ambulatory Clinic Building the Lowry and Peggy Mays Clinic. With its eight floors and 710,000 square feet of space, 590 new patient examination rooms and seventy-five-bed ambulatory treatment center, the clinic became home to the Nellie B. Connally Breast Center, the Laura Lee Blanton Gynecologic Oncology Center, and the Genitourinary Center. The Mays Clinic was a monument to ambulatory care and, according to Michael S. Ewer, assistant to the vice president for medical affairs at M. D. Anderson, "perhaps one of the benefits of managed care, in that it has challenged healthcare providers to find safe and effective ways to deliver care on an outpatient basis, which is clearly a benefit to our patients." Surely the most stunning example is the twenty-three-hour mastectomy. According to surgeon Raphael Pollock, "Twenty years ago, if you were having a mastectomy, you were an inpatient for eight days, and now it's routinely done on an outpatient basis."[21]

At the Red and Charline McCombs Institute, the metastasis center already existed. Metastasis is cancer's weapon of mass destruction, and so complex is the process that fewer than one hundred laboratories around the world deal with it. At M. D. Anderson, Isaiah J. Fidler had managed one such laboratory for more than twenty years. Were cancer to remain in place, its damage could be confined locally to the tissues and organs of origin, its effects not nearly as lethal, and the outcome of surgery and radiotherapy often curative. Primary tumors account for less than 10 percent of cancer deaths. Metastasis explains the rest.

From a solid tumor two-fifths of an inch in length, an estimated 1 million cancer cells per day sluff off into the circulatory system, and no more than one in hundreds of millions will survive the raging turbulence of the swim and escape prowling macrophages and killer cells. To make the initial escape, the cell must enjoy several peculiar properties—the ability to sever its bonds to the tumor, to dissolve the mortar of connective tissue, to alter its own shape, to grow legs like stilts to traverse the sticky, densely packed tumor cells, to insert itself into a capillary, and to thread its way through the circulatory system. Upon arriving at a distant site, the errant cell must find another capillary connected to a receptive organ.

Oncologists had long considered metastasis a random event, mysteriously intertwined with the type of disease and physiological characteristics in the patient. Fidler hypothesized that metastasis is a nonrandom process, and he carried out investigations into methods for learning the biochemical elements critical to metastasis.

Whether an organ is receptive to metastasis depends upon an intricate web of biological processes involving not only the malignant cell and its properties but the molecular characteristics of cells in the new microenvironment. Metastasizing cancer cells have special affinities for certain organs. Malignant colon cancer cells seem drawn to the liver, while prostate cancer cells prefer the bones. Lung tissues make homes for sarcoma cells, while breast cancer cells head for the liver, brain, bones, and lungs. Some oncologists likened the process to planting a seed in the garden. Success depends on the cancer cell and its microenvironment. The presence of the cancer cell prompts cells in the receptive organ to relax their cellular architecture and create space for the malignant cells to plant and proliferate.

At the genetic level, the migrating malignant cell and the cells of the receiving organ must enjoy a commonality allowing both to interact, like old relatives meeting after a long separation. Working from the Cancer Metastasis Research Center at M. D. Anderson in March 2007, Isaiah Fidler said, "We have reported that cancer cells growing in specific organ microenvironments express and release increased levels of [certain] proteins. . . . These ligands can bind to specific receptors on the tumor cells (autocrine) and tumor-associated organ-specific endothelial cells (paracrine). The activation of these receptors leads to stimulation of cell division and survival." Fidler continued, "The microenvironments of different organs are biologically unique. Endothelial cells [which pave] the vasculature of different organs express different cell-surface receptors and growth factors that influence the phenotype of metastases that develop there. The outcome of metastases depends on multiple interactions ("cross-talk") of metastasizing cells with homeostatic mechanisms."[22]

Some scientists speculate that a dangerous tumor cell triggers an embryonic program to direct its relocation. Others hypothesize that some cancer cells reignite a long dormant capacity, like stem cells, to become different types of tissue or to coexist with different types of tissues in different organs. Many oncologists surmise that millions of cells are able to reach distant organs, but only those with stem cell properties can trigger a new tumor.

Interrupting angiogenesis, the process by which metastatic tumor cells generate their own blood supply after relocating to a new site, has become a favorite target. On May 3, 1998, a media storm erupted when James Watson, the codiscoverer of DNA and early mentor of John Mendelsohn, allegedly told the *New York Times*, "Judah [Folkman] is going to cure cancer in two years." Watson later denied making that claim. Judah Folkman, at the eye of the media storm, was a sixty-seven-year-old oncologist at Children's Hospital in Boston. Folkman had spent years investigating angiogenesis—how rapidly growing tumors hungry for oxygen and nutrients draw to them, like magnets pulling in iron flakes, new capillaries that then inspire neighboring epithelial cells to divide more rapidly and migrate toward the tumor, where they change into new blood vessels.

The principle of antiangiogenesis spiked the curiosity of oncologists everywhere. "If we could slow tumor growth and prolong life with a drug that had little toxicity," said Lee Ellis, professor of surgery and cancer biology at M. D. Anderson, "we'd be doing far better than we are today." Ellis also suspected that angiogenesis drugs would fit well into the combination chemotherapy model with other anticancer drugs. "The future, I think," said the pathologist Frederick F. Becker in 2004, "will be in antiangiogenesis. Tumor cells can adapt in survival-of-the-fittest tradition to chemotherapy, radiation, and even targeted therapies, but they must have a blood supply. If we can block the formation of blood vessels, perhaps we can starve the tumor. A tumor that cannot vascularize itself cannot grow. It can survive, but it can't grow." On January 26, 2006, for example, the FDA approved the drug Sutent for the treatment of renal cell carcinoma. Kidney cancer had always exhibited resistance to radiotherapy and drug therapies, but Sutent inhibits a multitargeted tyrosine kinase and has antiangiogenic properties.[23]

In the spring of 2007, ground was broken for another of the centers of excellence—the Center for Advanced Biomedical Imaging Research. The state's

Texas Enterprise Fund provided a $25 million grant, and M. D. Anderson and the UT Health Science Center together supplied another $25 million. The UT System offered up $5 million, and GE Healthcare supplied $30 million in equipment and the expertise to run and maintain it. Mendelsohn had recruited Juri G. Gelovani from Memorial Sloan-Kettering to direct the center. An immigrant from Estonia, Gelovani was a leading figure in the effort to develop specific chemical, genetic, and cellular markers for imaging tumor-specific targets.

At the Center for Advanced Biomedical Imaging Research, Gelovani urged clinicians, chemists, and physicists to push beyond anatomical imaging to imaging techniques that measure metabolic activity as well. Traditional x-rays, CT scans, and MRIs, for all of their technological sophistication, display anatomical structures, while new technologies like MRI spectroscopies and PET scans expose what is occurring in those structures at a chemical and metabolic level. Malignant cells display metabolic activity distinct from that of normal cells.[24]

Gelovani soon demonstrated the promise of molecular imaging. He developed an imaging technique for identifying the EGF receptor tyrosine kinase, a molecule overexpressed in many human cancers. The drug Iressa (gefitinib) was designed to inhibit the molecule. The drug is of no use to patients whose cancers do not possess the overactive molecule. Iressa's side effects include diarrhea, acne, rash, dry skin, nausea, vomiting, pruritus (itching), anorexia, asthenia, and weight loss, certainly risks worth taking for a treatment capable of slowing the progress of cancer but not to be endured if therapeutic effect is impossible. By identifying or excluding the presence of the molecule, Gelovani could possibly assist medical oncologists in deciding whether a patient should take Iressa. Although the technique has not yet been shown effective in humans, its potential is enormous.

Molecular imaging allows oncologists to measure the progress of therapy more accurately. "Today, to assess the progress of chemotherapy," said Mendelsohn, "we take a CT scan of it at the outset of treatment, another one into the treatment, and with a ruler measure whether or not it is shrinking. This typically takes a few months." A molecular image, on the other hand, reveals the metabolic activity of a tumor, and within days of the delivery of the first dose of chemotherapy, if the drug is having an effect, change can be detected metabolically, even if the gross size of the tumor seems stable. If metabolic changes do not appear early in the molecular images of a woman receiving Adriamycin for breast cancer, for example, her oncologist, knowing the drug is ineffective, can terminate the treatment before she becomes vulnerable to its full side effects. Or instead of having to biopsy multiple metastatic lesions, a doctor can use molecular images to measure metabolic activity in all of them and identify a site of drug resistance.

In addition to the Cancer Metastasis Research Center and the Center for Advanced Biomedical Imaging Research, the McCombs Institute includes a Center for Cancer Immunology Research under the direction of Yong-Jun Liu, a newly recruited expert. Oncologists have long speculated about why the immune system is not more efficient in destroying cancer cells that so clearly pose a threat to the organism. Perhaps because the cancer cell possesses almost the same genetic code as the body's other cells, it sneaks under the immunological radar screen. Perhaps the process of natural selection in human evolution failed to eliminate cancer efficiently because, as a disease generally of older people, it does not stymie reproduction. At the heart of cancer immunology, for Liu, is the reality that "the immune system is capable of attacking self-tissues" as in such autoimmune illness as rheumatoid arthritis or lupus; "the trick is learning how to get it to attack malignant tissue instead." Still, Liu had a sober understanding of the challenges. "There can be up to 100,000 different proteins and protein variants getting turned over at different times in a cell, so trying to identify which ones the T-cell actually 'sees' is kind of like finding a needle in the haystack."[25]

A $5 million gift from the Robert J. Kleberg Jr. and Helen C. Kleberg Foundation set up Mendelsohn's fourth center of excellence—the Robert J. Kleberg Jr. and Helen C. Kleberg Center for Molecular Markers. Markers, like the prostate specific antigen test for prostate cancer, can signal that cancer may be present. Heirs to the fortune of the fabled King Ranch of South Texas, the Klebergs endowed the foundation and dedicated some of its resources to medical research. Under the direction of physician and scientist Gordon B. Mills, the center's goal is "to treat each patient's tumor based on what is happening with the biology of that person's cancer. We want the patient to 'teach' us what is important by looking at the genes that nature targeted for cancer in a particular patient. If we know what genes and their proteins are altered when cancer cells divide and multiply, we can better determine how to treat those proteins to stop cancer growth." One project at the Kleberg Center, for example, has isolated a novel predictive marker set for estrogen receptor positive breast cancer, the most common form of the disease, which may eventually assist physicians in its early diagnosis and treatment.

Investigations of the series of genetic and biochemical changes taking place within tumors and the relationships between biomarkers and the development and progression of cancer offer the promise of earlier detection and targeted therapy. Investigators at the Kleberg Center, using serum or tissue samples from patients, will then subdivide these classifications, searching for the molecular markers that dictate cellular proliferation and explain why some cancers originating in the same tissues behave so differently. Once those sets and subsets of markers are identified and classified, new diagnostic technologies can be developed to detect them before a patient is symptomatic and new treatments can be designed. This research dovetails closely with activities in the Center for Advanced Biomedical Imaging Research.

The Center for Targeted Therapy, under the direction of Garth Powis, serves as the fifth center of excellence. Targeted therapies, rather than killing cancer and normal cells, as does chemotherapy, employ molecules designed to interfere with specific molecules essential to the proliferation of cancer cells. A complex network and sequence of chemical and molecular signals instructs cells on growth, division, and apoptosis. Genetic changes can disrupt the signaling pathways and lead to cells that divide too rapidly or fail to die on schedule. Sometimes genetic alterations produce oncogenes that promote excessive cell division, or they incapacitate tumor suppressor genes, whose normal job is to slow down or stall cell division. In a variety of ways and at different stages of the mitotic process, targeted therapies interfere with the growth, division, and spread of cancer cells.

One class of targeted therapies includes small-molecule drugs that block specific enzymes and growth factor receptors required for cancer cell growth, an area of research pioneered by Mendelsohn. Cell biologists and oncologists also employ the term signal-transduction inhibitors to describe them. The drug Gleevec (also known as STI-571 or imatinib mesylate) is now used to treat gastrointestinal stromal tumor and certain forms of chronic myeloid leukemia. It targets abnormal proteins, or enzymes, that sometimes develop inside cancer cells and prompt unregulated growth. Iressa is a signal-transduction inhibitor.

Some targeted therapies interfere with the proteins involved in apoptosis, or cell death. The drug Velcade (bortezomib) can treat resistant cases of multiple myeloma. Velcade kills cancer cells by blocking enzymes known as proteasomes, which assist in the regulation of cell function and growth. The drug Genasense (oblimersen sodium) blocks a protein (Bcl-2) known to promote the longevity of tumor cells and may prove effective in treating leukemia, non-Hodgkin's lymphoma, and some solid tumors.[26]

Antiangiogenesis drugs strive to interfere with a tumor's effort to generate its own network of blood vessels. Avastin in 2005 was demonstrated to be effective in some colon cancer patients. Those who received it in combination with two chemotherapy drugs lived a little more than a year, compared to ten months for those receiving chemotherapy alone. In 2004, 173,000 Americans were diagnosed with lung cancer. Non small-cell cancer accounts for 80 percent of lung cancer. Avastin takes aim at molecules that help the cancer cells survive or grow. It is the first successful antiangiogenesis drug.

Arimidex became the number one drug for postmenopausal breast cancer patients. Better than tamoxifen, it blocks the aromatase enzyme and compresses the levels of estrogen in the body and inside the malignant cells. Arimidex reduces the rate of breast cancer reappearance by 70 to 80 percent, compared to tamoxifen's 50 percent. "Arimidex is a more effective treatment. It is a better drug," said Aman Buzdar, professor of breast medical oncology at M. D. Anderson. Buzdar thought all postmenopausal breast cancer patients should get the drug.[27]

The Proton Therapy Center was the sixth initiative. In 1988, James Cox had arrived at M. D. Anderson as vice president for patient affairs. Although he riled up some division heads by suggesting administrative reorganization, he "landed," in the words of biomathematician Stuart Zimmerman, "on his feet, in the job for which he was eminently qualified and at which he has excelled." When Lester J. Peters stepped down to return to Australia, Cox became head of the Division of Radiation Oncology, and from that perch he promoted more effective targeting of radiation therapy and proton therapy. Traditional radiotherapy shot photons— which many physicists describe as discrete particles having no mass and no electrical charge—at tumor sites to disturb metabolic processes within malignant cells. Because of their light weight, however, photons tend to scatter, elevating the risk of damage to normal cells. The holy grail of radiation oncology is conformational treatment in which the beams concentrate their energy to the shape of the tumor, without assaulting normal tissues.

During the late 1980s and early 1990s, two developments in radiation oncology advanced the frontier of conformational radiotherapy: three dimensional conformational radiotherapy (3DRT) and intensity modulated radiation therapy (IMRT). As 3DRT and IMRT improved, the advantages of proton therapy came into sharper focus.[28]

Employing CT scans and MRIs, radiation oncologists enhanced their powers of virtual simulation, which makes for greater accuracy in defining precisely the contours of the organs to be irradiated and the healthy organs to be avoided. With that precise, three-dimensional image, the oncologist can design the ballistics of the beam or beams and their intensity. The use of 3DRT permitted increased dosages to tumors while at the same time reducing dosages to neighboring healthy tissues. IMRT had similar advantages. IMRT directs several beams of photons at the tumor from different angles. They intersect at the tumor site, producing a three-dimensional treatment field, providing maximum doses to the tumor and smaller doses to surrounding tissue. Both 3DRT and IMRT proved to be clinically effective.

With both technologies, however, multiple radiation beams enter the radiation field, hit the tumor, and exit the field; the healthy tissues lying between the tumor and the entry and exit boundaries of the radiation field still absorb doses that can have long-term, deleterious consequences. The beauty of proton beam therapy is its capacity to unload radioactive particles exclusively at the tumor site. The beam stops at the distal margin of the tumor without exiting through normal tissues. Because proton therapy had advantages over other technologies, Cox became an advocate and convinced Mendelsohn of its virtues. With great fanfare, M. D. Anderson in 2000 announced its entrance into the new field. A group of private investors and companies set out to build a proton therapy center operated by M. D. Anderson radiation oncologists and physicists. All patients treated at the facility were placed on clinical research protocols to evaluate the efficacy of the new form of therapy.[29]

Before the Proton Therapy Center delivered its first beam, however, critics raised economic, medical, and ethical concerns. Some pointed out that 3DRT and IMRT could achieve the gold standard of conformational therapy at less than half the cost. Advocates of proton therapy countered that proton therapy might be superior to IMRT by significantly increasing dosages to the tumor site while reducing dosages to surrounding tissues. In tumors of the base of the skull, eye, spine, and brain, where critical structures needed protection, proton therapy allows maximum dose and maximum precision. Children with radiation-treated cancers often develop physical and mental deficits as well as new cancers. The more accurate the radiation beams, the less the exposure of neighboring tissues and the less severe the long-term side effects.[30]

Critics neglected to mention that the merits and demerits of proton therapy could not really be established until a large number of patients had been treated by it. With its gilt-edged tradition in radiation oncology, M. D. Anderson yearned to employ the latest technological innovation in radiotherapy, just as Gilbert

Fletcher in the 1950s and 1960s had turned away from the old x-ray technologies in favor of the cobalt-60 and the betatron. Fletcher's results made history. Proton therapy might be similarly successful. Or it could yield less than expected, such as the failed cyclotron experiments at M. D. Anderson in the early 1980s. M. D. Anderson intended to be on the front lines of investigating the therapy. "I'm convinced," Mendelsohn iterated, "that proton therapy is at least as good as standard radiation therapy, and there are preclinical and clinical data that it very possibly will be better."

Critics also talked money. Proton therapy requires powerful, expensive particle accelerators usually confined to advanced physics laboratories. At $125 million, the center represented a huge capital investment, and securing sufficient return would rest on the shoulders of patients and insurance companies. M. D. Anderson estimated, for example, that each proton treatment for prostate cancer would cost $850, compared to $308 for an IMRT treatment and $150 for standard radiation. Total average cost to treat a patient with protons would be $37,000, compared to $29,000 for IMRT and $21,000 for regular photon therapy. M. D. Anderson officials countered that every major innovation in the history of cancer medicine had initially attracted scientific skeptics and concerns embedded in medical economics. James D. Cox insisted that proton therapy offered more precision than IMRT and therefore more benefit to a wider range of patients. "I went from being a skeptic fifteen years ago," Cox was reported to have said, "to letting the science drive my decisions."

Some critics seemed blind to a fundamental reality of medical economics. American medicine in the 2000s, especially academic medicine, had become a highly competitive sector of the economy, where maintaining or increasing market share could guarantee the future of an institution or spell its decline. If economic history had proven anything, it had demonstrated the value of exploiting technological innovations. Only then could efficiencies be realized and market shares maintained.

Many regional oncologists feared that proton therapy might shrink their patient pool. Americans often purchase medical care like they buy the latest laptop, spending money on gadgets not really needed. If M. D. Anderson marketed proton therapy as unconditionally superior, patients might flee local radiation treatment facilities and put up with traffic and parking problems in the Texas Medical Center. M. D. Anderson oncologists, so the rant went, would promote protons for medical reasons, while M. D. Anderson accountants might do the same for business reasons. One Houston physician worried, "They may be able to take advantage of economies of scale. If the proton center gets enough volume, M. D. Anderson may [be] able to reduce the cost of protons to levels competitive with my charges for IMRT and x-rays. They will have a monopoly against which none of us can compete." Such an argument, of course, undermined those questioning it. As more and more patients were treated, the cost per treatment would fall when economies of scale kicked in. Mendelsohn and Cox believed that proton therapy had an excellent chance of proving economically and medically advantageous. In 2006, the Proton Therapy Center began treating patients.[31]

The six centers of excellence in the McCombs Institute were a natural outgrowth of an emphasis on collaborative clinical trials that began at M. D. Anderson in the 1980s. It really started with the early decision to physically organize specialty clinics around each type of cancer (breast cancer, for example) rather than around physicians' particular specialty, such as surgery. Surgeons, radiotherapists, medical oncologists, and others teamed up together to apply their specialized expertise to patients with the type of cancer they knew best, which also meant increased opportunities to develop collaborative clinical research protocols for patients with all stages of that type of cancer. Under Mendelsohn, the process of creating the multidisciplinary care centers was completed. It required redesigning most of the institution's clinics, no small undertaking.

The next stage involved expanding this model by focusing translational laboratory research on specific cancers. The NCI had encouraged this for many years, and in 2000 it announced a new type of research grant, which funded specialized programs of research excellence (SPOREs) designed to develop new targeted anticancer agents and bring them into clinical trials. M. D. Anderson competed successfully for ten of the nation's sixty SPOREs.

In 2007, Mendelsohn came up with plans to expand on the successful collaborative research model for the McCombs Institute by creating additional Centers of Excellence, many of them virtual rather than requiring a new research building, but in all cases promoting collaboration across traditional departmental structures. Most of M. D. Anderson's silos have been broken down.

Whether the next generation of M. D. Anderson's contributions to the war on cancer will come primarily from centers of excellence or from the likes of geneticist Benoit de Crombrugghe, molecular pathologist Ralph B. Arlinghaus, and biochemist William Klein—stellar scientists with highly competitive, investigator-initiated projects—is a question to be answered by a future historian of M. D. Anderson. Mendelsohn is betting on both approaches and points out that today many investigators at M. D. Anderson participate in collaborative, targeted projects while they also pursue their own favorite projects. "This is the trend in academic medicine," Mendelsohn believes. "My philosophy is that each member of our faculty is committed to both the institution's mission and their own personal missions, and we will achieve greater success if we support and encourage both missions."

14

Tipping History?

We may need a pharmacy with five hundred drawers to it, complete with RNA molecules, low molecular weight compounds, antibodies, replacement genes, and vaccines. We will type the cancer cell, identify its genetic abnormalities, and concoct a cocktail of therapies targeted at the malfunctions. Genes, not tissues, will steer the treatments.

—JOHN MENDELSOHN, 2006

Lightning struck the Stringer family twice. Both of their young sons were diagnosed with acute lymphocytic leukemia, Garrett in 2002 at the age of seven and Gatlin in 2004 at the age of three. Garrett responded quickly to the chemotherapy and went into remission, but getting Gatlin there proved more difficult, though he too eventually entered the promised land. Adjuvant therapy continued for both boys. Had it been 1957, before the curative work of Emil J Freireich, Emil Frei, and Don Pinkel, they would have gone the way of Robin Bush, leaving behind parents, grandparents, and siblings before anyone could catch a breath. Instead, the Stringers turned to their God and to M. D. Anderson, praying that earthly experts would use every tool to redeem the boys and that in heaven the Lord would unleash spiritual technologies of his own. Gatlin and Garrett survived and thrived, two rough-and-tumble East Texas boys caught, but not trapped, in a cancer blizzard. In 2006, they graced the cover of the Children's Art Project catalog, with family and physicians anticipating normal lives for them. In January 2008, Garrett finished his treatments, the first time since 2002 that a Stringer had not been on chemotherapy.[1]

—⟨⟨⟨⟨∫⟩⟩⟩⟩—

In 2004, as Mendelsohn proceeded with the development of the McCombs Institute and the Mays Clinic, he accelerated an earlier initiative. The cancer prevention program now had a symbol worthy of LeMaistre's original intentions. In January 2005, Mendelsohn unveiled the 384,000-square-foot Cancer Prevention Building, the new home to M. D. Anderson's Division of Cancer Prevention and Population Sciences, including the departments of Clinical Cancer Prevention, Behavioral Science, Epidemiology, and Health Disparities Research. The fledgling effort of 1979 under the leadership of Guy Newell and then the expansive efforts of his successor Bernard Levin had earned M. D. Anderson an aisle seat in the front row of cancer prevention.[2]

During the preceding decade, Bernard Levin had taken the lead, not only at M. D. Anderson but nationally as well. His group had revealed a great deal about inheritance patterns in colorectal cancer and about familial adenomatous polyposis (FAP), a genetic syndrome in which thousands of polyps appear in the colon and rectum, some of which eventually progress into malignant tumors. FAP is associated with the overexpression of the Cox-2 enzyme, and in collaboration with other investigators, Levin demonstrated the efficacy of nonsteroidal inflammatory drugs, particularly Celebrex, in inhibiting the enzyme and attenuating the progression of the polyps. The discovery provided a potent new technology for chemoprevention, and in 1999, the FDA approved it for the treatment of FAP. Levin suffered a setback in 2004, however, when Merck pulled its Cox-2 drug Vioxx from the market because of cardiac risks. Mendelsohn observed, "We always knew that putting a patient on a drug for many years entailed the risk of unforseen consequences. Our people will work it out. The more we learn about the circuitry of a cancer cell, the more able we become to prevent it."[3]

As an administrator, Levin had implemented a comprehensive vision of cancer prevention that included, in his words, "clinical practice and research in many sites of epithelial neoplasia; the molecular study of neoplasia, risk, and drug activity; classical and molecular epidemiology; the biology of tobacco- and obesity-related neoplasia; and the behavioral and nutritional sciences." Between 1995 and 2005, the number of faculty working in the field at M. D. Anderson had increased from nine to forty-three and the number of research projects from

23 to 140. Levin was no longer the chair of a small department but the vice president and head of a division. Cancer "prevention," Levin argues, "is very broad. It is not just prevention of cancer development but includes advances in diagnosis and treatment that reduce suffering from cancer by controlling pain and meeting psychosocial needs."[4]

Infection control and immunology constituted a new prevention frontier. The search for viral causes of cancer had finally yielded important information. The hepatitis B virus plays a direct role in many liver cancers, and public health campaigns in Taiwan to vaccinate children against it had yielded dramatic declines in mortality. Treating hepatitis B aggressively did the same. The human papilloma virus causes many cancers of the cervix, anus, throat, penis, vulva, vagina, and nasopharyngeal tissues, and the development of a vaccine against the virus promises to reduce the damage and the number of deaths from these lesions. Management of AIDS through protease inhibitors reduces the number of cases of Kaposi's sarcoma, AIDS-related lymphomas, and cancers of the anus and penis. Stomach cancers could be reduced through more effective treatment of the *H. pylori* virus.

Levin also cites wider acceptance of "practical cancer prevention," especially procedures targeting cancer precursors, such as colonoscopies and polypectomies to remove precancerous polyps and stall colon cancer, and prophylactic organ resection among high risk patients, such as FAP patients and women expressing the BRCA1 genetic mutation, a reliable harbinger of breast and ovarian cancer. Prophylactic mastectomies, hysterectomies, and colostomies became more common among high-risk individuals. Between 1998 and 2003, for example, the number of women with cancer in one breast opting for double mastectomies increased by 150 percent in the United States, even in the absence of data that such drastic preventive strategies extended life span.[5]

Cancer prevention also embraced molecular medicine, not only in the early detection of cancer but in the employment of targeted therapies. Clinical trials have established the efficacy in cancer prevention of tamoxifen, which takes aim at estrogen receptors in the breast; celecoxib, which targets cyclooxygenase-2 in the rectum and colon; and finasteride, which homes in on 5α-reductase in the prostate. For Levin and his colleague Scott Lippman, the next generation of targeted therapies will be useful for both prevention and treatment because fundamental to cancer prevention is the principle that cancer is "both a multistep process, which involves genetic and epigenetic alterations driven by genomic instability ultimately to cancer development, and a multifocal process,

which involves field carcinogenesis and intraepithelial clonal spread." Molecular pharmaceutical development has, therefore, "blurred the distinction between cancer prevention and [cancer] therapy."[6]

Finally, Levin built one of the nation's strongest programs in disparities research to explore why minority groups display higher cancer incidence and mortality rates. Between 1998 and 2002, for example, the annual age-adjusted death
rate from prostate cancer was 67.1 per 100,000 men among blacks compared
to 27.1 in whites. For more than two decades at M. D. Anderson, Lovell Jones, a
basic research scientist exploring endocrinology and tumor biology, had raised
the issue of cancer's impact on minority groups, where racism, poverty, and
lack of medical insurance limit access to good treatment. Under Levin, the Division of Cancer Prevention had come to include the Department of Health
Disparities Research, which dedicated itself to research on how to improve access.

That was also a major concern of the American Cancer Society (ACS), which
in 2007 began to throw millions of dollars into a public health campaign addressing the plight of Americans without health insurance. Although the ACS
tried to maintain an apolitical posture, the commercials on television, radio,
and the Internet and in the print media left little doubt that lack of health insurance for 47 million Americans posed the major barrier to continued progress
against the disease. "Reducing suffering and death from cancer," said Richard
C. Wender of the American Cancer Society, "may only truly be possible if all
Americans are able to visit their doctor . . . for quality cancer treatment if and
when they need it."[7]

On September 21, 2006, Mendelsohn delivered a state-of-the-institution address
to employees. The growth in ten years had been phenomenal. Since 1996, M. D.
Anderson revenues of $683 million had surged to $2.24 billion. Philanthropic
gifts jumped from $22.3 million in 1996 to $134 million, with a total of $878 mil-
lion for the decade. The research budget swelled from $121 million to $342 million, and the square footage of the physical plant, from more than 3.362 million
in 1996 to over 8.77 million in 2006. In 1996, M. D. Anderson employed 7,919
people; in 2006, 15,957 people. The number of faculty members rose from 694 to
1,321. The number of patient registrations for therapeutic clinical trials climbed
from 3,466 to 9,865, and the number of patients in all types of clinical trials from
5,087 to 23,158, giving M. D. Anderson the largest clinical cancer research program in the nation. M. D. Anderson served a total of 45,465 individual patients

in 1996 and 73,933 in 2006. External federal research funding had risen from $42.8 million to $161 million, and M. D. Anderson secured more grants and awards from NCI than any other institution. Mendelsohn's primary role has been to lead M. D. Anderson during an explosion in scientific knowledge and the greatest growth in its history. At the same time, he continues to lecture around the world on targeted therapy with tyrosine kinase inhibitors and writes extensively on the subject, which has been his central research interest for so long.[8]

In 2007, *U.S. News & World Report* released its annual rankings of the best hospitals in the United States. Between 2000 and 2006, M. D. Anderson had been rated number one on four occasions, and in 2007 again edged out Memorial Sloan-Kettering for the top spot. Somewhere in the heavens, R. Lee Clark must have been celebrating. Back in 1948, he had told Theophilus Painter, "Memorial Hospital is the only one of its kind in this country. While there are other cancer hospitals, their progress is limited in some of the fields. Memorial is the only one that has a complete program of Prevention, Research, Education, and Treatment." In 2007, so did M. D. Anderson.[9]

Mendelsohn is proudest of the tremendous growth in clinical trials to investigate potential new therapies for cancer. "What is most unusual about M. D. Anderson," he says, "what we are acknowledged to lead the world in, is translating scientific discoveries into the clinic for the benefit of patients. The expansion has involved a commitment by the entire faculty." Most trials with new drugs and new combinations are led by clinical investigators in the Division of Cancer Medicine, a division Mendelsohn created in 1998, splitting off a separate Division of Internal Medicine. To lead Cancer Medicine, he enlisted the energy and vision of Waun Ki Hong, a clinical scientist already on the M. D. Anderson faculty, after a nationwide search had confirmed Hong as the best candidate. Under Hong's leadership, clinical research and the infrastructure to support it grew rapidly, mirrored by increased institutional investment in supporting clinical trials and biostatistics.[10]

The earliest clinical studies of new therapies are called phase 1 trials, and Hong asked Razelle Kurzrock to spearhead expansion of these trials in the new Department of Investigational Cancer Therapeutics. In 2006–2007, the department carried out sixty-eight phase 1 trials, one-third involving drugs from pharmaceutical companies that were being administered to humans for the first time. And M. D. Anderson in 2007 had twenty-three anticancer agents invented by its own scientists in phase 1 trials or waiting for FDA approval to begin trials, with over a dozen additional drugs under development in its research laboratories. Mendelsohn points to this as a massive collaborative effort, involving the Center for Targeted Therapy and investigators from multiple departments and multidisciplinary care centers.

M. D. Anderson had also become a degree-granting institution. As far back as the mid-1950s, to put M. D. Anderson on a par with the medical schools, Lee Clark had longed for the day when M. D. Anderson, as a free-standing unit of the University of Texas, might add to its educational mission the awarding of formal degrees. Since 1963, M. D. Anderson faculty had taught with faculty from the nearby UT Health Science Center at Houston at the Graduate School of the Biomedical Sciences (GSBS) as full faculty members, but when students received their master's and doctoral degrees, the diplomas did not bear the M. D. Anderson imprint.

Mendelsohn began to lobby in the late 1990s for the ability to offer bachelor's degrees. Cytologist Michael J. Ahearn had spent most of his career at M. D. Anderson deeply involved in educational programs, and in 1998 he was named as the first dean of Allied Health Programs, which consolidated existing training programs in cytotechnology, cytogenetic technology, medical dosimetry, medical technology, and radiation therapy. In June 2000, the Texas legislature enabled M. D. Anderson to award baccalaureate degrees in those five fields. Within one year, there were four applicants for each of the forty-five slots, and at commencement in 2001, thirteen students received bachelor's degrees, with the diplomas bearing the name of M. D. Anderson and the signature of John Mendelsohn. At the same time, M. D. Anderson secured from the Texas Higher Education Coordinating Board permission, in conjunction with the UT Health Science Center, to award master's and Ph.D. degrees in the biomedical sciences. By 2006, when it won accreditation from the Southern Association of Colleges and Schools (SACS), the M. D. Anderson School of Health Sciences offered bachelor's degrees in eight programs, and students were also studying for advanced degrees in twenty-one areas. "SACS accreditation means that M. D. Anderson is now recognized as a true academic institution in the eyes of the state, the University of Texas System and colleges across the nation," noted Stephen P. Tomasovic, senior vice president for academic affairs.[11]

M. D. Anderson added other trappings of a full-blown university in 2004 with the dedication of the George and Cynthia Mitchell Basic Sciences Research Building. George Mitchell, the oil and real estate tycoon and founding father of the Woodlands, and his wife, Cynthia, had donated $20 million toward the construction of the seventeen-story basic science research edifice. Cynthia

Mitchell's brother Alando J. Ballantyne, the highly skilled head and neck surgeon and the first surgical resident in M. D. Anderson history, had, in the early 1960s, planted in the Mitchells a reverence for M. D. Anderson. The administrative staff and classrooms of the GSBS were then relocated to M. D. Anderson's George and Cynthia Mitchell Basic Sciences Research Building. The GSBS was a shared enterprise of M. D. Anderson and the University of Texas Health Science Center at Houston. The relocation brought along five hundred graduate students affiliated with the two institutions and added visibility to M. D. Anderson's efforts in the basic sciences. Now, M. D. Anderson was home to a university that granted bachelor's, master's, and doctoral degrees.

The remarkable growth of M. D. Anderson was fueled by funds that Mendelsohn gathered together from a variety of sources. A major contribution came from the margins earned by patient care reimbursement. The UT System and the state government provided some dollars for construction. The business group led by Leon Leach obtained UT regents' approval to incur debt by borrowing from banks. And philanthropic contributions grew substantially. Critical to the expansion of gifts to M. D. Anderson were successful efforts to broaden the size and scope of the board of visitors. Lee Clark had formed this group of community members interested in supporting M. D. Anderson. Mickey LeMaistre had nurtured it. Mendelsohn and Patrick Mulvey, vice president for development, expanded the vision for the board of visitors to go beyond the boundaries of Texas. Among a series of very effective and dedicated board chairs was former president George H. W. Bush. His seventy-fifth and eightieth birthday celebrations provided the opportunity to raise most of the funds for the George and Barbara Bush Endowment for Innovative Cancer Research, which supports pioneering research and helps create attractive recruitment packages for new faculty.

When the board of visitors approached two hundred members, such old-timers as Thomas Anderson, a nephew of Monroe D. Anderson and longtime member of the board, felt some loss of intimacy in the expansion. In 2006, he joked, "Pretty soon we will have to hold the meetings of the board in the Astrodome, but Mendelsohn's results speak for themselves. He's really done marvelous things."[12]

Mendelsohn decided that in a world that was becoming flat, R. Lee Clark's vision of a global impact for M. D. Anderson needed rekindling. After seven years of visits to Madrid by dozens of M. D. Anderson physicians, nurses, and administrators, M. D. Anderson International–España began to emerge as a strong presence in Spanish oncology. Putting emphasis on the mission statement "to eliminate cancer in Texas, the nation and the world," and drawing on the interests of dozens of faculty born or trained abroad, Mendelsohn persuaded management and faculty leaders of the importance of establishing academic collaborations with major cancer centers worldwide and of developing a named presence for M. D. Anderson at a few selected facilities on other continents. Sister institution agreements were signed with a series of cancer programs, and M. D. Anderson Global was formed to explore opportunities for a presence in countries seeking help in building a modern cancer hospital.

A series of meetings involving faculty, administrators, and employees worked diligently to create an institutional strategic plan for 2000–2005. Its update for 2005–2010 lists seven strategic goals centered around patient care, research, education, prevention, being an employer of choice, mission-driven collaborations, and stewardship of resources. The executive team paid particular attention to the fifth goal, advancing M. D. Anderson as an employer of choice. New seminars on leadership training, career counseling, courses for personal skill advancement, mentoring, and an employee health initiative all were designed to improve interactions between employees and to promote career pathways for each individual working at M. D. Anderson.[13]

———

In 2006, a stunning announcement from the American Cancer Society encouraged beleaguered oncologists. In 2003, for the first time since the government began to collect such statistics, the absolute number of Americans dying of cancer declined, even though the general population had increased, and the baby boomers had continued to age. The decline was modest to say the least, from 557,271 people in 2002 to 556,902 in 2003, a total of only 369, hardly much to cheer about and small enough to elicit concerns about statistical anomalies. In 2004, however, the trend continued, with 3,000 fewer deaths, a number less likely to be a methodological quirk. The decline was in actual deaths, not merely in the mortality or the age-adjusted rates. "This is an important achievement and milestone," said Elizabeth Ward of the American Cancer Society. "Two successive declines in deaths, the latter ten times more than the first, gives us confidence the trend is real." At M. D. Anderson, Maurie Markman, vice president for clinical research, hailed the report: "There's a real curve at work—the trend is unmistakable."[14] Mendelsohn attributed the ten-year decline in cancer death rates per 100,000 and the more recent decline in the absolute number of deaths to a combination of three factors: reduction in smoking by over 50 percent along with other cancer prevention measures, improved technologies which permit earlier diagnosis and better staging of the extent of disease, and development of new therapies.[15]

Charles A. LeMaistre expressed cautious optimism, much like a general engaged in a complicated battle. "It appears to be good news," he said, "but it's too soon to draw firm conclusions. We have to wait a while to see if these apparent trends hold." Caution had always been the best policy in making predictions about cancer.[16]

Incontrovertible, however, were declining death rates for many cancers. Among men, mortality rates had fallen for cancers of the pancreas, prostate, lung, brain, bladder, kidney, lung, stomach, and oral cavity, as well as for multiple myeloma, non-Hodgkin's lymphoma, and leukemia. Among women, mortality rates for systemic cancers had fallen as well for malignant tumors of the breast, colon, rectum, cervix, stomach, and kidney. During the 1990s, overall cancer mortality rates had dropped by an average of 1.1 percent a year, a number that accelerated to 2.1 percent annually between 2002 and 2004. Julie Gerberding, head of the Centers for Disease Control and Prevention, considered the data as certification that "important progress" was being made against cancer, and her colleague David Espey asserted, "We should expect to see continuing declines."[17]

Those pressed to decipher the decline turned to the four major killers—lung cancer, colon cancer, breast cancer, and prostate cancer—which accounted for most of the slide, along with less significant but very real drops in the number of deaths from head and neck cancers.

For lung cancer and cancers of the head and neck, declines in tobacco use shaved mortality. Between 1963 and 1998, the annual per capita consumption of cigarettes in the United States fell from 4,345 to 2,261. The Americans for Nonsmokers' Rights credited the campaign to ban smoking in the workplace. Science confirmed the claim. In 2006, U.S. Surgeon General Richard H. Carmona reported that levels of cotinine, a biological marker for secondhand smoke exposure among nonsmokers, had fallen 70 percent since the late 1980s and that the proportion of nonsmokers with detectable levels had been halved.[18]

Multiple factors explained the erosion in breast cancer deaths. Expanded use of mammograms had produced earlier diagnoses and earlier treatments. The identification of women with the BRCA1 and BRCA2 genes had improved surveillance. The wider use of such targeted therapies as tamoxifen and Herceptin, in combination with more traditional chemotherapy regimens, had postponed recurrences in women whose cancer cells exhibited estrogen receptors. According to Mark Pegram of the UCLA Jonsson Comprehensive Cancer Center, Herceptin heralded a new day in treatment. "In the past," he noted, "we used toxic drugs that kill any rapidly dividing cells, good or bad. The next step was drugs such as tamoxifen, which target estrogen receptors. The new generation, like Herceptin, targets actual flaws in cancer cells." For breast cancer patients, the proliferation of anticancer hormonal therapies helped. To tamoxifen has been added the aromatase inhibitors, which extend the lives of many women. Aromatase is an enzyme involved in the production of estrogen, and the drug reduces the amount of estrogen circulating in the body and inside the cancer cell. A host of drugs now target aromatase, including trioxifene, toremifene, droloxifene, and 4–hydroxyandrostenedione, and they extend disease-free survival. Such new hormonal treatments as leuprolide, buserelin, triptorelin, and goserelin help as well.[19]

The abandonment of hormone replacement therapy (HRT) may have accounted for part of the decline. In 2002, the Women's Health Initiative (WHI) study was halted when data indicated that more women using estrogen/ progesterone developed breast cancer than those not on HRT, and the news spread through virtually every media outlet. Large numbers of women abandoned HRT, and some epidemiologists speculated that pulling the hormones caused small tumors to shrink or even to disappear. Within a year, the incidence of breast cancer had fallen 7.2 percent, from an expected 200,000 new cases to approximately 186,000. Especially pronounced was the decline in women over the age of fifty, those most likely to have tumors with estrogen receptors. The news was first reported on December 14, 2006, at the San Antonio Breast Cancer Symposium. Peter Ravdin, a research professor and biostatistician at M. D. Anderson, remarked, "It is the largest single drop in breast cancer I am aware of. Something went right in 2003, and it seems that it was the decrease in the use of hormone therapy, but from the data we used we can only indirectly infer that is the case." Donald Berry, the M. D. Anderson biostatistician responsible for the study, expressed unfettered optimism: "[The] incidence of breast cancer had been increasing in the 20 or so years prior to July 2002, and this increase was over and above the known role of screening mammography. HRT had been proposed as a possible factor, although the magnitude of any HRT effect was not known. Now the possibility that the effect is much greater than originally thought all along is plausible, and that is a remarkable finding."[20]

In 2003, colorectal cancer deaths dropped by 1,110 in men and 1,094 in women.

"Screening has made a very important difference in breast cancer and colorectal cancer," said Elizabeth Ward of the American Cancer Society. Others cited
the so-called Katie Couric effect. On March 7, 2000, Katie Couric, the popular
cohost of the NBC *Today Show*, underwent a colonoscopy. Two years earlier,
colon cancer had killed her husband, Jay Monahan. Colon cancer is preventable if caught early through a colonoscopy, in which a physician probes the large
intestine with a scope. Couric made control of colon cancer a personal crusade,
and to demystify it, she underwent the procedure in front of a national audience. Because colon cancer takes many years to develop, Couric's demonstration could not explain the 2003 drop in deaths. But in the wake of the broadcast, the number of colonoscopies increased by 20 percent in the United States.[21]

To account for the drop in prostate cancer deaths, most experts looked to
improvements in surveillance and early diagnosis, especially PSA tests. Controversy still burned about the PSA, but to explain a raw decline in the number of deaths in the absence of any dramatic new treatments and against the
backdrop of a rapidly growing and rapidly aging American population, earlier
treatment carried the weight of logic. Emil Freireich and some others had doubts.
"We are operating under an enormous delusion that early detection will improve the cure rate," he remarked. "Most cancers are systemic from the very
beginning, and only systemic treatments are going to cure them. Early detection usually discovers cancers that won't kill anyway." Other oncologists, however, insist that early detection and early treatment, even if they do not lead to
cures, do lengthen survival time, and if hundreds of thousands of cancer patients are surviving for longer periods of time, the number of deaths in any given
year is bound to decline.[22]

————

John Mendelsohn looks optimistically to the future of oncology. Controlling
cancer is in the offing, not today or tomorrow but certainly within the century.
When R. Lee Clark in 1971 marked 1976 on his calendar as the year to cure cancer,
he would have scoffed at the notion that the real date might be closer to 2076,
the nation's tricentennial. For Mendelsohn, cancer is no more likely to be banished from the planet than pneumonia or tuberculosis, but it will be controlled
with an arsenal of therapies targeting not just the tissues in which a cancer orig-
inates but the underlying genetic abnormalities, perhaps a half dozen or more,
for each person's cancer, that give rise to uncontrolled cellular proliferation and
spread. Abnormalities in those genes and other products can express themselves

in a variety of ways. Take glioblastoma multiforme (GBM), brain cancer's death sentence. A team of M. D. Anderson neuro-oncologists led by Peter A. Steck in 1997 learned that more than 90 percent of people with GBM possess a gene designated MAC1, which has coding region mutations, suggesting a strong candidate for a tumor suppressor gene at chromosome 10q23.3. The gene's loss of function seems related to the oncogenesis of GBM. The team also learned of its presence in some kidney, breast, and prostate cancers.[23]

The p53 tumor suppressor gene, mapped to chromosome 17, is one of the most studied in biomedicine. In the early 1990s, Louise C. Strong at M. D. Anderson demonstrated that changes in the p53 tumor suppressor gene occur as acquired mutations in many patients as well as inherited mutations in cancer-prone families, and in February 1995, M. D. Anderson's Jack A. Roth carried out the first successful correction of a p53 gene. Since p53 is expressed in cancers of the lung, head and neck, bones, prostate, bladder, sarcomas, and brain, careful investigation of its signaling pathways, Mendelsohn believes, may lead to progress.

John Mendelsohn casts a long look back to the past and an equally long view into the future. When R. Lee Clark studied at the Pasteur Institute in Paris in the late 1930s, scientists were beginning to ferret out the pathogenesis of many infectious diseases, and as that process accelerated, an increasingly complex variety of drugs were synthesized to treat them. Mendelsohn notes that in a clinical laboratory today, a technician determines the exact nature of an infectious disease, and the clinician chooses, from an array of antibiotics or chemotherapies, the one or the combination most likely to arrest and to squelch the malady. Sometimes a cocktail is required for the most virulent infections, but resting in the drawers of the pharmacy are the remedies. Clinicians open the drawers, select the proven drugs, and cure most patients.

Mendelsohn anticipates a similar future for oncology. Until recently, he says, "Progress in oncology has generally been empirical in nature. Don't misunderstand me. I'll take all the good empirical data I can get, but we now know enough about uncontrolled cellular proliferation to design targeted therapies." With thirty thousand genes in the human genome, Mendelsohn sees only four to five hundred that are relevant to cancer, and perhaps only fifty to one hundred crucial to it.[24]

Much of his confidence arises from the work of cell and molecular biologists investigating the mutations that lead to the production of oncogenes, which gain abnormal functions, and suppressor genes, which experience loss of function. Mendelsohn agrees with Douglas Hanahan of the University of California, San Francisco, and Robert Weinberg of the Whitehead Institute for Biomedical Research at MIT, who suggest that "research over the past decades has revealed a small number of molecular, biochemical, and cellular traits—acquired capabilities—shared by most and perhaps all types of human cancers. . . . Virtually all mammalian cells carry a similar molecular machinery regulating their proliferation, differentiation, and death. . . . We foresee cancer research developing into a logical science, where the complexities of the disease . . . will become understandable in terms of a small number of underlying principles."

Cancer cells express six properties essential to their proliferation—(1) the acquired growth signal autonomy, which allows them to generate their own growth signals and reduces the need for stimulation from the outside environment; (2) an acquired insensitivity to the antigrowth signals present in normal cells; (3) an acquired resistance to the normal processes of apoptosis, which inhibits the rate of cell attrition; (4) the detachment of a cell's growth program from the normal signals in its environment; (5) the acquired ability to sustain the generation of new blood vessels to keep the growing tumor nourished; and (6) the acquired talent to exit the primary mass and make room for themselves in a new location. Success at metastasis relies on all six of the irregularities as well as other common characteristics, including communication between the cancer cell and the cells of the microenvironment it invades.[25]

Jack A. Roth, former chair of the Department of Thoracic and Cardiovascular Surgery and a cell biologist at M. D. Anderson, likens a cancer cell to the circuit board of a computer. "If you look at an integrated circuit on a circuit board for a computer, that gives you some understanding of how these various pathways interact with each other." Cancer develops resistance to all drug therapies that block only one pathway because the cells enjoy redundant pathways to achieve the same function and continue the process of proliferation. "It's probably going to turn out to be not just one pathway and one target," Roth continues, "but various groups of pathways or targets that need to be interrupted before we are going to make a difference therapeutically."[26]

During the next century, pharmacologists will synthesize a host of new drugs to treat cancer. Hundreds of magic bullets will have to be designed selectively to target the hundreds of gene products controlling cell proliferation, so that the proper cocktail of six or so therapeutic bullets can be assembled to attack the cancer in a particular patient. Mendelsohn foresees the day when oncologists will possess hundreds of targeted therapies that home in on the genetic abnormalities of an individual's cancer and the protein products of these genes. "We may need a pharmacy with five hundred drawers to it, complete with RNA molecules, low molecular weight compounds, antibodies, replacement genes, and vaccines. We will type the cancer cell, identify its genetic abnormalities, and concoct a cocktail of therapies targeted at the malfunctions. Genes, not tissues, will steer the treatments."[27]

To some, Mendelsohn's vision of the future of oncology, with its hundreds of targeted therapies, might seem ambitious until considered in the light of history. When Sidney Farber and Cornelius Rhoads laid the footings of medical oncology just after World War II, they possessed only one anticancer therapy—nitrogen mustard. Today, a vast array of anticancer drugs, approved or under study, stack the shelves and fill the drawers of medical oncologists, including the traditional chemotherapy agents and the new targeted therapies. Were we to kidnap from the last century the first medical oncologists and take them into the M. D. Anderson pharmacies today, they would find the pharmacopeia of modern oncology astonishing. The next century will vastly expand the arsenal. And as in the past, new therapies will join with surgery, radiation therapy, and chemotherapy in multidisciplinary combinations.

All around him today, Mendelsohn sees the future unfolding. The case of Samuel Hassenbusch, a neurosurgeon at M. D. Anderson, drew media attention from around the world. The story is simply irresistible. Headaches, as they do for most brain surgeons, prompted concerns about a brain tumor, and mostly to self-demonstrate what he thought to be a case of hypochondria, Hassenbusch in May 2005 underwent an MRI. He was in the x-ray reading room awaiting the report on his MRI when he glanced at the view box and said, "That patient has a large glioma." The MRI revealed a tumor half the size of a banana in the right temporal lobe, and it displayed the angry features of a deadly glioblastoma multiforme (GBM). On closer examination, Hassenbusch discovered that his name was on the film. "The median survival is one year," Hassenbusch recalled. "My life flashed in front of me."

After surgery to remove as much of the tumor as possible, Hassenbusch heard from colleagues in the Department of Neurosurgery. They posited two options, a standard treatment of radiation and chemotherapy using the drug Temodar, or radiation followed up by immunotherapy employing an experimental vaccine protocol. In serum tests, a telltale tumor marker showed up in his blood; Temodar works best on GBM patients expressing the marker. Hassenbusch went with both of the possible options, a decision that required a priori approval from an M. D. Anderson Institutional Review Board and FDA. After the radiation, he took large monthly doses of Temodar; when his white blood cell count cratered after twenty-one days, he switched to the vaccine, which restored his immune system. Month after month, the tumor melted in size. Two years later, when other GBM contemporaries were already dead, Hassenbusch was still performing brain surgery and riding his motorcycle. Within three years, the cancer had recurred in his brain, and he died in 2008. But he had defied the GBM odds.[28]

Mendelsohn uses the words "unprecedented" and "spectacular" for Jeffrey Molldrem's work in the Department of Stem Cell Transplantation and Cellular Therapy. In a phase 1 clinical trial in 2004, Molldrem had administered three subcutaneous injections of a 9-amino acid peptide antigen to forty-five patients with myeloid leukemia. All standard therapies had failed, and each patient had a life expectancy of less than a year. Eleven of the patients experienced objective clinical responses, and four enjoyed complete molecular remissions. "We were startled," Molldrem remembers. "Initially we were just trying to see if we could boost immunity to the antigen we had identified—we didn't expect molecular remissions, especially in a phase I trial and in such a refractory group. That's never been described for *any* vaccine." The vaccine has moved into large-scale clinical trials.[29]

Recent immunological advances in managing metastatic melanoma hold out hope. Patients with metastatic melanomas involving major organs have a survival rate of less than seven years, and the only traditional chemotherapy treatment—the drug dacarbazine—produces responses in only 10 percent of cases. In 2004, however, medical oncologist Patrick Hwu came to M. D. Anderson as chair of the Department of Melanoma Medical Oncology, and with him came a new technology developed at NCI. Interleukin-2 (IL-2), a derivative of interferon, stimulates killer T cells to attack melanoma. Fifteen to 20 percent of patients respond to IL-2. Eight percent of those enjoy a durable response. According to Kevin Kim of the Department of Melanoma Medical Oncology, "We can essentially cure some of the patients with advanced disease, but only a small minority have this remarkable response."[30]

The new discipline of epigenetics offers novel approaches to cancer. It seems apparent, in some cancers, that the typical genetic mutations are absent. Instead, an external molecule attaches to the DNA inside a cell, a process labeled methylation, which alters the normal functioning of the gene. The most typical is a hydrocarbon molecule, but others exist, and they appear to be connected to environmental factors, including smoking, diet, and aging. In the Department of Leukemia at M. D. Anderson, hematologists Jean-Pierre Issa and Hagop Kantarjian have demonstrated some success with demethylation, or stripping away the external molecule. Traditional cancer genetics has invested enormous resources in identifying the genetic causes of cancer and then trying to alter the function of the products of these genes.

Issa and Kantarjian, for example, treated seventy-one-year-old Nancy Stanley for myelodysplastic syndrome, a lethal harbinger of leukemia. The two doctors decided to administer decitabine, a drug already at use in an M. D. Anderson clinical trial. Because of its toxicity, the drug had been shelved in the 1970s; Kantarjian and Issa, however, recognized its epigenetic potential—it strips away the DNA methyl tags when administered in much lower doses over a longer period of time. Stanley went into sustained remission. "This is the beginning of a new era of research," says Kantarjian. "Epigenetic therapy is going to have a very large role over the next five to ten years in a large number of cancers." Issa compared epigenetic therapy to old-fashioned chemotherapy: "It's like the difference between war and politics. Instead of bombing the cancer cells, we're trying diplomacy."[31]

In November 2004, the FDA approved the drug Tarceva, an oral tyrosine kinase inhibitor that targeted the same EGF receptor Mendelsohn's research had pinpointed with monoclonal antibody 225 nearly two decades earlier. Charles Gibson, a truck driver and a former smoker, embodied the dream of transforming cancer from an acute to a chronic disease. In 2001, he was diagnosed with inoperable stage 4 adenocarcinoma of the lung. His physician informed him that 42 percent of lung cancer patients die within one year of diagnosis, and that only 15 percent survive for five years. "I will never forget it," Gibson later recalled. "They said I had a lot of cancer cells and it was inoperable." A local oncologist put him on a standard, two-drug chemotherapy regimen, but Gibson failed to respond. Referred to M. D. Anderson, he reported to Roy S. Herbst, chief of thoracic medical oncology. "He came in three years ago," Herbst told a reporter. "He looked just awful." Herbst put Gibson in a clinical trial involving Tarceva, which disrupts cancer cell to cancer cell signals that stimulate growth. In 8 to 10 percent of lung cancer patients with non–small cell tumors, the drug shrinks tumors by 50 percent or more. Gibson was lucky. He experienced a near 90 percent reduction in the size of his tumors, with only some chest acne and skin rashes as side effects. "He had one of those gee-whiz type of responses," said Herbst. "He's an example of how the new wave of therapies can help someone live longer. He's not cured by any means, but he's living with stable lung disease. He will take one pill of Tarceva a day indefinitely."[32]

Perhaps developments in prostate cancer research best illustrate Mendelsohn's outlook on the future. Until just a few years ago, medical oncologists offered few treatments to men with advanced prostate cancer, including chemical castration to reduce testosterone levels or surgical castration— treatments not much different from what in 1953 had earned Charles B. Huggins the Bertner Award.

In 1993, Michael Milken, the former junk bond guru, was diagnosed with an aggressive prostate cancer. His lymph nodes were thick with disease. Physicians gave him twelve to eighteen months. Milken threw himself into an effort to save his own life and the lives of other men. He established the Prostate Cancer Foundation to raise money for research and the Dendreon Corporation to bring new treatments to the market in accelerated fashion. Milken also arrived at M. D. Anderson to go under the care of Christopher Logothetis, a leading expert in prostate cancer.

Under Andrew von Eschenbach and then Christopher Logothetis at M. D. Anderson and at other research centers, oncologists had acquired a better understanding of the natural history of prostate cancer and the pattern of its responses to therapy. Because in 70 percent of autopsies carried out on men over the age of seventy the presence of prostate cancer is evident, many oncologists now see lower grades of prostate cancer in older men as a chronic disease that should go untreated, and new therapies may be converting some aggressive prostate cancers into a low-grade, chronic disease.

Clinical trials of dozens of targeted therapies were under way around the nation; Milken's Prostate Cancer Foundation funded several. Logothetis tailored his treatments to the biology of Milken's cancer, using molecular markers as a kind of pharmacological road map. At Memorial Sloan-Kettering, medical oncologist Howard Scher recalled the time when he had little to offer men with advanced disease. "Now I can talk to them about eight or ten different options. I'm successfully treating patients with advanced disease for four, five, six, eight, or even nine years. That just didn't happen before." Logothetis did the same for Milken. Eighteen months had ballooned into fourteen years, and he had made good use of the time, lighting a fire under prostate cancer research. The targeted therapies drugs were designed to block the signaling pathways that allow prostate cancer cells to metastasize to bones, to send killer T cells on a search-and-destroy mission against malignant cells, and to prod those cells into apoptosis. For the time being at least, Logothetis transformed Milken's prostate cancer from an acute to a chronic disease.[33]

Not everyone in cancer medicine feels that Mendelsohn has accurately divined the future. In fact, some argue that the entire paradigm of searching for genetic mutations and malfunctions and targeting them with smart bombs may be fundamentally flawed. Biochemist Wallace L. McKeehan of Texas A&M University, for example, insists that genetic mutations may in the long run provide few clues to the explanation of cancer and to the design of effective treatments. The entire cancer establishment today, as far as McKeehan is concerned, worships at the altar of genetic mutations, but "there are just a mind-boggling number of mutations associated with cancer. We need some new ideas." Australian geneticist George Miklos echoes such concerns, as does William R. Brinkley, a senior vice president at the Baylor College of Medicine. A cell biologist, Brinkley trained at M. D. Anderson and for years worked there with T. C. Hsu. For Brinkley, future therapies patterned on the gene mutation model are likely to affect only a minority of patients and to affect them only temporarily. "If it could have happened," claimed Brinkley, "it would have already happened." Just how much control genetic mutations exercise in tumor development, critics argue, remains a mystery. McKeehan worries that the entire field of targeted therapies will go the way of the NCI's ill-fated virology program, spending tens of millions of dollars with, in the end, little to show for it. Mendelsohn is no ideologue viewing oncology as an either-or contest of competing theories. "If I were designing [a new NCI research program]," he told journalist Eric Berger, "I'd probably take a holistic approach. For example, let's take a look at DNA methylation."[34]

So is John Mendelsohn an emissary from the future our descendants will inhabit? Or does Charles Darwin still speak from the dust? Will targeted therapies eventually transform the disease by curing even more patients and converting cancer from an acute to a chronic disease for some others, even most? Or is cancer, as Frederick F. Becker remarked, the "Houdini of diseases," still floating atop a reservoir of untapped survival strategies, forever to evoke natural selection and its derivatives of adaptation and replication?[35]

Mendelsohn does not think so, nor does Gloria Belsha (Robertson). In 1963, when she reported to M. D. Anderson with an osteosarcoma of the leg and metastases to her lungs, Gloria was as good as dead, given the state of medical oncology. To the astonishment of R. Lee Clark, however, surgery and experimental chemotherapy removed and dissolved every trace of the disease. In 1976, on a return visit to Houston, she let Clark cuddle her new baby boy. She later went to work at M. D. Anderson. In 2007, Belsha retired from M. D. Anderson having lived a very full life as a cancer survivor of forty-four years. Hundreds of thousands of patients have similar stories to tell, and another generation of faculty and staff is carrying on M. D. Anderson's culture and crusade to make cancer history.

Epilogue

In some ways, being an oncologist must be like spending a lifetime waiting to exhale, anxiously anticipating the outcomes of the latest clinical trial, wondering whether the research, the money, and the treatments will make a difference and the rhetoric of translational research be fully realized. Patients hold their breath too.

At first glance, the Pink Palace hardly inspired in me the hope and sunrise R. Lee Clark envisioned in 1953 when construction workers attached to the new hospital's exterior walls the great slabs of rosy Georgia Etowah marble. Late in 1980, I felt a lump on my left hand, not a mass with the smooth shape of a small marble but irregular and knotty, like a piece of gristle on a cheap cut of beef. The mass could be moved from the right to left and left to right, but not up or down along the line of the tendon, to which it seemed attached. For two months, I casually tugged at the knot like a baby discovering his big toe. It did not seem to change in size, but the "seven warning signs of cancer" reverberated in my head. A general practitioner referred me to Paul Vilardi, an orthopedic surgeon with the bedside manner of a sumo wrestler, who in January 1981 removed the tumor.

The pathologist diagnosed a benign, giant-cell tumor of the tendon sheath. For some reason, the word *benign* brought little reassurance, and I requested a second opinion. Vilardi had trained at the Roswell Park Institute and sent the slides there. Pathologists at Roswell Park confirmed the diagnosis. Still uneasy, I requested a third opinion from M. D. Anderson and learned the hard way of the rivalries in cancer medicine. Vilardi's mood darkened quickly from irritation to agitation and then agitation to anger; he blurted that Roswell Park was better than M. D. Anderson and that only an alarmist would insist on a third opinion. "Listen," I said, "I'm not getting into an argument with you about the relative merits of M. D. Anderson and Roswell Park. It's my hand. I want a third opinion."

Two weeks later, he called for a meeting. A bit sheepishly, Vilardi passed on the diagnosis from M. D. Anderson—a malignant epithelioid sarcoma—and then asked a bit sarcastically whether I wanted a fourth opinion, this time from the Armed Forces Institute of Pathology, apparently the final arbiter in the close calls of pathology. I did. Until then, cancer had always seemed a straightforward disease, dangerous but never equivocal. As a professor of history, not of medicine, I found it unsettling that doctors might argue as much about medicine as historians contested the past. From the institute came the coup de grâce: M. D. Anderson had it right, Roswell Park wrong. I called M. D. Anderson to set up an appointment, only to learn that because of the original legislation founding the hospital, a physician referral was needed. By telephone I asked Vilardi for a referral. He insisted on another visit to his office. Truculent and belligerent, he all but demanded an immediate amputation and warned that M. D. Anderson would recommend radiation. "If you listen to them it might very well cost you your life. Don't be foolish." The badgering concluded, he gave me the referral.

I registered in 1981 as a patient. Nearly 160,000 others had preceded me in M. D. Anderson history. A friend and survivor of embryonal testicular cancer drove me to Houston. Brent promised to help me navigate the hospital's labyrinthine corridors and windowless basement, which he compared to the catacombs of Rome. We drove south along I-45 to downtown, switched to 59, and exited on Fannin. Flower vendors lined the east side of the street. I hardly noticed. We continued south into the cavernous Texas Medical Center. Brent turned left at what was then M. D. Anderson Boulevard and inched through traffic toward the fountain in front of Baylor College of Medicine. At the stop sign, he suppressed a dry heave and explained that for several years after his treatments, upon seeing the glistening marble walls, in Pavlovian style he had vomited onto the street. At Station 19 in the late 1970s, Brent had endured prodigious doses of bleomycin administered according to prescriptions developed by "Megadose" Melvin Samuels. Even years later, the pink exterior evoked not images of hope and sunrise but of the modern art he splattered on the asphalt.

My first glimpse of the Pink Palace was less dramatic but equally vivid. In anticipation of the first visit, and anxious to learn more about the tumor, I had called ahead and talked with pathologist John M. Lukeman, who agreed to examine the tissue samples Vilardi had forwarded. Lukeman, confirmed the epithelioid sarcoma diagnosis and described the tumor as "dark gray and pitted" in texture. When Brent dropped me out on the ramp in front of the hospital, I got an up-close look at the pink marble, pieced tightly together in square slabs resembling giant shower tiles. Streaks of gray cut through the surface of the marble like river rapids tumbling toward the sea. Small black dots pitted the gloomy smears. The sinister image melted quickly into memories of a college bacteriology class, with its microscope and glass slides. The marble slabs morphed into giant slides, the rosy hues representing healthy cells and the pitted grey and dark spots the nasty sarcoma cells that might already be marching out of my hand to other parts of my body. At the time, the Etowah marble betokened death, not life and hope.

All cancer survivors interpret their own disease through the multiple lenses of grade, stage, faith, and temperament, with optimists taking courage amid the direst prognoses and pessimists bemoaning even the best of outcomes. Emotional ligaments connect gut to throat. For weeks, thoughts of death and disability accompanied me everywhere like a stubborn toothache. Several local surgeons in Houston, still wired into the old paradigm of cancer as a local disease and thinking along the lines of William Stewart Halsted, pushed for amputation immediately. To wait another day was to tempt the demons of metastasis.

On that first day at M. D. Anderson, after a morning of paperwork, bureaucratic tangles, and misplaced medical reports, of walking the catacombs, of posing for chest x-rays and nuclear scans of liver and bones, of enduring multiple sticks from phlebotomists, I ended up at Station 80 of the Clark Clinic, surrounded by patients with sarcomas, melanomas, and breast cancers, the waiting room televisions tuned to the favorite soaps of the receptionists. At the time, surgeons stood as gatekeepers at M. D. Anderson, and from their offices patients scattered to the other medical specialties. Now just three years on the job, LeMaistre had not yet worked out the scheduling kinks that had accumulated during the end stages of Clark's tenure. Schedules imploded that first day. The two-hour wait to see surgeon Richard Martin seemed interminable. Then, a nurse led my wife, Judy, and me to the inner sanctum.

As with Freddie Steinmark twelve years earlier, Martin greeted me warmly and immediately put to rest concerns about any tumor deposits in the lungs, bones, or liver. While inquiring about professional and family matters and taking the pulse of my psyche, he determined that my lymph nodes—from groin to neck and everywhere in between—also seemed to be free of disease. Unlike the other surgeons, he expressed no inclination to rush to the scalpel and suggested radiotherapy first, reserving radical surgery only for the worst cases; he assured me that I was not a worst case. Because most cancers are systemic, he explained, the long-term survival was the same for each treatment approach. A surgeon who recommended radiation first! At that moment, R. Lee Clark's legacy of multidisciplinary treatment and salaried physicians embraced me.

At first I assumed that to radiate or to amputate was a life-and-death decision and that a wrong choice portended my demise. Like Freddie, I wanted to keep my limb, but not at the price of my life. Back in 1971, I had read *I Play to Win*. At least I had a choice about radiotherapy or amputation. Freddie had to choose between amputation and nothing. Now, his surgeon was my surgeon, surely a random coincidence but proffering an irrational sense of comfort. After all, Freddie had survived for less than a year under Martin's care. I raised the case of Steinmark, and Martin explained that Ewing's sarcoma was more deadly than my soft-tissue sarcoma.

Martin's nurse made an appointment with radiotherapy, and two days later I sat in an amphitheater listening to Gilbert Fletcher discuss my case with the assembled residents and fellows. Seated in the front row were Robert Lindberg, who ran the soft tissue sarcoma clinic at M. D. Anderson, and Marsha McNeese, a young radiotherapist working with Fletcher. In the early 1980s, the waiting room at each clinic offered brochures describing the cancers being treated there, and I had casually picked up the pamphlet labeled "osteosarcoma." For me, a sarcoma was a sarcoma, but Fletcher quickly set me straight, brusquely snatching it away. "You have a soft-tissue sarcoma, not an osteosarcoma," he announced. "They're different. You're going to be fine. I will cure you." So unambiguous. He then explained to the audience, not to me, the size of the radiation field, the doses of electrons to be programmed into the betatron, and the schedule of fractionated treatments to be delivered. The case demanded exact calibration, enough to wipe out the residual sarcoma cells while avoiding any scarring of the skin that might create adhesions to the tendons below and limit the mobility of my fingers. At the time, the importance of the setting did not register. The audience listened to Fletcher in rapt attention, with respect obvious in their demeanor. With dumb luck, I had drawn the long straw, and now enjoyed, as my physician, the father of modern radiotherapy. Had I known of Fletcher's stature, I might have taken comfort in his assurances.

Selecting radiotherapy over amputation seemed a foregone conclusion. Who wouldn't? But what if Vilardi had it right? Did a wrong choice bode death? I had yet to absorb fully Martin's explanation about the systemic nature of solid tumors. Surgery and radiotherapy might wipe out the tumor at the site of its origin, but the cancer cells already floating in the circulatory system would either fall victim to the immune system or return with a vengeance.

Fletcher handed me over to McNeese, who repeated Martin's logic, reiterating that the five-year survival rates for patients who received radiation mimicked those of patients undergoing amputation. Still, I was skeptical, and she grew frustrated. With her red hair and Celtic roots, McNeese had a short fuse, or perhaps my stubbornness had simply lit it at the stem. She left the examination room and exploded to a resident, "I can't believe it! Does he just want us to cut off his arm?" I was out of earshot, but Judy happened to be in range and immediately made common cause with McNeese. I surrendered. Two days later, a nurse drew with a purple Magic Marker a square target on the top of my hand. Treatments began a few days later—fractionated daily doses to be delivered over the course of six weeks. On many of those days, when head and neck patients gathered in the waiting room, the place resembled the bar scene in *Star Wars*, full of maimed faces decorated in purple lines.

On that first day, as I entered the room housing the betatron, I walked slowly; it hovered over me like a great brown grizzly bear, presiding with an authority not to be brooked. When the technician announced by intercom the commencement of the treatment, I braced for the unknown, only to hear several minutes of anticlimactic buzzing as the electrons attacked. The hand felt nothing, nor did it sense anything until the last two weeks, when the two-inch square of skin first assumed the appearance of a modest sunburn, then the fiery red of a serious sunburn, and finally the darkness of a second degree burn. The hand emitted a slight odor.

Several days into the treatment, Mickey LeMaistre appeared in the waiting room. He casually introduced himself, and inquired about our welfare, also asking each of us how long we had been waiting for treatment. It happened to be a bad day; the betatron had malfunctioned. The patients would be shifted for one day to the cobalt-60. LeMaistre apologized for the delay and went about his business. When the technicians positioned me for the treatment, the cobalt-60 seemed aesthetically primitive compared to the sleek betatron, and it conjured up memories of old dental drills, with their exposed cables and screeching, grinding sounds. After the treatment, I found Fletcher in his office, and asked whether the archaic cobalt-60 was as efficacious as the betatron. He scoffed, "Of course," and dismissed me with a wave of the hand. Taking Fletcher at his word, late in March 1981, I departed M. D. Anderson for good, except for follow-ups with Martin—or so I thought.

Martin saw me at the clinic every three months for a routine chest x-ray one day and a clinic visit the next. Overnight a radiologist, always invisible to me, and in my mind a furtive nocturnal creature adorned in the white lab coat of a physician and peering at x-ray films while most people slept, read the chest x-rays and rendered a diagnosis. An inverse relationship governs anxiety in the M. D. Anderson clinics: stress levels decline according to the amount of time since the most recent recurrence. By 1984, the clinic visits, though never routine, no longer seemed quite so potentially ominous.

While waiting for an appointment, I met R. Lee Clark. A volunteer had pushed the Jolly Trolley, a rolling dolly packed with coffee, soft drinks, and snacks, near the elevators at Station 80. When appointment times fell behind, patients sometimes lingered in the waiting room for hours without eating, unwilling to get a meal in the cafeteria, eight stories below, for fear of being absent just when the receptionist called their names. Some experienced patients brought a sack lunch to the clinic, just in case the delay stretched into hours. As I paid for a soda, the elevator door opened and out walked Clark, wearing a smile and a white lab coat. I recognized him from a portrait on the first floor. Noticing that I had noticed, he extended his hand. Clark inquired about my physician. "Dr. Martin is the best," he assured me. "You're in good hands."

A few minutes later, in the examining room, just before Martin opened the door, I detected a small, hard lump on the inside of the left forearm, about eight inches above the wrist. Martin never wasted time with good news, and as he went to the sink to wash his hands before probing my armpit for suspicious lumps, he intoned, "Your chest is clear."

In clinic visits, such an announcement usually eased my distress like air hissing from a deflating balloon. Never, however, had relief been so short lived. "There's a new lump in my left arm," I blurted. "Right here." Osteoporosis had already rounded Martin's shoulders, but as he hovered over me, he seemed visibly shorter and older. He handled the arm carefully, almost tenderly, touching the lump and pinching the surrounding tissues. Martin pushed his fingers into the axilla under my arm. Finding nothing of concern, he still said, "The lymph nodes seem to be fine, but we'd better take out the lump." He told the nurse to bring a surgical kit. When she returned, Martin injected a local anesthetic, chatted for a few minutes, and then excised the tumor, lifting it out with a large set of tweezers. The specimen he placed in a stainless steel bowl. "We'll send the tissue to pathology and call you tomorrow with the results. It doesn't look suspicious, but epithelioids can be tricky. We need to be sure."

In the morning at work, each ring of the phone startled me. Finally, a woman with a faint voice indicated that she needed to talk to "Dr. Olson." At M. D. Anderson, owning a doctorate endows special status; a doctor is a doctor and a title acknowledged at every turn. The voice paused for a moment and announced. "Dr. Olson, we have the pathology report, and it's a lipoma." For cancer patients unfamiliar with the entire lexicon of neoplastic formations, "oma" is the suffix of death—sarcomas, carcinomas, gliomas, lymphomas, meningiomas, melanomas, myelomas, seminomas, neuroblastomas, retinoblastomas, astrocytomas, and dozens of others. Thoughts of a new cancer popped up like ads on today's computer screen. Catching my breath and composing my thoughts, I asked, "What do I need to do to schedule treatments?" "What treatments?" she replied. "The cancer treatments," I responded. "You don't have cancer," she insisted. "You have a lipoma, a benign fatty tumor." At my audible groan she replied, "Dr. Olson. Are you OK?" "Yes, I'm fine." Plaintively, however, I asked, "May I offer a suggestion?" She said, "Of course." "When you talk to a patient, the word 'benign' always needs to precede 'oma.'"

She then announced the graduation from quarterly to semiannual checkups, a sign that a valedictory might come before the eulogy to my life. Instead of a diploma, I took away certification that the battle with cancer was proceeding well, that maybe Gilbert Fletcher had indeed effected a cure.

I spoke to myself too soon. Six months later, in 1984, it seemed clear that Fletcher might have exaggerated his powers. A hard irregular knot had arisen within the radiated field of the hand. The sarcoma still had legs. Martin operated again, and the pathologists called it a recurrence. He suggested more radiation. Judy and I rode the elevator from Station 80 to the second floor, traipsed the long corridor, and then descended the stairs to Fletcher's basement office, hoping to steal a moment. He had retired! For cancer patients, losing a trusted oncologist to retirement is troubling. To Marsha McNeese fell the chore of directing a second course of radiotherapy. Before undergoing the first of the thirty-five treatments. I secured a second opinion from sarcoma expert Robert Lindberg, now on the faculty of the University of Louisville. By telephone, he concurred with the recommendation.

McNeese seemed more pleasant this time, but probably I was less knotted up emotionally. In 1981, I had stubbornly resisted radiation at the very moment when she was cutting her teeth on Gilbert Fletcher's legacy—the demise of radical surgery. She prevailed; I kept the arm. The cancer was back now, but not in the lungs. McNeese had been right. Over the next six weeks, we talked several times. She had just purchased a Betamax, the latest rage in technology, and dutifully recorded the ABC miniseries *Masada*, about the ancient Jewish redoubt in Palestine. Somewhat melodramatically, though privately, I wondered if the sarcoma would overcome me, just as the Roman legion had overwhelmed the Jews. After the six weeks, when I asked McNeese to assume the mantle of Fletcher's promise for a cure, she played her cards close to the stethoscope.

With good reason. In 1985, the tumor showed up again, like a greedy relative twice-banned from the family. So that Martin could take another whack at the tumor, I entered the hospital and enjoyed the comforts of a lovely room in the Lutheran Pavilion thanks to the beneficence of Marshall and Lillie Johnson. An agitated patient, however, complicated the stay. With oat cell tumors thriving in both lungs, he had latched on to interferon as his salvation, the drug that might allow him someday to go deer hunting with his young son. Like an autistic child, the man pummeled every nurse with the same request for interferon, and each advised, "You need to discuss that matter with your physician." He spoke to anybody wearing white. The next morning, Martin revealed that my sarcoma cells were still on the offensive. When I inquired about the possibility of more radiation, he demurred. The hand had absorbed its fill. Martin recommended wait-and-see. Amputation was still an option should the tumor flare up again, and apparently the sarcoma cells had an aversion to my lungs. Quarterly checkups resumed.

I waited and waited, my optimism waxing with each uneventful clinic visit. In 1987, Martin again traded three-month checkups for six months. Six years a cancer survivor, I prepared to exit a dark cloud. In 1981, 1984, and 1985, I had outlasted the disease, but cancer has no regard for calendars. Late in 1987, Freddie Steinmark's dilemma became my own. The tumor seemed to sprout overnight, like the beanstalk of my daughter Susan's junior high school science project. I asked a physician friend to take a look at it. Al Gebert knew Anderson well, having completed part of an internship there that included the memorable duty of sticking maggots into the gaping, pelvic wounds of women who had absorbed too much radiation in the days of Fletcher's oxygen-rich hyperbaric chamber treatments. The oily little creatures fattened up on the rotten tissues. "You better get back to Anderson," he urged.

One week before Christmas in 1987, in an examining room at Station 80, Martin, using a local anesthetic, excised the tumor. "I'll send it to pathology for a frozen section," he explained. "Why don't you get some lunch and come back in a few hours." I wondered silently whether another round of radiotherapy might be in order. Martin intuited the question before its posting. "If it is malignant, you will have a decision to make. Two times the tumor has survived full blasts of radiation, and it might now be more aggressive. We might try radiation again. We'll talk about it later today."

Dear friends accompanied Judy and me to Ninfa's, a Mexican restaurant, and we ate tacos with our hands, all thinking but none voicing the obvious. Later in the afternoon, back at Station 80, Martin explained the pathology report and Solzhenitsyn echoed, "to amputate or not to amputate. To amputate or not to amputate." We drove home to four children awaiting the diagnosis. Brad, our eight-year-old, pushed to the point immediately. "Is it cancer?" I nodded affirmatively. Knowing its consequences, he burst into tears, scrambled up the stairs, and slammed and locked the door to his bedroom. I followed him and gently knocked. Sobbing himself into giant hiccups of grief, he opened the door. We sat together on the bed for thirty minutes and talked. Brad calmed down. "Are you OK?" I inquired. "Yeah," he reassured me. I left his room. While I walked downstairs, he said, "Dad. After the operation, will you be handicapped?" I responded, "Yes, I guess so." With the reductionism only children muster, he responded, "From now on, can we park anywhere we want to?"

With the banality of getting a cavity filled, M. D. Anderson scheduled an amputation. On December 28, 1987, Judy and I arose very early. Six years in the M. D. Anderson clinics had taught the virtues of punctuality. A crescent moon as thin as a cut fingernail hung in the sky. To save money, the hospital had abandoned overnight stays on the eve of some surgeries. For me, the overnight stay felt more comfortable, more transitional, like a drivers' training course at the age of fifteen before taking the wheel solo. "Same-day amputation" had the thud of a tennis racket with loose strings. Promptly at 8:00 a.m., a perky nurse appeared in the waiting room, praising the weather, with its cloudless sky, moderate temperature, and low humidity. In silence, I thought, "Yes, it's a pretty day for an amputation." Things got worse before they got better. Looking at the chart, she amiably remarked, "July 15. What's your sign?" "Cancer."

The woman, however, did have her wits about her. Oncology nurses are the first responders of cancer medicine, straddling the divide between patients and the surgeons wielding sharp scalpels, the radiotherapists aiming the photons, electrons, and protons, and the medical oncologists injecting the old-fashioned poisons and newfangled molecules. After escorting us through an Alice in Wonderland succession of progressively shrinking waiting rooms, she paused and knowingly asked, "Would the two of you like to be alone for a few minutes before I take you to surgery?" Judy silently nodded. The nurse took us to a room barely large enough for the sofa it contained and quickly disappeared. Judy started to cry, not stricken sobs of grief but the less wrenching tears of losing the important but not the essential. Clueless, I resorted to male platitudes, suggesting that she "not worry too much," promising that "everything will be OK," and certifying that "we'll be fine." With a forefinger to her lips, she hushed me. "Shush." Glistening eyes conveyed the layered affections of a long marriage. Cradling the hand, she gently tugged the wedding band from a ring finger thickened over twenty-one years. "This hand," she whispered, "has made love to me and comforted my children. It is my flesh. Let me mourn it." And she kissed the hand farewell.

Then I changed to a hospital gown, and the gurney arrived. The nurse pushed out into a queue of gurneys transporting other men and women about to lose a body part. A polite orderly rolled me into a surgical suite and a masked anesthesiologist introduced himself. Richard Martin in scrubs materialized. In a few minutes, with a scalpel and a Gigli saw, he severed the arm.

One week later, we saw him again. Martin needed to check the surgical wound and discuss the pathology reports on the lymph nodes excised from the axilla. No evidence of disease had been found in the nodes. Still thinking like William Stewart Halsted, I asked, "Does that mean that we caught it in time?" Again a gentle lecture on the new paradigm in cancer medicine, a discussion of the systemic nature of most solid tumors. "The absence of positive lymph nodes does not really mean that the previous surgeries and radiotherapy treatments have killed the cancer," Martin explained. "More likely, your body is handling them on its own." To drain the fluids causing the lymphedema on my left side, he drew from the drawer what looked to be a syringe more suited to a horse than to a human. Judy gasped and Martin smiled, "Don't worry. He won't feel a thing. The nerves were all severed in the operation." In went a four-inch, large gauge needle with neither twinge nor tickle, and out came several ounces of cloudy fluid.

Two weeks later, a well-meaning breast cancer survivor called with suggestions for easing the lymphedema. "I did exercises," she instructed, "by standing next to a wall and repeatedly marching my fingers as far up the wall as I could." I listened with silent bemusement. "Just walk your fingers as far as you can stretch. The swelling will go down." A smile creased my face. "Do it four or five times at day." Then silence. I stuffed laughter, hoping she would break in before my explosion. She did. "Oh Jim! You don't have any fingers. I'm so sorry."

One year after the amputation, I traveled to West Lafayette, Indiana, to deliver a lecture at Purdue University. Unknown to me, the PGA was staging at Indianapolis that week the national amputee golf championship, the "U.S. Open" championship for scratch golfers with stumps. I played some golf but with little skill; in fact, my scores improved *after* the amputation. I approached the Hertz desk, and the agent politely said, "Welcome to Indianapolis, Mr. Olson, what's your handicap." Perhaps he had not noticed the hook at the end of my prosthesis. "What?" I replied. "What's your handicap?" he repeated. "What?" I said, some irritation sneaking into my voice. "What's your handicap?" Eyeballing him, I lifted my prosthesis and said, "I have only one arm for heaven's sake." He exclaimed, "Do you mean you're not here for the amputee golf championship?" "No," I said. Cancer has its moments.

In 1995, after eight years of quarterly and semiannual clinic appointments— a total of twenty appointments since the amputation in 1987 and fifty-four since the first encounter with Richard Martin—I went for another, only to be greeted by a very young surgeon, one too young for my taste. Although forty-nine, I still expected physicians to be older than me. Raphael Pollock introduced himself and stated that Richard Martin had retired. So many times during my cancer odyssey, Martin had extended a steady hand and been the fulcrum, balancing my fear with hope and the realistic with the impossible. And now he had gone the way of Gilbert Fletcher. Pollock, however, quickly earned my confidence. We both shared a history major as undergraduates. That he had trained in the humanities and the sciences erased the stigma of youth. Pollock displayed an easy congeniality and measured temperament, a fitting successor to Martin. No sooner had I warmed to him than he announced that I was cured of the sarcoma and no longer in need of M. D. Anderson. Annual chest x-rays in my own community would suffice.

I skipped out of Station 80, descended via elevator to the main entrance. Saying a permanent good bye to M. D. Anderson gave me pause, so I sat down to savor a sweet juncture in life. And at that very moment, my colleague Joan Coffey and her husband, Ed, walked through the waiting room deep in thought. They had just left Raymond Alexanian's office with the dismal diagnosis of Waldenström's lymphoma. We greeted each other cordially, but when I related my own good news, survivor's guilt throbbed. At M. D. Anderson, the possibility of death always plays subtext to the rhetoric of hope. As I left the world of oncology, Joan entered.

Or had I left? In 1996, a surgeon skillfully excised from my face a basal cell carcinoma, the carcinogenic legacy of a childhood drenched in the ultraviolet glow of southern California. The lesion had evolved over the course of several years from a blemish to a small tumor, but it rested just above a forehead wrinkle, and the surgeon skillfully concealed the scar within the fissure of age. Compared to the epithelioid sarcoma, the basal cell cancer seemed a trifle.

Four years later, I reentered Joan's world. A severe, head-on collision had deposited me in an emergency room near Seattle, Washington. We had just attended my son's wedding and were driving to the reception. After a CAT scan, the attending physician confided, "Mr. Olson, there appear to be no injuries from the accident, but do you know that there is a mass in your brain?" I had no idea and no symptoms but recounted my odyssey with M. D. Anderson. He prescribed an antiseizure medication and suggested that I return without delay to the hospital. Because of the legislation LeMaistre had secured in 1995, I needed no referral from a physician, but my family physician, David Prier, arranged it anyway.

Early in January 2002, fourteen years after the amputation, Judy, my daughter Karin, and I sat anxiously in the reception area of the Brain and Spine Center at M. D. Anderson, waiting to meet neurosurgeon Jeffrey Weinberg, who would soon go spelunking through the caverns of my brain. Like my son Brad, Weinberg had a gregarious nature and an undergraduate degree from the University of Pennsylvania. The seventh-floor waiting room of the Clark Clinic rested one floor under the old Station 80. I could almost feel the benign ghost of Richard Martin floating in the corridors. Two days later, Weinberg carved a horseshoe opening over the right parietal lobe and pulled out a tumor the size of a ping pong ball. The next day, in a room the Alkek family had financed, Weinberg arrived on his rounds. I was sitting up in bed reading the *Houston Chronicle*. "That's a good sign," he remarked. I posed the inevitable Halstedesque question, "Did you get it all?" The reply guaranteed a lifetime with M. D. Anderson. "No."

The pathology report tendered sobering news within a larger context of "not as bad as it might have been." My skull harbored an incurable glioma, though one not as swiftly lethal as glioblastoma multiforme or the somewhat less aggressive but equally lethal astrocytoma. I owned the "good glioma," a low-grade oligodendroglioma, not quite as complicated to pronounce as Joan Coffey's Waldenström's macroglobulinemia, but difficult enough to require practice before perfecting.

Vijay Puduvalli, my neuro-oncologist, was an emigrant from Kerala state in India, one of a steady stream of well-educated and highly skilled ethnic Mayalam streaming into the Texas Medical Center. He had the soul of Richard Martin and the heart of Emil Freireich, whose long reach now touched me. Freireich, in the 1950s, had spawned the era of combination chemotherapy, a fundamental principle that stood to characterize oncology into the distant future. Puduvalli prescribed for me a triple punch—six rounds of vincristine, lomustine, and procarbazine, delivered in a finely honed schedule over the course of six months. I braced myself for the wrath of chemo and the misery so many cancer patients had experienced, but my treatments took place after the advent of the antiemetic drugs. During the entire time, I experienced a persistent, low-grade nausea but vomited only once. The other side effects were minimal—a modest bout with pneumonia, some weight loss, and fatigue not serious enough to keep me from work.

And so, in the fall of 2002, I stood cancer watch again, now on two-month follow-ups as Puduvalli monitored the tumor to see if it remained true to its grade 2 origins. Standard MRIs exposed its gross dimensions and MRI spectroscopy its metabolic activity. Puduvalli assured me that it was behaving indolently, and I responded, "I prefer a lazy tumor to a hardworking one."

In 2004, however, it acquired some ambition, and the tumor watch gave way to a tumor warning. Puduvalli wanted a biopsy to determine whether other than oligo cells now inhabited the tumor, and Weinberg operated again, screwing to my skull a heavy cage apparatus resembling the headgear of early deep sea divers. He needed the head fixed in place to lend more precision to the probe. Weinberg poked the probe deep to the center of my brain and then managed to drag its tip through several sites in the tumor. The pathology reports were as good as could be expected—still a low-grade oligodendroglioma. Two days after the operation, I went to my barber. He noted the presence of two stitches and observed, "That cut must have been deep. Doctors usually use tape these days, not stitches." "Oh it was deep all right," I replied. "Really, really deep."

The tumor remained true to its oligo heritage but had also grown a bit uppity; Puduvalli felt that the time was at hand for brain radiation. I blanched at first. During a chance encounter in the Woodlands with Pamela Schlembach, a radiation oncologist at M. D. Anderson, we had talked casually about my condition and about the long-term side effects of brain radiation. I recalled the appearance of my hand in 1981 after the first full round of treatments—a rectangular square that had reddened and then blackened. I realized that the dosages to my brain would be nowhere near as powerful as those to my hand, but brain radiation created images of senility, of voluntarily leaping toward dementia. I felt trapped and paralyzed, unable to make the decision, and from indecision I descended into clinical depression.

The descent, however, also took me into a world fashioned a decade earlier by Charles LeMaistre. In his campaign to streamline patient services, LeMaistre had decided gradually to cluster physicians according to organ site, not just to department. Patients usually had to crisscross the campus many times in order to meet with the specialists providing their care. Some spent nearly as much time getting to the appointment as they did at the appointment. Pulling it off proved more difficult than LeMaistre anticipated, but at the Brain and Spine Center, it worked.

One morning in the midst of my troubles, I had a planning appointment with Puduvalli. The oncology nurse Karen Baumgartner, who usually preceded him into the examining room to conduct an initial evaluation, noticed my funk. "You don't seem to be yourself, Dr. Olson," she inquired. "Are you OK?" I confessed to feeling somewhat depressed, to which my daughter and wife simultaneously chimed, "Really depressed!" During the next three hours, I sat in the same chair and greeted a succession of consults—neurosurgeon Jeffrey Weinberg, medical oncologist Vijay Puduvalli, radiation oncologist Eric Chang, and psychiatrist Alan Valentine. With their help over the course of several weeks, I climbed from the pit, and in September 2005, when the treatments were about to begin, felt better, not recovered yet but markedly improved. The old self chased away the new.

In the midst of my recurrence, I learned that Lynn Bull, a dear friend from San Antonio, needed to come to M. D. Anderson because of breast cancer. I offered to drive her to Houston and introduce her to the institution. Lynn too then experienced the organ site program at its best—the Nellie B. Connally Breast

Center, named after the former first lady of Texas, a breast cancer survivor and breast cancer advocate. The Nellie B. Connally Breast Center at M. D. Anderson is home to twenty-one medical breast oncologists, eleven surgical oncologists, seven radiation oncologists, and five specialists in blood and bone marrow transplantation, to say nothing of radiological equipment and laboratory medicine. In an extraordinarily effective example of one-stop medicine, the Nellie B. Connally Breast Center centralized in one place the diagnostic and treatment technologies that Lynn needed to recover. It changed the way breast cancer patients are treated today and has been widely adopted at other cancer centers. The primary site today for the treatment of breast cancer is the outpatient clinic.

Puduvalli raised the issue of proton therapy. Mendelsohn's Proton Therapy Center, however, was not yet online, and to secure such treatment I would have to travel either to Loma Linda University in California or Massachusetts General in Boston. Neither coast appealed to me, especially after Eric Chang explained IMRT—intensity modulated radiation therapy, in which multiple beams of energy hit the tumor from different angles. At the point of their conjunction at the tumor site, the dosage reaches a maximum, but at the point of their entrance and exit from the skull, the dosages amount to much less.

Back in 1981 and 1984, when the technicians set my hand up for each radiation treatment, they positioned it carefully according to a lighted image emanating from the machine. They then left the room, reminding me not to move the hand. In 2005, with multiple beams angling into my brain and destined for a single site in the middle of it all, a reliable, fixed target was at even more of a premium. To fix my head exactly in place, the same way every day for six weeks, the radiation oncologists made a mold of my face and fashioned from it a plastic mask, a contraption resembling the headgear of a fencer or of Hannibal Lecter in *Silence of the Lambs*.

Positioning me on the treatment table, the nurse bolted the head mask tightly into place. "Mr. Olson," she said. "I'm Jeannette Ards. I gave the treatments to your hand twenty-six years ago." I strained to see through the mesh of the mask and recognized a familiar face and a friendly countenance. Ards administered the IMRT treatments daily in September and October 2005.

Today, in the spring of 2008, I am still alive, 158,788th of M. D. Anderson's more than 700,000 patients. Writing about cancer survivors, oneself or others, is not like describing the cycles of the moon, as predictable tomorrow as they were yesterday. Cancer survivorship more resembles a meteor shower, spectacular and inspiring at first but subject to swift disappearance into the void. The futures of cancer and cancer survivors join hands in ambiguity. For an expansive John Mendelsohn, the future of oncology glows with hope; new therapy upon new therapy will attack cancer at the level of its genetic abnormalities, steadily disrupting and disabling the cell's legion of pathways until the disease, like tuberculosis and pneumonia, shrinks into a manageable malady. Whether Mendelsohn is an emissary from the future remains to be seen.

For me today, Richard Martin whispers from a past that I once inhabited: "In medicine . . . we don't know the end from the beginning. Just live your life."

NOTES

CHAPTER 1: R. Lee Clark, History, and the Dread Disease

1. Steinmark, *I Play to Win*, 64–65, 127, 135; Charles A. LeMaistre interview.
2. The author, over the course of sixteen years, was treated for a soft tissue sarcoma by Richard G. Martin, meeting with him dozens of times. The treatments included a limb amputation. The author was not present during Steinmark's meetings with Martin. The description in the text is extrapolated from the author's own experience.
3. Richard G. Martin interview.
4. *Messenger* 8 (Aug. 1979): 9.
5. *Dallas Morning News*, Jan. 2, 1970; Steinmark, *I Play to Win*, 64–65, 127, 135; Charles A. LeMaistre interview.
6. *Houston Chronicle*, Dec. 7–8, 1969; Steinmark, *I Play to Win*, 64–65, 127, 135, 237–38; Frei, *Horns, Hogs, and Nixon Coming*, 277–80, 287–91.
7. Steinmark, *I Play to Win*, 270; Richard G. Martin interview.
8. Steinmark, *I Play to Win*, 254; *Cancer Chemotherapy Reports* 52 (June 1968): 485–87.
9. Patterson, *Dread Disease*, 248–49; *Dallas Morning News*, June 23, 1971; Norman Jaffe interview; Ray Dowdy interview. In 1971, the major scoreboard in Memorial Stadium (now Darrell K. Royal Memorial Stadium) was named for Freddie. It reads: "Dedicated to the memory of Fred Steinmark, 1949–1971, defensive back of the Texas Longhorns National football champions of 1969, whose courageous fight against savage odds transcended the locker room, the playing field, the campus, the nation itself. The indelible memory of his indomitable spirit will provide an inspiration to those who play the game of life."
10. Eleanor Macdonald interview.
11. *Houston Chronicle*, Feb. 3, 1974; Eleanor Macdonald interview; Rosemary Cumley interview; Marion Wall (Lowrey) interview.
12. Swaim, *Walking TCU*, 1–58; Clark, *Thank God*, 437–51, 470; Macon, *Clark and the Anderson*, 50, 69. Add-Ran College became Texas Christian University, and Wichita Falls Junior College became Midwestern State University.
13. Macon, *Clark and the Anderson*, 91.
14. Ibid., 63.
15. Ibid., 69. See Schaffer, *Daniel H. Burnham*; Eleanor Macdonald interview.
16. Quoted in Nuland, *Doctors*, 409–10; Eleanor Macdonald interview.

17. Pernick, *A Calculus of Suffering*, 42–59, 148–57.

18. Eleanor Macdonald interview; Macon, *Clark and the Anderson*, 99–100; Aronowitz, *Unnatural History*, 86–89.

19. J. Walter Wilson, "Virchow's Contribution to the Cell Theory," *Journal of the History of Medicine* 2 (Spring 1947): 163–78.

20. William Seybold interview; Richard G. Martin interview; Eleanor Macdonald interview.

21. H. M. Said and H. M. Barakati, "Cancer: The Last Two and a Half Millennia of Aetiology and Cure," *Hamdard* 21 (July–Sept. 1978): 28–47.

22. Betty Galluci, "Selected Concepts of Cancer as a Disease: From the Greeks to 1900," *Oncology Nursing Forum* 12 (July–Aug. 1985): 67–71.

23. D. G. Lytton and L. M. Resuhr, "Galen on Abnormal Swellings," *Journal of the History of Medicine and the Allied Sciences* 33 (Oct. 1978): 531–49.

24. Elizabeth C. Miller and James A. Miller, "Milestones in Chemical Carcinogenesis," *Seminars in Oncology* 6 (Dec. 1979): 445–56; Benjamin Rush, "An Account of the Late Dr. Hugh Martin's Cancer Powder, with Brief Observations on Cancers," American Philosophical Society, *Transactions* 11 (1786): 212–17; Rush, *Letters*, 1:251 and 2:1104.

25. A. Harvey McGehee, "Early Contributions to the Surgery of Cancer: William S. Halsted, Hugh H. Young, and John G. Clark," *Johns Hopkins Medical Journal* 135 (Dec. 1974): 399–417; Wilson I. B. Onuigbo, "The Paradox of Virchow's Views on Cancer Metastasis," *Bulletin of the History of Medicine* 6 (Sept.–Oct. 1962): 444–49.

26. Patterson, *Dread Disease*, 1–37; E. S. Judd, "Tumors of the Breast, with Special Reference for Obtaining Better Results in Malignant Cases," in *Collected Papers by the Staff of St. Mary's Hospital and Mayo Clinic, 1905–1909* (1919), 368.

27. Starr, *Social Transformation of American Medicine*, 150–51.

28. Eleanor Macdonald interview; Patterson, *The Dread Disease*, 74; Samuel Hopkins Adams, "What Can We Do about Cancer? The Most Vital and Insistent Question in the Medical World," *Ladies Home Journal*, May 1913, 21–22.

29. Virginia Gardner, "Vanity, Modesty, and Cancer," *Hygeia* 2 (Apr. 1933): 300–302; Tobey, *Cancer*, 3–16; Gardner, *Early Detection*, 53–88; *New York Times*, Nov. 3, 1913.

30. See Wheatley, *Politics of Philanthropy*; Ludmerer, *Learning to Heal*; Rothstein, *American Medical Schools*; *Cancer Research* 29 (1969): 1615–40; *Journal of the National Cancer Institute* 59 (Aug. 1977): 551–58.

31. Macon, *Clark and the Anderson*, 107–13.

32. See Reynolds, *How Pasteur Changed History*.

33. See Giroud, *Marie Curie*.

34. *Cancer Bulletin* 8 (Jan.–Feb. 1956): 8–12.

35. Reginald Murley, "The Treatment of Breast Cancer: A Study in Evolution," *Annals of the Royal College of Surgeons of England* 69 (Sept. 1987): 212–15.

36. www.mayoclinic.org/tradition-heritage/model-care.html.

37. Macon, *Clark and the Anderson*, 117–18.

38. See *Fortune*, Mar. 1937, 112; *Life*, Mar. 1, 1937, 36.

39. Henderson, *Maury Maverick*, 1–10; *New York Times*, July 24 and Aug. 6, 1937, and June 2, 1938.

40. Quotes are from Hilts, *Protecting America's Health*, 78–95; Lebineau, *Medical Science and Medical Industry*; www.harvardsquarelibrary.org/unitarians/cabot.html; *New York Times*, June 24–25 and July 15–18, 1938.

CHAPTER 2: Present at the Creation

1. Macon, *Clark and the Anderson*, 123; Richard G. Martin interview; Walter Pagel interview. Clark's tastes later ran to Porsches and Citroens.
2. For a look at Monroe Anderson, see Macon, *Monroe Dunaway Anderson*.
3. *Houston Press*, June 24, 1958.
4. See Sibley, *The Port of Houston*, 3–23.
5. Eleanor Macdonald interview; William Seybold interview; Thomas D. Anderson interview; *Messenger* 9 (Aug. 1980): 1, 3; *Houston Chronicle*, Dec. 2, 1979; Fleming, *Growth of the Business*.
6. Kelsey, *Doctoring in Houston*, 9–10, 22–23; Mavis P. Kelsey interview; William Seybold interview; *Houston Chronicle*, July 19–22, 1950.
7. Eleanor Macdonald interview; Mavis P. Kelsey interview; Karl John Karnaky to E. W. Bertner, Sept. 20, 1946, Folder 1, Box 1, Ernst W. Bertner Papers; Kelsey, *Doctoring in Houston*, 9–10, 22–23; *Houston Chronicle*, July 19–22, 1950.
8. See clippings in Folder 1, Box 1, Bertner Papers.
9. Karl John Karnaky to E. W. Bertner, Sept. 20, 1946, Folder 1, Box 1, Bertner Papers; *Houston Chronicle*, Nov. 22, 1932; *Texas State Journal of Medicine* 46 (Sept. 1950): 728–29; *Houston Action* 5 (Aug. 5, 1950): 1, 3; Eleanor Macdonald interview.
10. Eleanor Macdonald interview; Thomas D. Anderson interview.
11. Macon, *Mr. John H. Freeman*, 20–22; *Messenger* 30 (June 2001): 6; *Messenger* 9 (Aug. 1980): 1, 3.
12. Thomas D. Anderson interview; Bryant Boutwell, "Two Bachelors, a Vision, and the Texas Medical Center," *Houston Review of History and Culture* 2 (Fall 2004): 8–14.
13. Eleanor Macdonald interview; William Seybold interview; Thomas D. Anderson interview.
14. Lester Clark to Charles A. LeMaistre, June 29, 1987, MDA, 1D, Development, General; James Cato interview; Bettie Belcher interview. (Bettie Belcher is Arthur Cato's daughter.) Bettie Lillian Cato later died from bone cancer.
15. R. Lee Clark interview.
16. *Houston Chronicle*, June 16–25, 1941; *Houston Post*, June 25, 1941; Bettie Belcher interview; William T. Jackson interview.
17. Macon, *South from Flower Mountain*, 26–27; "Agreement between M. D. Anderson Foundation Trustees and Board of Regents of the University of Texas," MDA, RLC, Org. Func. and Hist., MDAH 1947–1954, Early Administration, ADM, General; William T. Jackson interview; *Houston Post*, June 25, 1941.
18. Macon, *Clark and the Anderson*, 135; Eleanor Macdonald interview; Mavis P. Kelsey interview; *The First Twenty Years*, 17–22.
19. John Musgrove interview; Eleanor Macdonald interview; *The First Twenty Years*, 24–28; Ernst Bertner to Homer P. Rainey, Nov. 25, 1942, MDA, RLC, Board of Regents, Misc. Correspondence, Inactive, the University of Texas, 1942–1951.
20. *The First Twenty Years*, 28–29; John Musgrove interview; Ernst Bertner to Homer P. Rainey, Nov. 25, 1942, MDA, RLC, Board of Regents, Misc. Correspondence, Inactive, University of Texas, 1942–1951.
21. Richard G. Martin interview; Marion Wall (Lowrey) interview; John Musgrove interview; Eleanor Macdonald interview; *The First Twenty Years*, 24–26.

22. John Musgrove interview; Eleanor Macdonald interview.

23. Steve Stuyck interview; *Houston Chronicle*, Feb. 18–19, 1944.

24. Macon, *Clark and the Anderson*, 123–25; Eleanor Macdonald interview.

25. Mavis P. Kelsey interview; Macon, *Clark and the Anderson*, 128–32; Marion Wall (Lowrey) interview; Eleanor Macdonald interview.

26. R. Lee Clark interview; Joseph T. Painter interview.

CHAPTER 3: Designing a Dream, 1946–1950

1. R. Lee Clark interview; Lee Clark to Ernst Bertner, Oct. 12, 1949, Folder 6, Bertner Papers; Kelsey, *Doctoring in Houston*, 26, 296.

2. Cornelius Rhoads to Ernst Bertner, May 22, 1950, and Ernst Bertner to Cornelius Rhoads, May 26, 1950, Bertner Papers, Folder 6; *Houston Chronicle*, July 2, 1950.

3. Mary Schiflett, "The Second Downtown," *Houston Review of History and Culture* 2 (Fall 2004): 2–7.

4. R. Lee Clark interview.

5. Smith, *Controversy*, 4–12.

6. *Houston Post*, Dec. 13–15, 1943.

7. In 1958, the state legislature changed its name to the University of Texas Dental Branch; *Houston Post*, Dec. 13, 1943, and Mar. 1, 1946; Bryant Boutwell, "Two Bachelors, a Vision, and the Texas Medical Center," *Houston Review of History and Culture* 2 (Fall 2004): 12–13, 56–57; Carroll D. Simmons to C. M. Sparenberg, Oct. 29, 1951, RLC, UT, TMC, 1947–1951.

8. William Seybold interview; R. Lee Clark interview.

9. *Science* 138 (1972): 233–37; Daniel M. Fox, "The Politics of the NIH Extramural Program, 1937–1950," *Journal of the History of Medicine and the Allied Sciences* 42 (1987): 447–66; Geiger, *To Advance Knowledge*.

10. R. Lee Clark interview.

11. Ludmerer, *Time to Heal*, 117–19, 141–44.

12. Ibid.; Charles A. LeMaistre interview; R. Lee Clark interview.

13. R. Lee Clark to Ben R. Barbee, Nov. 6, 1959, MDA, RLC, Special Projects, Nuclear Reactor.

14. Charles A. LeMaistre interview.

15. Jerzy Einhorn, "Nitrogen Mustard: The Origin of Chemotherapy for Cancer," *International Journal of Radiation, Oncology, Biology, and Physics* 11 (July 1985): 1375–78.

16. Jane C. Wright, "Cancer Chemotherapy: Past, Present, and Future," *Journal of the National Medical Association* 76 (Aug. 1984): 773–84; Carl G. Kardinal, "Cancer Chemotherapy: Historical Aspects and Future Considerations," *Postgraduate Medicine* 77 (May 1, 1985): 165–74.

17. R. Lee Clark to Theophilus Painter, May 14, 1948, MDA, RLC, Series 1, Board of Regents, Misc. Correspondence, University of Texas, 1942–1951. Memorial Sloan-Kettering Cancer Center was founded in 1884 as the New York Cancer Hospital and then evolved into the General Memorial Hospital for Cancer and Allied Diseases. In 1902, railroad magnate Collis Huntington gave the hospital one hundred thousand dollars for cancer research. Sixteen years later, James Douglas of Phelps Dodge Corporation donated six hundred thousand dollars for the treatment of cancer

patients, insisting, however, that the word "General" be dropped from the hospital's name. In 1945, Alfred P. Sloan and Charles F. Kettering of General Motors founded the Sloan-Kettering Institute for Cancer Research, which in 1960 merged with Memorial Hospital.

18. Shaughnessy, "The Story of the American Cancer Society," 221–27, 257–59.

19. E. V. Cowdry Papers, Bernard Becker Library, Washington University School of Medicine, Boxes 77:4; 71:17; 78:16; 78:25; 78:32; 78:36.

20. Carroll D. Simmons to R. Lee Clark, Feb. 29, 1952; "Confidential Memo," Jan. 15, 1952; Dr. Heflebower to Dr. Clark, Dec. 13, 1951; MDA, RLC, Business Office, Administration; www.koreanwar.org/html/units/usaf/90bs.htm; Minutes of the Meeting of the Coordinating Committee, Oct. 24, 1949, MDA, RLC, Coordinating Committee Minutes. John Musgrove interview; Eleanor Macdonald interview; Richard G. Martin interview.

21. Arthur Kleifgen Memo, Apr. 12, 1952, MDA, RLC, Business Office, Administration; Minutes, Coordinating Committee, Sept. 28, 1959, Coordinating Committee Minutes.

22. R. Lee Clark to Mr. Kleifgen, Nov. 17, 1950; Marion Wall to Mr. Kleifgen, Sept. 11, 1951, MDA, RLC, Business Office, Administration; Allan Shivers to R. Lee Clark, Oct. 12, 1950; Anna W. Hanselman to R. Lee Clark, July 1, 1949, MDA, RLC, Blood Donor Service, Special Projects; Eleanor Macdonald interview; Coordinating Committee Minutes, Sept. 24, 1951, MDA. RLC, Coordinating Committee Minutes; *The First Twenty Years*, 52–55; Arthur Kleifgen to R. Lee Clark, Aug. 9, 1949, MDA, RLC, Business Office, Administration.

23. *U.S. News & World Report*, Apr. 2, 1950, 13.

24. Macon, *Clark and the Anderson*, 187–90; Peter Almond interview.

25. Peter Almond interview; Eleanor Macdonald interview; Marion Wall (Lowrey) interview; Macon, *Clark and the Anderson*, 189–92.

26. "Leonard G. Grimmett," *British Journal of Radiology* (Sept. 1951): 92–93; Eleanor Macdonald interview; Peter Almond interview; John Musgrove interview.

27. Gilbert Fletcher to R. Lee Clark, May 21, 1950, MDA, RLC, OP-M-RLC, Admin. Radiology 1956.

28. Leonard G. Grimmett et al., "Design and Construction of a Multicurie Cobalt Teletherapy Unit," *Radiology* 59 (July 1952): 1929; *Houston Chronicle*, Sept. 28, 1952.

29. William O. Russell, Biography, William O. Russell Papers; Marion Wall (Lowrey) interview; Eleanor Macdonald interview; John Skarstad interview.

30. John Aikin interview; *Marshall News-Messenger*, Nov. 7, 2005; *Houston Chronicle*, Mar. 29, 1953.

31. Roy C. Heflebower to department heads, June 14, 1949, MDA, RLC, Series 1, Early Administration, Assistant Director, Early Administration.

32. *Houston Post*, Oct. 28 and Dec. 21, 1950, and Sept. 28, 1952; *Houston Chronicle*, Mar. 30, 31, and May 5, 1949, Oct. 28, Nov. 23, and Dec. 21, 1950; R. Lee Clark interview; *Austin-American Statesman*, Apr. 12, 1950; Allan Shivers interview; American Cancer Society Broadcast, Apr. 2, 1951, Allan Shivers Papers; R. Lee Clark interview.

33. R. Lee Clark interview; *Cancer Bulletin* 2 (Sept.–Oct. 1950): 113, and 2 (July–Aug. 1950): 74–75.

34. *Texas Triangle* 1 (Sept. 1950): 1; Kelsey, *Doctoring in Houston*, 296, 305–6.

CHAPTER 4: The Pink Palace, 1950–1955

1. Bush, *Barbara Bush*, 39–44; Bush, *All the Best*, 76.
2. Eleanor Macdonald interview; *Houston Chronicle*, May 28, 1951.
3. Eleanor Macdonald interview; Marion Wall (Lowrey) interview; R. Lee Clark interview.
4. Excerpt from the Minutes of the Board of Regents of the University of Texas on Feb. 6, 1953; Lee Clark to James P. Hart, Jan. 30, 1953, Organization and Functions, Early Administration, Admin. General, RLC Papers; *Houston Chronicle*, Mar. 18–21, 1953.
5. R. Lee Clark interview; *Houston Chronicle*, Jan. 22 and Feb. 1, 7, 14, and 25; May 21, 1953; Aug. 11–13, 1953; *Houston Post*, Sept. 29, and Dec. 16, 1952; Jan. 8 and 11, May 16, Sept. 15, and Nov. 4, 1953; Minutes of the TMC Council, Feb. 10, 1954, MDA, RLC, Texas Medical Center Council.
6. A. Clark Griffin to Stanley W. Olson, Oct. 4, 1955; A. Clark Griffin to R. Lee Clark, Mar. 24, 1955, MDA, RLC, Department of Biochemistry, Administration.
7. W. K. Sinclair to Dr. Clark, Oct. 25, 1954; W. K. Sinclair, "Nuclear Reactor Project," Jan. 1955, MDA, RLC, Special Projects, Nuclear Reactor.
8. John C. Burgher to R. Lee Clark, June 30, 1958; Austin M. Brues to R. Lee Clark, Nov. 4, 1958; W. K. Sinclair, "Summary Report," May 5, 1958; and R. J. Shalek to R. Lee Clark and W. K. Sinclair, Feb. 11, 1959, MDA, RLC, Special Projects, Nuclear Reactor; Edward Munger interview; W. K. Sinclair to R. Lee Clark, Aug. 9, 1957; Minutes of the Radiation Health and Hazards Panel, July 25, 1957, MDA, RLC, Radiation Health and Hazard Panel (1957).
9. W. K. Sinclair, "Memorandum," Feb. 17, 1959; John C. Burgher to R. Lee Clark, June 30, 1958; Austin M. Brues to R. Lee Clark, Nov. 4, 1958; W. K. Sinclair, "Summary Report," May 5, 1958; and R. J. Shalek to R. Lee Clark and W. K. Sinclair, Feb. 11, 1959, MDA, RLC, Special Projects, Nuclear Reactor; Edward Munger interview; W. K. Sinclair to R. Lee Clark, Aug. 9, 1957; Minutes of the Radiation Health and Hazards Panel, July 25, 1957, MDA, RLC, Radiation Health and Hazard Panel (1957).
10. Felix Haas Biography, Manuscript Collection No. 27, Felix Haas Papers, 1; Marion Wall (Lowrey) interview.
11. T. C. Hsu, "Confidential," typescript, UTMDA, T. C. Hsu Papers, unprocessed.
12. T. C. Hsu interview; Charles A. LeMaistre interview; Andrew DeWees interview; *Houston Chronicle*, July 13, 2003; S. Pathak, "T. C. Hsu: In Memory of a Rare Scientist," *Cytogenetic Genome Research* 105 (2004): 1–3; Marion Wall (Lowrey) interview.
13. *Houston Chronicle*, Sept. 15, 1960; "Termination Report, NCI C-1751"; William O. Russell to Osmond Molarsky, Aug. 16, 1962; Paul Swaffer to William O. Russell, Feb. 12, 1960, MDA, RLC, Research, Study Section on Cancer Eye.
14. Emil Freireich, "Min Chiu Li: A Perspective in Cancer Therapy," *Clinical Cancer Research* 8 (Sept. 2002): 2764–65; Min Chiu Li, "The Historical Background of Successful Chemotherapy for Advanced Gestational Trophoblastic Tumors," *American Journal of Gynecology and Obstetrics* 135 (Sept. 15, 1979): 266–72; *Newsletter* 2 (Sept. 1957).
15. Emil Freireich interview.
16. N. S. R. Maluf, "The History of Blood Transfusion," *Journal of the History of Medicine and Allied Sciences* 9 (Jan. 1954): 59–107; Böttcher, *Wonder Drugs*; Richard G. Martin interview.

17. Marion Wall (Lowrey) interview; Eleanor Macdonald interview; Richard G. Martin interview; Mary Lou Rogge interview; "Alexander Brunschwig (1901–1969)," *CA-A Cancer Journal for Clinicians* 24 (Nov.–Dec. 1974): 361–62; *Cancer* 5 (Sept. 1952): 992–1008.

18. Taylor, *Remembrances and Reflections*, 139–50, 181–83; Eleanor Macdonald interview; "Biography," Wataru W. Sutow Papers.

19. *Houston Chronicle*, Apr. 6, 2008; Eleanor Macdonald interview.

20. Eleanor Macdonald interview; John Musgrove interview; George Beto interview.

21. Edna Wagner to R. Lee Clark, Sept. 15, 1955; R. Lee Clark to Members of the Emergency Relief Committee for Cancer Patients' Aid, Oct. 25, 1955; Carroll Simmons to R. Lee Clark, May 6, 1960, Special Projects, Part 1, Series 1; Eleanor Macdonald interview; *Houston Chronicle*, Dec. 9, 1951.

22. Carroll D. Simmons to W. Leland Anderson, May 6, 1959; Edna Wagner to R. Lee Clark, Sept. 15, 1955; R. Lee Clark to Members of the Emergency Relief Committee for Cancer Patients' Aid, Oct. 25, 1955; Minutes of Meeting, Sept. 27, 1955; Edna Wagner to R. Lee Clark, Sept. 16, 1955, MDA, RLC, Part 1; Special Projects, La Posada; Wagner refused to breach patient confidentiality. Albino Torres to Lee Clark, May 31, 1955, MDA, RLC, Social Services.

23. Olson, *Bathsheba's Breast*, 10–12, 14–26.

24. Eleanor Macdonald interview; Richard G. Martin interview; William Seybold interview; *Cancer Bulletin* 4 (Mar.–Apr. 1952): back cover.

25. *Time* (Apr. 10, 1953); *New York Times*, Apr. 12, 1953.

26. Cayleff, *Babe*, 217–42; *New York Times*, Apr. 6, 1953, and Sept. 28, 1956; Richard G. Martin interview; Eleanor Macdonald interview; Marion Wall (Lowrey) interview.

CHAPTER 5: Changing Paradigms, 1956–1963

1. *Anderson Messenger*, 7 (Mar. 1978): 4; Wheatley, *Politics of Philanthropy*.

2. *Newsletter* 2 (June 1957): 1; *Newsletter* 3 (Sept. 1958): 3.

3. Marion Wall (Lowrey) interview; Vincent Guinee interview; *Newsletter* 1 (Dec. 1956): 1; *Newsletter* 2 (Dec. 1957): 4; *Newsletter* 3 (Dec. 1958): 4; *Newsletter* 4 (May 1959): 4; *Newsletter* 4 (Oct. 1959): 1.

4. Minutes, Board of Regents Meeting, Sept. 27–29, 1963; R. Lee Clark to Harry Ransom, Apr. 23, 1963; Murray Copeland to R. Lee Clark, Sept. 19, 1963; Robert Hickey to R. Lee Clark, Sept. 12, 1963; "Agenda Item for Board of Regents Meeting," Sept. 27 and 28, 1963, RLC, MDA, Graduate School of Biomedical Sciences; *Newsletter* 3 (May 1958): 1; *Newsletter* 8 (July 1963): 2–3; James P. Hart to Board of Regents, Nov. 17, 1953; R. Lee Clark to Judge D. K. Woodward Jr., Feb. 15, 1952; MDA, RLC, Postgraduate School of Medicine; R. Lee Clark to L. D. Haskew, Apr. 11, 1959, MDA, RLC, Org. Func. And Hist. MDAH-1955, Early Administration, ADM, General; R. Lee Clark to Logan Wilson, Oct. 10, 1957; R. Lee Clark to Logan Wilson, Sept. 26, 1957; R. Lee Clark to Harry Ransom, May 23, 1961; Minutes, Jan. 10, 1957, MDA, RLC, Medical Affairs Council.

5. Gerald Dodd interview; *Pitt Med* (Apr. 2001): 31–33; *American Journal of Radiology* 150 (Mar. 1988): 493–98; *Newsletter* 5 (Oct. 1960): 1; *Newsletter* 6 (Aug. 1961): 1–2; Dodd, *A History*, 27–29.

6. *Newsletter* 4 (Dec. 1959): 1; *Newsletter* 6 (Aug. 1961): 2; *Newsletter* 7 (Sept. 1962): 2–4; Gilbert Fletcher to Head and Neck Service, Jan. 1966, MDA, RLC, Department of Radiotherapy; *Newsletter* 3 (Dec. 1958): 3.

7. *Messenger* 10 (Feb. 1981): 8; Dodd, *A History*, 29, 117–18. By 2007, more than one-third of M. D. Anderson's fifty-seven radiation oncologists were women.

8. www.vanderbilt.edu/alumni/publications/pr/reflector_su04_2.pdf; MDA, RLC, Coordinating Committee Minutes, Feb. 23, 1959; Renilda Hilkemeyer interview; Charles A. LeMaistre interview; Joyce Alt interview; *Newsletter* 1 (Apr. 1956): 4.

9. Richard C. Hay, Takeshi Yonezawa, and W. S. Derrick, "The Control of Intractable Pain in Advanced Cancer by Subarachnoid Alcohol Block," *Journal of the American Medical Association* 169 (Mar. 21, 1959): 1315–20; Mary Lou Rogge interview.

10. J. P. Chang, W. O. Russell, E. B. Moore, and W. K. Sinclair, "A New Cryostat for Frozen Section Technic," *American Journal of Clinical Pathology* 35 (Jan. 1961): 14–19; *Newsletter* 2 (June 1957): 1.

11. J. S. Stehlin and R. Lee Clark, "Perfusion of the Extremities. Techniques, Perfusion Conference," *Cancer Chemotherapy Reports* 10 (Dec. 1960): 3–4.

12. Helmuth H. Goepfert interview; H. J. C. Goldstein and G. A. Sisson Sr., "The History of Head and Neck Surgery," *Otolaryngology Head and Neck Surgery* 115 (Nov. 1996): 379–85; *American Journal of Surgery* 136 (Oct. 1978): 414; *Newsletter* 4 (Aug. 1959): 1.

13. *Newsletter* 1 (May 1956): 4; *Newsletter* 1 (Dec. 1956): 1–2; Richard G. Martin interview.

14. Crile, *Cancer and Common Sense*; *Life*, Oct. 31, 1955, 126–31.

15. Richard G. Martin interview; Gilbert Fletcher interview; Lerner, *Breast Cancer Wars*, 94–95, 109–15; Olson, *Bathsheba's Breast*, 103–5.

16. Ralph Freedman interview; "Cumulative Percent Survivals for Total Amputations by Age Groups," M. D. Anderson Hospital; MDA, RLC, Rehabilitation Symposium (New York Academy of Sciences).

17. Philip A. Kreiger to R. Lee Clark, June 10, 1968; Wilfred D. Keith to R. Lee Clark, May 9, 1968; MDA, RLC, Rehabilitation Symposium (New York Academy of Sciences).

18. Felix Rutledge to R. Lee Clark, June 15, 1972, MDA, RLC, Gynecology, Department of, Administration; *New York Times*, June 27, 1997; Grant Taylor to R. Lee Clark, June 25, 1962, MDA, RLC, Director Advisory Committee Minutes.

19. R. Lee Clark interview, 18; *Texas Times* 5 (Mar. 1977): 12.

20. Minutes, Coordinating Committee, Dec. 23, 1957, MDA, RLC, Minutes of the Coordinating Committee, Coordinating Committee Minutes; Report to the Research Committee, Oct. 25, 1961, MDA, RLC, Director Advisory Committee Minutes. Clifton F. Mountain to Robert C. Hickey, May 31, 1963, MDA, RLC, Biomathematics and Computer Science Committees.

21. A. Clark Griffin to R. Lee Clark, John Stehlin, William O. Russell, and Darrell N. Ward, Oct. 17, 1963, MDA, RLC, Department of Biochemistry, Administration; "A. Clark Griffin," *Cancer Letters* 20 (1983): 247–48; *Newsletter* 5 (Dec. 1960): 1.

22. G. D. Adams, "Formation and Early Years of the AAPM," *Medical Physics* 5 (July–Aug.1978): 290–96; *New York Times*, Apr. 15, 1964; Hynes, *The Recurring Silent Spring*, 2–4, 16–20, 30–34; McCay, *Rachel Carson*, 2–5; 40–44.

23. Leon Dmochowski, "Viruses and Tumors in the Light of Electron Microscopic Studies: A Review," *Cancer Research* 20 (Aug. 1960): 993–1003; *Newsletter* 7 (Dec. 1962): 2.

24. R. Lee Clark to Harry H. Ransom, Aug. 19, 1961; J. N. P. Davies to Leon Dmochowski, June 12, 1961; Leon Dmochowski to John F. Dominick, June 20, 1961; Leon Dmochowski to R. Lee Clark, Apr. 20, 1961, MDA, RLC, African Research Foundation, Special Projects, 1961–1962; *Time*, July 18, 1960, 45.

25. Spangenburg and Moser, *Disease Fighters*; www.whonamedit.com/doctor.cfm/2199.html.

26. Sharif, *Fury on Earth*, 298–303; *Science Newsletter* 63 (June 13, 1953): 366–68; *Psychoanalytic Review* 42 (1955): 217–27; *New York Times*, Nov. 3, 1957; *Time*, Aug. 16, 1954, 40; Edna Wagner to R. Lee Clark, Mar. 25, 1952; Clifton Read to Beatrix Cobb, June 15, 1953, RLC Papers, MDA, AF, MP.

27. Edna Wagner to R. Lee Clark, Mar. 26, 1952; Heflebower memo, Sept. 21, 1954; "Advisory Council Meeting for the Medical Psychology Training Program," Apr. 9, 1953; "Psychosocial Factors and Neoplasia Panel," Feb. 1954 MDA, AF, SS.

28. Department of Surgery, "Psychiatric and Psychological Service," Oct. 2, 1959; Beatrix Cobb to R. Lee Clark, Aug. 27, 1958; Beatrix Cobb to Joe E. Boyd, July 8, 1958; R. Lee Clark to Gilbert H. Fletcher, Clifton D. Howe, and E. C. White, Aug. 10, 1959, MDA, RLC, Part 1, Series 2, Administration, Medical Psychology. Two decades later, medical psychologists and physicians would again study the relationship between personality, depression, and treatment outcomes, but notions of carcinogenic personalities and psychosexual etiologies of cancer remained highly controversial. Charles A. LeMaistre, who would succeed Lee Clark at M. D. Anderson, said he is "certain that this episode . . . biased a number of faculty against further support of psychological and psychiatric services." Charles A. LeMaistre to James S. Olson, Apr. 2007.

29. Gerald Dodd interview; Eleanor Macdonald interview.

30. James Bowen interview; Marion Wall (Lowrey) interview; Teaching Activities Program. Report of Associate Director, Office of Education (Oct. 1964), MDA, RLC, Office of Education, Administration, 1971; *Newsletter* 5 (Sept. 1960): 1–2.

31. In English the International Union against Cancer.

32. *New York Times*, Oct. 8–11, 1957, and May 26, 1961.

33. Murray M. Copeland, "Contacts with Russian Physicians and Scientists at Georgetown University Medical School," Oct. 18, 1957; Murray M. Copeland, "Contacts with Russian Physicians and Scientists while a Member of the TNM Committee on Clinical Staging," MDA, RLC, Soviet Interaction, 1973; Eleanor Macdonald interview.

34. *Newsletter* 6 (Aug. 1961): 8; *Newsletter* 8 (Oct. 1963) 1–2; Richard G. Martin interview.

CHAPTER 6: M. D. Anderson and the Rise of Medical Oncology, 1964–1969

1. *Journal of the American Medical Association* 263 (1985): 2995–97.

2. *New York Times*, Jan. 12–14, 1964. For a history of the Surgeon General's Advisory Committee, see R. Kluger, *Ashes to Ashes*, 223–61; Charles A. LeMaistre to James S. Olson, Apr. 2007, James S. Olson File.

3. Roberts and Olson, *John Wayne*, 513–15.

4. Patti Watson LeClear interview.

5. Dorothy Howard to Renilda Hilkemeyer, May 4–5, 1964, MDA, RLC, Part 1, Series 2, Administration, Nursing, Department of, 1973.

6. Gilbert Fletcher to Richard H. Jesse Jr., Sept. 18, 1968; Gilbert Fletcher to John Bardwil, Mar. 24, 1967; Gilbert Fletcher to Richard Jesse, Dec. 28, 1967; Gilbert Fletcher to Robert Hickey, Jan. 13, 1969; Gilbert Fletcher to Head and Neck Service, Jan. 1966, MDA, RLC, Department of Radiotherapy, Administration.

7. Gilbert Fletcher to William S. MacComb, Jan. 5, 1967; William S. MacComb to

R. Lee Clark, Dec. 13, 1966; Gilbert Fletcher to Robert Hickey, Sept. 12, 1968 and Jan. 13, 1969; Gilbert H. Fletcher to A. J. Ballantyne, Jan. 5, 1967; Gilbert Fletcher to R. Lee Clark, Apr. 3, 1968; Gilbert Fletcher to Alando Ballantyne, Oct. 3, 1968; R. Lee Clark Memo, Jan. 26, 1969; MDA, RLC, Department of Radiotherapy, Administration.

8. Gilbert Fletcher to Robert Hickey, Sept. 18, 1968; R. Lee Clark Memo, Jan. 26, 1969; Gilbert Fletcher to Head and Neck Service, Jan. 1966; Gilbert Fletcher to Robert Hickey, Jan. 13, 1969, and Nov. 16, 1970; Gilbert Fletcher to Robert Moreton, Nov. 16, 1970; Gilbert Fletcher to William S. MacComb, Alando J. Ballantyne, Richard H. Jesse Jr., and John Bardwil, July 21, 1966; Gilbert H. Fletcher to R. Lee Clark, Dec. 6, 1966; Gilbert Fletcher to William S. MacComb, Jan. 5, 1967; William S. MacComb to R. Lee Clark, Dec. 13, 1966; Gilbert Fletcher to Alando Ballantyne, Jan. 5, 1967, MDA, RLC, Department of Radiotherapy, Administration.

9. Gilbert Fletcher, "Progress Report: Jan. 1, 1962–Nov. 1, 1967," 4, 5, 11–15; Gilbert Fletcher to Head and Neck Service, Jan. 1966, MDA, RLC, Part 1, Department of Radiotherapy, Administration.

10. "Progress Report," 18–22, 27, 74–75.

11. Ibid., 18–22, 74–75; H. D. Kerman to Gilbert Fletcher, Dec. 10, 1970, MDA, RLC, Department of Radiotherapy, Administration.

12. Dodd, *A History*, 117–18.

13. Marion Wall to Joe E. Boyd Jr., Aug. 9, 1966; R. Lee Clark to section chiefs, July 11, 1966, MDA, RLC, Department of Radiotherapy, Administration; Gerald Dodd interviews (UTMDA and author); Dodd, *A History*, 23–25.

14. Lazlo, *The Cure*, 205–6.

15. Ibid., 93–107.

16. Ibid., 156–89; Emil Frei interview; Emil Freireich interviews (UTMDA and author).

17. Lazlo, *The Cure*, 108.

18. Ibid., 108–55; Emil Frei interview; Emil Freireich interviews (UTMDA and author).

19. *Cancer Research* 26 (1966): 1284–89; Emil Freireich interview; Edmund Gehan interview; Lazlo, *The Cure*, 179; Gerald Bodey interview.

20. ASCO, *40 Years of Quality Care*, 5–15.

21. Emil Frei interview; Emil Freireich interviews (UTMDA and author); Gerald Bodey interview.

22. Kenneth Endicott to Lee Clark, Jan. 13, 1965, MDA, RLC, Regional Medical Program, University of Texas, 1966; Kardinal, "Cancer Chemotherapy," 165–74; R. Lee Clark to Secretary Cohen, July 1, 1968, LBJ Papers, Lyndon B. Johnson Presidential Library, White House Central File, C Box 251, R. Lee Clark.

23. *Biometrika* 52 (1965): 203–23.

24. *Cancer* 32 (1973): 1150–53; Richard G. Martin interview; Eleanor Macdonald interview; William Seybold interview; Norman Jaffe interview; *Newsletter* 11 (Aug. 1966): 3; *Newsletter* 10 (Dec. 1965): 1; *Newsletter* 11 (Aug. 1966): 3.

25. *Newsletter* 12 (Sept. 1967): 1; Emil Freireich interview; Gerald Bodey interview; Lazlo, *The Cure*, 190–96.

26. Ralph Freedman interview.

27. *Messenger* (Sept. 1972): 3; *Medical Tribune*, Dec. 7, 1966; "Cancer Workshops, Princeton," July 10, 1968; Philip A. Kreiger to June 10, 1968; Wilfred D. Keith to R. Lee

Clark, May 9, 1968; R. Lee Clark to Participants of the Second Conference on the Rehabilitation of the Cancer Patient, Apr. 22, 1968; Minutes, "Symposia on Rehabilitation of the Cancer Patient," June 20, 1966; MDA, RLC, Rehabilitation Symposium (New York Academy of Sciences), 1966 Special Projects; Richard G. Martin interview.

28. *Journal of the National Cancer Institute* 60 (Mar. 1978): 545–71.

29. *Newsletter* 11 (Aug. 1966): 3; *Newsletter* 12 (Apr. 1967): 1.

30. *Newsletter* 12 (Apr. 1967): 1; *Houston Chronicle*, June 22, 1969.

31. Eleanor Macdonald interview; Gerald Dodd interview; James Bowen interview; Helmuth Goepfert interview; *Newsletter* 13 (May 1968): 2–3; *Newsletter* 14 (Sept. 1969): 3; *Newsletter* 9 (Feb. 1964): 3; *Newsletter* 9 (May 1964): 1–2; *Newsletter* 10 (Jan. 1965): 1–2; *Newsletter* 11 (Aug. 1966): 3; *Newsletter* 13 (May 1968): 2–3; *Newsletter* 13 (Sept. 1968): 8; R. Lee Clark to Department Heads, May 23, 1969, MDA, RLC, Memorandums, Administration 1971.

32. King, *Confessions of a White Racist*, 7, 12–13; Eleanor Macdonald interview; Martha Jones interview; Taylor, *Remembrances and Reflections*.

33. Mavis P. Kelsey interview; Martha Jones interview; Eleanor Macdonald interview; Taylor, *Remembrances and Reflections*, 184; Minutes, Education Committee, Nov. 5, 1949, MDA, RLC, Education Committee, Committees MDAH; Hesta Anderson interview; Rosie L. Baines interview; Blanche Polunsky interview; Bettie Aldridge interview; Armandina Blanco interview; Dodie Frost interview; Lovell Jones interview; "Minutes," Aug. 11, 1965, MDA, RLC, Administrative Committees, 1971.

34. Edward Munger interview; Martha Jones interview; Armandina Blanco interview; Rosie Baines interview; Bettie Aldridge interview; Hesta Anderson interview.

35. Eva Child to Arthur Kleifgen, Mar. 15, 1960; Glenn M. Johnson to Arthur F. Kleifgen, Nov. 16, 1964; Arthur F. Kleifgen to Madeline Long, Aug. 22, 1966; Minutes of the OES Board Members and M. D. Anderson Hospital Administrative Staff, July 9, 1964; Donald R. Olson OES Report, Feb. 21, 1964; R. Lee Clark to Mr. W. C. Long, Oct. 13, 1966. The program would later be revived and today remains available to patients of all races.

36. Martha Jones interview; Frances Coatney interview; Renilda Hilkemeyer interview; *New York Times*, Aug. 29–30, 1963; *Birmingham News*, July 16, 1967.

37. Oncology, 1970, 1–3; *Newsletter* 11 (Nov. 1966): 1; *Newsletter* 12 (Nov. 1967): 1; Murray Copeland to R. Lee Clark, Jan. 3, 1973, MDA, RLC, Soviet Interaction, Committees UT-HAC, 1973; *Newsletter* 11 (Aug. 1966): 1; *Newsletter* 12 (Dec. 1967): 1–4.

CHAPTER 7: The Summit, 1970–1971

1. *Houston Post*, May 22 and 27, 1970; *Newsletter* 11 (Dec. 1966): 1; *Newsletter* 12 (Nov. 1967): 1; *Newsletter* 13 (May 1968): 5–6; *Newsletter* 14 (June 1969): 5; *Newsletter* 15 (Oct. 1970): 1–5; Alan Scott to Murray Copeland, Feb. 18, 1970, MDA, RLC, Department of Personnel, Administration, Needs Group for 10th.

2. Marion Wall (Lowrey) interview; Eleanor Macdonald interview.

3. *Houston Post*, May 15, 1970.

4. *New York Times*, Apr. 23–24, 1970; *Houston Chronicle*, Apr. 23–24, 1970.

5. Brynner and Stephens, *Dark Remedy*.

6. *Houston Post*, June 10, 1970; *New York Times*, Apr. 4; May 6; and Oct. 18–21, 1969; Troetel, "Three-Part Disharmony," 3–23.

7. Arthur L. Herbst, Howard Ulfelder, and David C. Poskanzer, "Adenocarcinoma of the Vagina: Association of Maternal Stilbesterol Therapy with Tumor Appearances in Young Women," *New England Journal of Medicine* 284 (1971): 878–81.

8. Harry Ricketts interview; Randy Roberts interview; Marion Wall (Lowrey) interview; *Houston Post*, May 1, 1970.

9. *New York Times*, Apr. 1–4, 1970; *Houston Post*, May 23, 1970; Marion Wall (Lowrey) interview.

10. *Houston Chronicle*, May 22–25, 1970; *New York Times*, May 24, 1970; Harry Ricketts interview.

11. Harry Ricketts interview; *Oncology*, 1970, 4–5.

12. *Houston Chronicle*, Apr. 11–20, 1970.

13. *Houston Post*, May 27, 1970.

14. International Union against Cancer, *Abstracts*, 827, 204.

15. Ibid., 240.

16. Ibid., 596–97.

17. Ibid., 188–89, 818, 827, 841.

18. Ibid., 661–64, 818; *Houston Post*, May 24, 1970.

19. R. Alexanian, A. U. Khan, et al., "Treatment for Multiple Myeloma: Combination Chemotherapy with Different Melphalan Dose Regimens," *Journal of the American Medical Association* 208 (1969): 1680–85; Raymond Alexanian interview; International Union against Cancer, *Abstracts*, 327, 458–59, 488.

20. International Union against Cancer, *Abstracts*, 123, 150–51, 161, 164–68, 204–5, 208, 215, 511, 834; James Bowen interview.

21. *Anderson Messenger* 4 (Aug.–Sept. 1975): 5; Emil Freireich interview; Melvin Samuels interview.

22. Brent Ferrin interview; Melvin Samuels interview; Emil Freireich interview; Steve Stuyck interview; M. L. Samuels, P. Y. Holoye, and D. E. Johnson, "Bleomycin Combination Chemotherapy in the Management of Testicular Cancer," *Cancer* 36 (1975): 318–26.

23. Cox, *Ralph W. Yarborough*, 150–59, 214–19.

24. *Science* (Mar. 20, 1964): 216–27; R. Lee Clark to Edward Dempsey, Mar. 1, 1965, LBJ Papers, White House Central File, Name file, "C" Box 251, R. Lee Clark folder.

25. *New York Times*, Dec. 9, 1969.

26. Aaron Bendich to J. H. Burchenal, July 15, 1970; R. Lee Clark to Ralph W. Yarborough, June 1, 1970, MDA, RLC, Yarborough Panel of Consultants, June 1970.

27. *Wall Street Journal*, Aug. 26, 1970.

28. Cox, *Ralph W. Yarborough*, 239–47, 260–62.

29. *Drug Researcher Reporter* 13 (1970): 14–16.

30. W. K. Joklik to Joseph H. Burchenal, Aug. 6, 1970; Robert L. Sinsheimer to Mathilde Krim, July 13, 1970; Harry Rubin to Mathilde Krim, Aug. 12, 1970; Heinz Fraenkel-Conrat to Mathilde Krim, July 6, 1970; Benno Schmidt to R. Lee Clark, Sept. 24, 1970; "Report of Meeting on Cancer Research," Aug. 22, 1970; Presentation of James F. Holland, President, American Association for Cancer Research to Subcommittee on Labor, Health, Education and Welfare, Committee on Appropriations, House of Representatives, June 3, 1970; J. K. Joklik to Joseph H. Burchenal, Aug. 6, 1970, MDA, RLC, Part 1, Panel of Consultants, June 1970.

31. *The First Twenty Years*; *New York Times*, Dec. 5–7, 1970; R. Lee Clark interview.

32. R. Lee Clark, "M. D. Anderson's Legacy to Texas," *Texas Medicine* 64 (Oct. 1968): 85; *Washington Post*, Dec. 5–6, 1970; *Science*, 174 (Oct. 8, 1971): 127–31; *Science*, 176 (Apr. 28, 1972): 353–437.

33. James Ernstrom and Donald Austin, "Interpreting Cancer Survival Rates," *Science* 195 (Mar. 14, 1977): 847–51; *Newsweek*, Feb. 22, 1971, 84.

34. *New York Times*, Sept. 16–18, 1971; *Washington Post*, Sept. 15–18, 1971.

35. *Science* 174 (Dec. 24, 1971): 1306–11. Also see Beeman, "Robert J. Huebner, M.D.: A Virologist's Odyssey."

36. Fred Rapp, Renato Dulbecco, and J. Earle Officer, "Special Virus Cancer Program," *Science* 175 (Mar. 10, 1972): 1061–63; *Newsletter* 16 (Aug. 30, 1971): 1; *Nature* 232 (July 14, 1971): 61–62; Jane Brandenberger to R. Lee Clark, July 9, 1971; Richard Nixon to Dr. Priori, Aug. 4, 1971; R. Lee Clark to Richard M. Nixon, Aug. 25, 1971; MDA, RLC, Type C Virus (ESP-1), 6.30; *Science News* 100 (July 10, 1971): 21–22.

37. James Bowen interview.

38. B. J. Kolenda to Leon Dmochowski, Oct. 26, 1971, MDA, RLC, Administration, Virology, Type C; *Houston Post*, Sept. 11, 1971; *Medical World News*, Sept. 17, 1971, 31; Robert J. Huebner to Dr. Manaker, Aug. 17, 1971; Elizabeth S. Priori to R. Lee Clark, Aug. 27, 1971; "Girls Note Memo," Aug. 27, 1971; "JAH for RLC" Memo, Sept. 23, 1971; MDA, RLC, Type C. Virus, 6.30; *Nature* 232 (Sept. 10, 1971): 102–3.

39. *Washington Post*, Sept. 6, 1971; *New York Times*, Sept. 16, Nov. 16, Dec. 10–11, and 24, 1971.

40. *Newsletter* 17 (June 1972): 1; *Houston Post*, Dec. 14, 1971. For a history of the National Cancer Act of 1971, see Rettig, *Cancer Crusade*.

41. www.bearshistory.com/lore/brianpiccolo.aspx.

CHAPTER 8: Waging War and Fading Away, 1971–1977

1. *Messenger* 20 (Apr. 1975): 1–2; Richard Cording interview; Frances Cording interview.

2. Ralph Arlinghaus to James S. Olson, Jan. 10, 2008.

3. *JUMANA* 50 (2005): 62–65; Administrative Staff Meeting, Nov. 18, 1968, MDA, RLC, Administrative Staff Meeting, Committees, MDAH, 1971.

4. James Bowen interview; *Anderson Messenger* 1 (1972): 1.

5. *Washington Post*, May 1 and 12, 1977; Charles A. LeMaistre interview; Joseph Patrick Kennedy, "Environmental Science Park, Outline," Chapter 4, and John Jardine, "Section of Experimental Animals," Center for American History, University of Texas at Austin.

6. *Messenger* 6 (Apr. 1977): 12; *Messenger* 6 (Sept. 1977): 5.

7. Richard G. Martin interview; Marion Wall (Lowrey) interview; R. Lee Clark to Lynne Turnage, Nov. 18, 1975, MDA, RLC, Development Office (1975), 6.6.

8. Lutheran Memorial Hospital Brochure, MDA, RLC, Inquiries, General Florida Land, Special Projects.

9. *Anderson Messenger* 1 (Nov. 1972): 1.

10. Joseph L. Kunic to Elmer R. Gilley, Aug. 5, 1971; Joseph Kunic to R. Lee Clark, Sept. 27, 1971, MDA, RLC, Development Office of MDAH, 1971, Administration; *Houston Chronicle*, Dec. 14, 1970; *Houston Post*, Dec. 5, 1972; *Fort Worth Star-Telegram*, Feb. 2, 1973.

11. *New York Times*, Oct. 2, 1971. See Gardner, *Early Detection*, 132–39; *McCall's*, Feb. 1973, 82, 112–14; Norman Jaffe interview.

12. Steve Stuyck interview; Norman Jaffe interview; Minutes, Administrative Staff Meeting, June 5, 1972, MDA, RLC, Administrative Staff Mtg., Minutes, Committee, 1972.

13. http://cancercenters.cancer.gov/about/our-history/html; *Newsletter* 17 (June 1972): 1; *Houston Post*, Dec. 14, 1971; Paul Van Nevel interview; John S. Spratt to R. Lee Clark, Feb. 26, 1974, MDA, RLC, Cancer Centers, 1972–74, 17.2.

14. http://cancercenters.cancer.gov/about/our-history/html; *Newsletter* 17 (June 1972): 1. For a history of the National Cancer Act of 1971, see Rettig, *Cancer Crusade*; *Houston Post*, Dec. 14, 1971; Paul Van Nevel interview.

15. Frank J. Rauscher Jr. to Saul Rosenman, July 10, 1973 and June 29, 1973; Benno C. Schmidt to R. Lee Clark, May 15, 1975; Saul Rosenman to Frank J. Rauscher Jr., June 29, 1973; "Statement by Representatives of the Councils of the National Institutes of Health to the President's Biomedical Research Panel," Sept. 30, 1975, 17, MDA, RLC, General (1973–74), 2.1; *Wall Street Journal*, Feb. 3, 1975; Nathaniel Polster to R. Lee Clark, Aug. 14, 1975; R. Lee Clark, "Confidential Memo to Benno C. Schmidt, Ray D. Owen, and Frank. J. Rauscher," Sept. 10, 1974, MDA, RLC, General (1975), 6.1.

16. *Anderson Messenger* 1, no. 2, 5.

17. Arthur B. Pardee to Benno Schmidt, July 20, 1977; R. Lee Clark to Jimmy Carter, July 31, 1978; James F. Hagerty to R. Lee Clark, Aug. 24, 1977, MDA, RLC, General (1975), 6.1.

18. Frank P. Rauscher to Saul Rosenman, May 1, 1973; Rulon W. Rawson to Frank Rauscher Jr., Mar. 17, 1974; Saul Rosenman to Rulon W. Rawson, n.d., MDA, RLC, General (NCI) (1978), 4.1.

19. "Discussion with Dr. Jonathan Rhoads," Memorandum to the Files, Mar. 28, 1974, MDA, RLC, National Cancer Advisory Board, 7.0.

20. Thomas D. Anderson interview; Rulon W. Rawson, "Report on Cancer Activities of the University of Texas System"; Minutes, Administrative Staff Meeting, Oct. 16, 1972, MDA, RLC, General (1975).

21. R. Lee Clark to E. C. DeLand, May 2, 1969, and Sept. 24, 1970; Stuart O. Zimmerman to R. Lee Clark, Apr. 9, 1969, and Sept. 14, 1970; R. Lee Clark to J. R. Goldstein, May 30, 1969; Dick Goldstein to R. Lee Clark, Nov. 10, 1968, MDA, RLC, Rand Corporation, Biomathematics, ADM, 1969; "Informatics Meeting," Jan. 16–18, 1974, MDA, RLC, Informatics, Inc., (1973–74), 8.2; R. Lee Clark to Kenneth M. Endicott, Aug. 16, 1968, MDA, RLC, Biomathematics, Dept. Of Administration, 1971; Stuart O. Zimmerman interview.

22. Robert D. Moreton to Robert J. Turner, July 17, 1972, MDA, RLC, Medicine, Dept. of Administration; Joseph T. Ainsworth interview.

23. Steve Stuyck interview; Joseph T. Painter interview; Joseph T. Ainsworth interview; Vincent Guinee interview; *Anderson Messenger* 4 (Mar.–Apr. 1975): 1.

24. U.S. Public Health Service, "Investigations Involving Human Subjects," Dec. 12, 1966; "The Institutional Guide to DHEW Policy on Protection of Human Subjects," Dec. 1971; D. T. Chalkley to R. Lee Clark, Aug. 3, 1972, MDA, RLC, Surveillance Committee, Committees, MDAH, 1973–74.

25. Jones, *Bad Blood*; *Washington Star*, July 25, 1972.

26. James Bowen to Files, Aug. 24, 1976, MDA, RLC, General (1975), 6.1; Marion Wall (Lowrey) memo to R. Lee Clark, June 4, 1976, MDA, RLC, Information Office (1975),

6.12.1; *Houston Post*, May 3, 1976; "Statement of Emil J Freireich, Oct. 1, 1976," MDA, RLC, General (1975), 6.1.

27. "Activities Involving Human Subjects," 1974; Felix Haas to Raymond Alexanian, Oct. 28, 1974; Jeffrey Gottlieb to C. C. Shullenberger, Mar. 20, 1973; M. D. Anderson, "Activities Involving Human Subjects: Code and Methods of Procedure," all in MDA, RLC, Memorandums, Administration, 1971; Robert C. Hickey to Stuart O. Zimmerman, Sept. 20, 1972, MDA, RLC, Research Committee, Committees, MDAH 1971; Norman Jaffe interview.

28. Guither, *Animal Rights*, 1–34; also see Scully, *Dominion*.

29. Robert Hickey to Department Heads, Oct. 4, 1971, MDA, RLC, Memorandums, Administration, 1971; Minutes, Dec. 8, 1976, Animal Resources and Physical Facilities Committee, MDA, RLC, Animal Resources and Facilities, Advisory Committee (1975), 13.3.

30. R. Lee Clark to A. M. Aiken, Feb. 16, 1978, MDA, RLC, Legislature, Texas (1975), 17.22.1; R. Lee Clark to Mary Lasker, Apr. 19, 1977, MDA, RLC, 13.2; *Houston Chronicle*, July 19, 1978; R. Lee Clark to Dolph Briscoe, June 10, 1975, MDA, RLC, Legislative Biennial, 1973–1975, 26.1; Joyce Alt interview; Eleanor Macdonald interview.

31. *SAWDUST* Memo, Feb. 27, 1976, MDA, RLC, Information Office (1975), 6.12.1; Raphael Pollock interview; Marion Wall (Lowrey) interview; Charles A. LeMaistre interview; James Bowen interview.

32. Clifton D. Howe to Kenneth B. McCredie, Dec. 6, 1973; Clifton D. Howe to Jeffrey A. Gottlieb, Dec. 6, 1973; Clifton D. Howe to Jeane P. Hester, Dec. 6, 1973, MDA, RLC, Administration, Clinics; Clifford D. Howe to R. Lee Clark, July 14, 1970, MDA, RLC, Department Head Reports on Committee on Cancer; George R. Blumenschein to Robert C. Hickey, Aug. 11, 1976, MDA, RLC, Medicine (1975), 6.16; C. Stratton Hill interview.

33. T. C. Hsu to Robert C. Hickey and R. Lee Clark, Feb. 6, 1978; Arthur B. Pardee to Benno Schmidt, July 20, 1977; R. Lee Clark to Jimmy Carter, July 31, 1978, MDA, RLC, General (1975), 6; *Cytogenetic and Genome Research* 105 (2004): 1–3; Felix Haas to R. Lee Clark, July 8, 1970, MDA, RLC, Department Head Reports on Committee on Cancer; Frederick Becker interview; "2 PO1 CA 05831-15A1," 23, MDA, RLC, Cancer Clinical Research Grant CA 05831 (1975), 10.3.3; James Bowen interview.

34. *Anderson Messenger* 5 (Oct.–Nov. 1976): 3; John R. Bush to Lynne Turnage, Nov. 18, 1975, MDA, RLC, Development Office (1975), 6.6.

35. See press release, the University of Texas System, July 9, 1976, MDA, RLC, Development Office (1975), 6.6; *Anderson Messenger* 1 (1972): 5.

36. Richard G. Martin interview; George Blumenschein interview; Olson, *Bathsheba's Breast*, 127–42; Howard Skipper, "Some Thoughts on Cancer Research," typescript, Nov. 1972, MDA, RLC, Minutes, Agendas of NCAB, 1972, 7.3; *Oncology News* (May–June 1975): 4.

37. *Anderson Messenger* 6 (Apr. 1977): 11; Rough Draft, "Cancer Chemotherapy Reference," June 29, 1976, MDA, RLC, Pharmacy (1975), 2.1.

38. *Times-Herald* [Dallas], June 1, 1977; Robert C. Hickey to File, Nov. 13, 1978, MDA, RLC, General, Research, 1970–1979, 10.1; Minutes, Cancer Clinical Research Grant Advisory Committee, Nov. 17, 1975, MDA, RLC, Cancer Clinical Research Grant Advisory Committee (1975), 13.4; *Newsletter* 21 (Sept.–Oct. 1976): 7; *Messenger* 16 (Aug. 1987): 3.

39. Emil Freireich interview; Richard G. Martin interview; Melvin Samuels inter-

view; *Anderson Messenger* 4 (Aug.–Sept. 1975): 5; R. Lee Clark to Frederick Silber, July 22, 1975, MDA, RLC, General (1975) 18.1.

40. Richard G. Martin interview; Marion Wall (Lowrey) interview; Helmuth Goepfert interview.

41. Minutes, Administrative Council of the UTHSC/UTSCC, (1975), 1.1; *Anderson Messenger* 6 (Feb. 1977): 12.

42. "A Look at the M. D. Anderson Hospital and Tumor Institute," MDA, RLC, Development Office (1975), 6.6. Also see "Affiliation Agreements," MDA, RLC, Affiliation Agreements (Official Filed in JTP Area), 1976, 7.22.

43. Dodd, *A History*, 36–38, 74–77, 119–20; www.imaginis.com/ct-scan/history.asp.

44. L. C. Strong, W. R. Williams, M. A. Tainsky, "The Li-Fraumeni Syndrome," *American Journal of Epidemiology* 135 (1992): 190; D. E. Anderson, W. B. Falls, R. T. Davidson, "The Nevoid Basal Cell Carcinoma Syndrome," *American Journal of Human Genetics* 19 (1967): 12; *Messenger* 33 (June 2004): 6–7; Minutes, President's Advisory Council, Feb. 26, 1973, MDA, RLC, PAC, Minutes, 1972.

45. Frederick Becker interview; *Messenger* 6 (Sept. 1977): 3; Evan Hersh interview; Jordan Gutterman interview; Evan M. Hersh to R. Lee Clark, July 21, 1970, MDA, RLC, Department Head Reports on the Committee on Cancer.

46. Richard G. Martin interview; *Anderson Messenger* 6 (Sept. 1977): 6.

CHAPTER 9: Charles A. LeMaistre and the Consolidation of Excellence, 1978–1983

1. Raphael Pollock interview; Richard G. Martin interview.

2. See the case of Bruce Wood, MDA, CAL, 1I, Grievance Procedure, 83–84.

3. Charles A. LeMaistre interview; James Bowen interview; Steve Stuyck interview.

4. *Covington [Ala.] News*, June 27, 1929; *Andalusia [Ala.] Star News*, June 25, 1929; *Dothan [Ala.] Eagle*, June 11, 1909; Charles A. LeMaistre interview.

5. Mary Jane Schier, "Leaving a Legacy of Excellence," *Conquest* 11 (Summer 1996): 8–12; Charles A. LeMaistre interviews (UTMDA and author).

6. David S. Jones, "The Health Care Experiments at Many Farms: The Navajo, Tuberculosis, and the Limits of Modern Medicine, 1952–1962," *Bulletin of the History of Medicine* 76.4 (2002): 749–90; *Time*, Mar. 3 1952, 44; Charles A. LeMaistre interview.

7. Schier, "Leaving a Legacy of Excellence,"13–17; Charles A. LeMaistre interview.

8. Charles A. LeMaistre, "Welcoming Comments," Patient Advisory Committee, Mar. 25, 1988, CAL Personal Papers; Charles A. LeMaistre to James S. Olson, Apr. 2007, James S. Olson File.

9. Frederick Becker interview.

10. Frederick Becker interview.

11. Minutes, May 6, 1988, "Committee to Evaluate the Status of Minority and Women Faculty and Administrators; Faculty Analysis as of 2-29-88"; Margaret Kripke to Charles A. LeMaistre, Mar. 1, 1988; Charles A. LeMaistre to Margaret Kripke, Feb. 25, 1988; Margaret Kripke to Isaiah W. Dimery, Apr. 19, 1988, MDA, CAL, Committee to Evaluate the Status of Minority and Women Faculty, 1987–88; Elizabeth Travis interview; Charles A. LeMaistre interview; Lovell Jones interview.

12. Committee on Diet and Health, *Diet and Health*.

13. Bruce Ames, "Mother Nature Is Meaner Than You Think," *Science* 84 (July–

Aug. 1984): 98; Epstein, *Politics of Cancer*; Proctor, *Cancer Wars*, 57–64; Efron, *The Apocalyptics*.

14. 92 STAT. 3412, Public Law 95-622, Nov. 9, 1978.

15. *Messenger* 8 (Oct. 1979): 3; Peter W. A. Mansell interview (UTMDA).

16. *Messenger* 8 (Oct. 1979): 5; Charles A. LeMaistre to Department Heads, Section Chiefs, and Administrative Officers, Aug. 30, 1982; Frederick Becker interview.

17. *Anderson Messenger* 5 (May–June 1976): 10; James Bowen interview; Charles A. LeMaistre interview; Roger K. Schomburg to John Sheridan, Aug. 19, 1980, MDA, CAL, 1 J Nutrition and Food Service, 81–82.

18. Report on Cost Containment and Management Efficiency, Nov. 13, 1984, MDA, CAL, 1A. Public Information and Evaluation, 84–85.

19. *Washington Post*, Oct. 19, 1981.

20. "Response of the University of Texas System Cancer Center to the First Biennial Report of the President's Commission for the Study of Ethical Problems in Medicine and Biomedical Research," May 27, 1982, MDA, CAL, 14.NIH, 81–82; Ralph Freedman interview.

21. Irwin Krakoff interview; Charles A. LeMaistre to Ti Li Loo, June 30, 1981; Charles A. LeMaistre to Boh Seng Yap, June 30, 1981, MDA, CAM, 14, NCI; Emil J Freireich to Charles A. LeMaistre, Aug. 28, 1978, MDA, CAL, 10.1a, Food and Drug Administration.

22. Ralph Freedman to James S. Olson, Dec. 16, 2007, James S. Olson File.

23. Steve Stuyck interview; Charles A. LeMaistre interview.

24. Response of the University of Texas System Cancer Center to the First Biennial Report of the President's Commission for the Study of Ethical Problems in Medicine and Biomedical Research, May 27, 1982; Charles A. LeMaistre to Carol Young, May 27, 1982; MDA, CAL, 14.NIH, 81–82; *Newsweek*, Nov. 7, 1981; Charles A. LeMaistre interview; James Bowen interview; Steve Stuyck interview.

25. Frederick F. Becker to Charles A. LeMaistre, Oct. 8, 1982; Gerald P. Bodey to Frederick F. Becker, Oct. 8, 1982; Gerald P. Bodey to Emil J Freireich, Sept. 23, 1982, Charles A. LeMaistre to Staff Involved in Studies of Investigational Agents, Oct. 21, 1981; MDA, CAL, 1G-VP for Research, General, 81–82; MDA, CAL, 1J, Bone Marrow, 79–80.

26. Charles A. LeMaistre interview; James Bowen interview; Joseph T. Ainsworth interview; Walter Pagel interview; Frederick Becker interview; *Houston Chronicle*, Dec. 18–19, 1982. The institution implemented strict security measures requiring ID badges on every employee and passes for all vendors. These measures continue today.

27. Renilda Hilkemeyer to Janet Lunsford, Aug. 2, 1976; R. Lee Clark to Art Ulene, Nov. 24, 1978; Renilda Hilkemeyer interview; University of Chicago, Press Release, Dec. 19, 1975, MDA, RLC, Nursing (1975), 6.17; Joyce Alt interview.

28. Minutes, Administrative Committee, Dec. 10, 1968, MDA, RLC, Administrative Committee, Committees, MDAH, 1971; Renilda Hilkemeyer and Dorothy Rodriguez, "Development of an Enterostomal Therapy Education Program," *Nursing Clinics of North America* 11 (Sept. 1976): 469–78; Nevidjon, *Building a Legacy*; Dorothy Rodriguez to R. Lee Clark, May 7, 1976, MDA, RLC, Nursing (1975), 6.17.

29. Joyce Alt interview; Renilda Hilkemeyer interview.

30. "Nursing Services at the University of Texas M. D. Anderson Hospital and Cancer Institute," May 1979; Joyce Alt to Joseph Painter, Nov. 27, 1978; Renilda Hilkemeyer to R. Lee Clark, Robert Hickey, and Joseph Painter, Apr. 14, 1978, MDA, RLC, Nursing

(1975), 6.171; *Houston Post*, Sept. 2, 1978; Richard H. Jesse to Robert C. Hickey, June 17, 1981, MDA, CAL, 8. Nursing, 81–82.

31. Joyce Alt interview; Patricia Tedder interview; Renilda Hilkemeyer interview; Charles A. LeMaistre interview; "1986 Annual Report," *Conquest* (Winter 1987); "RN Appointments and Separations," 1975–1977; MDA, RLC, Nursing (1975), 6.17. MDA, RLC, Nursing (1975), 6.17; Joyce Alt interview.

32. James Bowen interview; *Messenger* 10 (Apr. 1981): 1–2; *Messenger* 10 (Nov. 1981): 12; Peter Almond interview.

33. James Bowen interview; Peter Almond interview.

34. *Newsletter* 25 (Nov.–Dec. 1980): 8; *Messenger* 11 (Feb. 1982): 9; *Messenger* 12 (Jan. 1983): 2; Melvin Samuels interview; "A. Clark Griffin (1917–82)," *Cancer Letters* 20 (1983): 247–48.

35. Ludmerer, *Time to Heal*, 148–52.

36. Darrell N. Ward to Charles A. LeMaistre, Nov. 25, 1980 and Mar. 22, 1982; William K. Plunkett to Charles A. LeMaistre, July 16, 1982; MDA, CAM, I.G., Basic and Clinical Oncology, 81–82; Nicole Kresge, Robert D. Samuel, Robert L. Hill, "The Biosynthesis of Membrane Glycoproteins: The Work of William J. Lennarz," *Journal of Biological Chemistry* 281.20 (July 21, 2006): 23.

37. Frederick Becker interview; James Bowen interview; Garth Nicolson to Charles A. LeMaistre, Jan. 6, 1984, MDA, CAL, 1G, Tumor Biology.

38. Frederick Becker interview; Charles A. LeMaistre interview; Felix Haas to Charles A. LeMaistre, Feb. 12, 1980, MDA, CLM, 1H, VP for Academic Affairs; Charles A. LeMaistre to Barbara Bynum, June 11, 1985, MDA, CAM, 1G, Basic Science and Clinical Oncology.

39. *Messenger* 11 (Sept. 1982): 5.

40. Frederick F. Becker to Judith Johns, Dec. 19, 1985; Judah Folkman to Frederick F. Becker, Mar. 10, 1985; David Patterson to Frederick F. Becker, Mar. 14, 1986, MDA, CAL, 1G, General, 85–86.

41. Grady F. Saunders to Charles A. LeMaistre, June 23, 1981; Frederick F. Becker to Charles A. LeMaistre, May 8, 1981; T. C. Hsu to Charles A. LeMaistre, May 12, 1981; Alfred G. Knudson Jr., Report of Genetics Advisory Committee, 11–12 June, 1982; MDA, CAL, 1G, Genetics.

42. Thomas E. Andreoli to Fred Conrad and Thomas P. Haynie, Oct. 10, 1979, MDA, CAM, 8, Medical School.

43. Charles A. LeMaistre to Eugene F. Frenkel, Dec. 28, 1978; Eugene F. Frenkel to Charles A. LeMaistre, Dec. 12, 1978, MDA, RLC, Nursing (1975), 6.17.

44. William B. Hood Jr., "Subdividing Departments of Medicine," *Annals of Internal Medicine* 119 (1993): 1225.

45. Charles A. LeMaistre to James Bowen, Mar. 31, 1982; Howard Skipper to Charles A. LeMaistre, Mar. 9, 1982; Gianni Bonadonna to Charles A. LeMaistre, Mar. 19, 1982; James F. Holland to Charles A. LeMaistre, Mar. 2, 1982; James F. Holland to Evan M. Hersh, Feb. 18, 1981; Emil Frei III to Charles A. LeMaistre, Feb. 26, 1982, MDA, CAL, Medicine Search.

46. *Messenger* 13 (Feb. 1984): 1–2.

47. G. Blumenschein et al., "FAC Chemotherapy for Breast Cancer," *Proceedings of the American Society of Clinical Oncology* 15 (1974): 193; Emil Freireich interview; George

Blumenschein interview; Irwin Krakoff interview; James Bowen interview; Charles A. LeMaistre interview.

48. Houston AIDS Working Group, "Acquired Immune Deficiency Syndrome: Criteria and Definition," MDA, CAL, RG 3-1J, Cancer Prevention, 83–83; "Acquired Immuno-Deficiency Syndrome (AIDS): A New Clinical Entity," MDA, CAL, RG3-1J, Cancer Prevention, 79–80; Debra Danburg to Charles A. LeMaistre, July 26, 1982, MDA, CAL, 1J, Cancer Prevention, 82–83.

49. Peter W. A. Mansell interview; Sue Cooper interview; Guy R. Newell to Robert C. Hickey, Mar. 7, 1983; Guy R. Newell to Irwin H. Krakoff, July 17, 1984; Peter Mansell to Eugene McKelvey, July 23, 1984, MDA, CAL, RG3-1J, Cancer Prevention, 81–82.

CHAPTER 10: Chasing the Devil, 1984–1988

1. T. C. Hsu to Charles A. LeMaistre, Dec. 30, 1985; Roger Hewitt to Charles A. LeMaistre, Dec. 3, 1985; William J. Lennarz to Frederick Becker, Aug. 9, 1985; Garth Nicolson to Charles A. LeMaistre, Aug. 5, 1985, MDA, CAL, 1G, Genetics, 84–85; www.hhmi .org/genesweshare/b220.html.

2. *Nature* 314 (1985): 1985–86; Frederick F. Becker to William J. Lennarz, July 18, 1986; Frederick Becker to Benoit de Crombrugghe, May 15, 1986; see LeMaistre's handwritten note on the document "Department of Genetics, Research and Educational Activities 1984–1985," MDA, CAL, 1G, Genetics, Search Committee; Charles A. LeMaistre to Frederick F. Becker, Mar. 1, 1985; Siciliano declared himself a candidate, Michael J. Siciliano to William J. Lennarz, May 9, 1984, MDA, CAL, 1G, Genetics, Search Committee.

3. Lazlo, *The Cure*, 230–38; Ralph Arlinghaus interview.

4. Ralph Arlinghaus interview; Moshe Talpaz, "Use of Interferon in the Treatment of Chronic Myelogenous Leukemia," *Seminars in Oncology* 21 (1994): 8–13; Charles A. LeMaistre interview.

5. Charles A. LeMaistre to George and Barbara Bush, Nov. 5, 1987, MDA, CAL, 13. City of Houston, Houston Economic Summit (1990); Irwin Krakoff to Charles A. LeMaistre, Dec. 18, 1987; Albert Deisseroth to Irwin Krakoff, Dec. 16, 1987, MDA, CAL, 1G., General, 85–86; Bush, *Barbara Bush*, 214; *Messenger* 16 (Nov. 1987): 2.

6. Charles A. LeMaistre to Chancellor E. D. Walker, Sept. 7, 1979, 4. Physicians Referral Service, 82–83; Charles A. LeMaistre to All Physician PRS Members, Mar. 3, 1980; MDA, CAL, 4. Physicians Referral Service, 82–83; Charles A. LeMaistre to Charles Balch, Mar. 4, 1985; Ben Love to Charles A. LeMaistre, Feb. 25, 1985; Charles A. LeMaistre to Members of the University Cancer Foundation Board of Visitors, Feb. 25, 1985; Charles Balch to Charles A. LeMaistre, Feb. 12, 1985; Marion J. McMurtrey et al. to Charles A. LeMaistre, July 18, 1983, MDA, CAL, 1J, Surgery Search; Richard G. Martin interview.

7. Charles A. LeMaistre interview; *Messenger* 14 (Dec. 1985): 5; Roger Anderson to Robert Moreton, Aug. 2, 1982, MDA, CAL, 1J, Pharmacy, 81–82; William J. Dana to Roger W. Anderson, Mar. 28, 1979; Warren Rutherford to Charles A. LeMaistre, Apr. 5, 1979, MDA, RLC, Pharmacy (1975), 6.21; James McKinley interview; Roger W. Anderson interview; William J. Dana to Roger W. Anderson, Mar. 28, 1979, MDA, RLC, Pharmacy (1975), 6.21.

8. Roger Anderson to James S. Olson, Nov. 13, 2007, James S. Olson File; *Messenger* 25 (Oct. 1996): 1, 6.

9. Boston Women's Health Book Collective, *Our Bodies, Ourselves,* 52–55; Olson, *Bathsheba's Breast,* 117–21; Ludmerer, *Time to Heal,* 267; Rothman, *Strangers at the Bedside.*

10. *Messenger* 13 (June 1984): 2; *Messenger* 13 (Aug. 1984): 6; H. S. Sahdev, "Patients Rights or Patient Neglect: The Impact of the Patients Rights Movement on Delivery Systems," *American Journal of Orthopsychiatry* 46 (Oct. 1976): 660–68.

11. *Messenger* 13 (June 1984): 2; *Messenger* 13 (Aug. 1984): 6; Charles A. LeMaistre interview.

12. Fred G. Conrad to record, June 24, 1980, MDA, CAL, 1J, Developmental Therapeutics, 79–80; Helmuth Goepfert to Charles A. LeMaistre, July 7, 1982, MDA CAL,1J, Thoracic Surgery, 81–82; Charles A. LeMaistre interview; *Messenger* 17 (Feb. 1988): 1, 3.

13. Charles A. LeMaistre, "Progress Report. University of Texas System Cancer Center, presented to the UT Board of Regents, Dec. 3, 1987"; Charles A. LeMaistre to Legislative Budget Board Staff, Sept. 13, 1987, MDA, LeMaistre's private papers.

14. Lowell Carlson to Charles A. LeMaistre, Dec. 30, 1981; Charles A. LeMaistre to Lowell Carlson, Jan. 6 and Mar. 30, 1982; Robert C. Hickey to File, May 4 and 26, 1982; Peter Almond to Robert C. Hickey, Apr. 13, 1982; Robert C. Hickey, "The Cyclotron Meeting," June 28, 1982, MDA, CAL, 1G, Physics, 80–81; *Wall Street Journal,* Feb. 3, 1982.

15. David Hussey to Charles A. LeMaistre, May 8, 1981; Charles A. LeMaistre to David Hussey, May 1, 1981, Peter R. Almond to Charles A. LeMaistre, Jan. 23, 1983; Lester J. Peters to Robert C. Hickey, Mar. 2, 1983, MDA, CAL, 1G, Physics, 80–81; Peter Almond interview.

16. *Messenger* 17 (Feb. 1988): 1–2; Karel Dicke, "Bone Marrow Transplantation," Oct. 7, 1981, MDA, CAL, 1D, Foundations (Private Grants); *Guardian,* Mar. 27, 2002.

17. Richard Bloch interview.

18. *Newsletter* 25 (Sept.–Oct. 1980): 1, 6; Charles A. LeMaistre interview.

19. American Cancer Society, "Background on Interferon," MDA, CAL, 1G, Tumor Biology; Evan Hersh, "Statement on Biological Therapy of Cancer Research at M. D. Anderson Hospital and Tumor Institute"; Charles A. LeMaistre interview.

20. Charles A. LeMaistre to James S. Olson, Apr. 2007.

21. John R. Bush to Robert Anderson, May 21, 1980, MDA, CAL, 1D (Private Grants); Charles A. LeMaistre to Leon Jaworski, Apr. 16, 1980; Charles A. LeMaistre to T. C. Campbell, Aug. 11, 1980; MDA, CAL, 1G, Tumor Biology, 83–84; Carroll G. Sunseri to Charles A. LeMaistre, Aug. 21, 1984, MDA, CAL, 1D, Donations, Gifts, Endowments, Trusts.

22. James S. Olson observation.

23. *Houston Chronicle,* Aug. 10, 1985.

24. Charles A. LeMaistre to Ray Weisberg, Mar. 13, 1985, MDA, CAL, 12. Ciglit, 84–85; Charles A. LeMaistre interview; *Messenger* 17 (Oct. 1988): 1–2; *Messenger* 18 (Jan. 1989): 1, 5; Joe E. Boyd to all Employees, June 20, 1975, MDA, RLC, Memorandums, 5.6, 1976.

25. *Messenger* 17 (Jan. 1988): 1; Steve Stuyck to James Olson, Dec. 17, 2007.

26. *Messenger* 17 (May 1988): 1; Charles A. LeMaistre to Vice Presidents; Clinical and Basic Science Division Heads and Department Chairmen, Feb. 26, 1988; Charles A. LeMaistre, "Institutional Identity Briefing," Mar. 7, 1988; Charles A. LeMaistre, "Setting the Stage: Issues Affecting Our Strategic Agenda," Nov., 1988, LeMaistre personal papers.

27. Annual Hospital Report, Aug. 31, 1984, 1985, 1986; Mark White to Jess Hay, Feb. 5, 1986, MDA, RLC, UT System. General Admin. and Policy, 81–8; Charles A.

LeMaistre to Joseph T. Painter, Dec. 23, 1986; Minutes, President's Executive Board, Aug. 19, 1987, MDA, CAL, 1E, Vice President for Physicians Referral Development and Extramural Programs 84–85, General.

28. *Messenger* 17 (Oct. 1988): 2.

29. *Messenger* 16 (June 1987): 1–2; *Messenger* 17 ; (Mar. 1988): 1–2; Steve Stuyck interview; Charles Balch interview; Richard G. Martin interview. See Kenney, *Biotechnology*.

30. Project Assessment, Southwest Institute of Immunological Diseases and Related Foundation, June 1, 1985; JoAnne Hale to File, Nov. 22, 1985; Richard D'Antoni to Joseph Ainsworth, Mar. 11, 1986, MDA, CAL, American Medical International, Inc., 84–85.

31. *Medical World News* (Aug. 26, 1985): 41–68; Eugene McKelvey to Frederick Becker, Aug. 28, 1985; Evan Hersh to Steve Schultz, July 17 and Aug. 5, 1985; Irwin Krakoff, Evan M. Hersh, Guy R. Newell, and Peter W. A. Mansell to Steve Schultz, Aug. 13, 1985; Irwin Krakoff to Charles A. LeMaistre, Aug. 30, 1985, and Feb. 10, 1986, MDA, CAL, American Medical International, Inc. (AMI), 84–85.

32. *Houston Chronicle*, Jan. 28 and Feb. 13, 1986; Roger Bulger to Charles A. LeMaistre, Feb. 11, 1986; Roger J. Bulger to John Whitmire, Feb. 11, 1986; John Whitmire to Charles A. LeMaistre, Feb. 3, 1986; Richard D'Antoni to Charles B. Mullins, Jan. 28, 1986; C. S. Roper to UT Board of Regents, Jan. 24, 1986; Minutes, President's Executive Board, Oct. 16, 1985, MDA, CAL, American Medical International, Inc. (AMI), 84–85.

33. Tom Sawicki to Charles B. Mullins, Aug. 5 and Oct. 2, 1987; AMI News Release, Aug. 6, 1987; Michael B. Wilson to Charles A. LeMaistre, May 28, 1987, MDA, CAL, American Medical International, Inc. (AMI), 84–85. Irwin Krakoff interview; Charles A. LeMaistre interview; Peter W. A. Mansell interview; Sue Cooper interview.

34. Charles A. LeMaistre to Irwin H. Krakoff, July 16, 1987; Pat Smith to Dan J. Oldani, July 16, 1987; Tom Sawicki to Charles B. Mullins, Aug. 5 and Oct. 2, 1987; AMI News Release, Aug. 6, 1987; Michael B. Wilson to Charles A. LeMaistre, May 28, 1987, MDA, CAL, American Medical International, Inc. (AMI), 84–85. Irwin Krakoff interview; Charles A. LeMaistre interview; Peter W. A. Mansell interview; Sue Cooper interview.

35. *Newsletter* 19 (Mar. 1974): 2; *Newsletter* 27 (Nov.–Dec. 1982): 6; C. Stratton Hill interview; *Drug Treatment of Cancer Pain*, 3; www.iasp-pain.org/AM/Template.cfm?Section=In_Memoriam1&Template=/CM/HTMLDisplay.cfm&Cont entID=1262.

36. www.aslme.org/research/mayday/24.4/24.4i.php; *Messenger* 17 (Mar. 1988): 5; *Messenger* 19 (Nov. 1990): 2; C. Stratton Hill interview.

CHAPTER 11: Victory, Defeat, and an Elusive Enemy, 1988–1996

1. Joan Coffey interview; Edward Coffey interview; Raymond Alexanian interview.

2. *Messenger* 14 (Jan. 1995): 1; *Messenger* 33 (Jan. 1994): 1; *Messenger* 19 (Jan. 1990): 1; *Messenger* 22 (Dec. 1993): 1, 2; "Progress Report: M. D. Anderson Cancer Center Outreach Corporation, Period Covered: Mar. 1, 1990 to June 30, 1991, 7–9.

3. *Messenger* 23 (Jan. 1994): 1; *Messenger* 24 (Feb. 1995): 1, 5; SHARP Report, *M. D. Anderson Cancer Center: Strategic Direction*, Aug. 11, 1995, 12; Charles A. LeMaistre to Joseph T. Painter, Sept. 11, 1986; Charles A. LeMaistre to Joseph T. Painter, Nov. 25, 1987; J. T. Painter to C. A. LeMaistre, Oct. 26, 1987, 1E, Vice President for Physicians Referral Development and Extramural Programs, General, 84–85. See John Iglehart, "The American Health Care System: Managed Care," *New England Journal of Medicine* 327 (1992): 742–47; Kenneth M. Ludmerer, *Time to Heal*, 266.

4. Charles A. LeMaistre interview; David Bachrach interview.

5. *Messenger* 23 (Mar. 1994): 1; *Messenger* 24 (Apr. 1995): 1, 4.

6. Minutes, President's Executive Board, Oct. 24, 1990, CAL, Series 4, PRS, 1990–91.

7. *Messenger* 23 (Mar. 1994): 1.

8. Margaret Meyer interview; Lois Woodard interview; *Messenger* 24 (Apr. 1995): 1, 4.

9. Charles A. LeMaistre interview; Harry Holmes interview.

10. Harry Holmes interview; *Wall Street Journal*, Aug. 29, 2000.

11. *Houston Post*, July 12, 1990; *Washington Post*, July 12, 1990; Barbara Bush to Mickey, July 17, 1990; Charles A. LeMaistre to Richard E. Wainardi, Mar. 22, 1990, MDA, CAL, 13; Houston, City of, Houston Economic Summit (1990).

12. *Messenger* 23 (June 1994): 4; Charles A. LeMaistre interview; James Bowen interview; Harry Holmes interview.

13. David C. Hohn, "Patient Care Summary, 1995," Jan. 12, 1996, MDA, CAL, Office of Patient Care; *Messenger* 25 (Oct. 1996): 1, 6.

14. *Messenger* 28 (July 1989): 5; *Messenger* 19 (Jan. 1990): 6; *Messenger* 19 (Nov. 1990): 6; *Messenger* 19 (Dec. 1990): 7; *Messenger* 20 (May 1991): 2; *Messenger* 21 (Feb. 1992): 2; *Messenger* 21 (Mar. 1992): 3. Although a number of important figures in M. D. Anderson history died of cancer, the incidence rate was not unusual because cancer at the time was the second leading cause of death in the United States and had an incidence rate higher among the old than the young,

15. Martha Jones interview; *Messenger* 33 (July 1994): 1.

16. Joyce Alt interview; Renilda Hilkemeyer interview; Charles A. LeMaistre interview; Randy Clark interview; James Bowen interview; *New York Times*, May 4, 1993.

17. Peter W. A. Mansell interview; Sue Cooper interview; Charles A. LeMaistre interview; Debra Danburg interview; James Bowen interview; *Messenger* 23 (Dec. 1994): 2.

18. *Messenger* 18 (June 1989): 1; *Messenger* 19 (Aug. 1990): 2; *Messenger* 19 (Oct. 1990): 5; *Messenger* 20 (Jan. 1991): 8; *Messenger* 20 (Feb. 1991): 2; *Messenger* 20 (Mar. 1991): 1; *Messenger* 21 (July 1992): 1; *Messenger* 22 (Feb. 1993): 2; *Messenger* 33 (Oct. 1994): 6; *Messenger* 24 (Apr. 1995): 2.

19. Joseph Painter to Charles A. LeMaistre, Apr. 8, 1980; Charles A. LeMaistre to Anderson Mayfair residents, Feb. 12, 1980; Police Report 84-426-184; Robert Moreton to Charles A. LeMaistre, Dec. 29, 1979, and July 31, 1984; Deed L. Vest, File Letter, Jan. 28, 1980; Robert D. Morton to C. E. Fennell, Feb. 28, 1980; William J. Schull to Frank Romanelli, Nov. 2, 1979; Minutes of President's Executive Board Minutes, June 20, 1987; MDA, 1I, Anderson-Mayfair.

20. *Messenger* 33 (Mar. 1994). Construction would be completed in Feb. 1993.

21. Gardner, *Early Detection*, 53–88; Charles A. LeMaistre, "Cancer Prevention: Realism or Idealism?" 23rd Annual Meeting, U.S. Public Health Service Professional Association.

22. Olson, *Bathsheba's Breast*, 231.

23. *New York Times*, May 5, 1993; *Houston Chronicle*, Sept. 14, 1994; B. Kushka, "BRC1 Alteration Found in Eastern European Jews," *Journal of the National Cancer Institute* 87 (Oct. 15, 1995): 1505; Olson, *Bathsheba's Breast*, 238.

24. Lisne Clorfene-Casten, "The Environmental Link to Breast Cancer," *Ms.* (May–June 1993): 52–54; Susan Rennie, "Breast Cancer Prevention: Diet v. Drugs," *Ms.* (May–June 1993): 38–51; Monte Paulsen, "The Cancer Business," *Mother Jones* (June 1994): 41; "Breast Cancer, Inc.," *Houston Press*, Aug. 12–18, 1993, 19–27.

25. Olson, *Bathsheba's Breast*, 182–84.

26. Karen Steinberg, Stephen B. Thaker, S. Jay Smith, et al., "A Meta-Analysis of the Effect of Estrogen Replacement Therapy on the Risk of Breast Cancer," *Journal of the American Medical Association* 265 (Apr. 17, 1991): 1985–90; "Prospective Study of Oral Contraceptive Use and the Risk of Breast Cancer," *Journal of the National Cancer Institute* 81 (Sept. 6, 1989): 1313–21.

27. "Interdepartmental Oversight Committee on Tobacco Control," Sept. 28, 1989, MDA, CLM, 1A.M, ACS Committees General, 1989–90; Charles A. LeMaistre interview.

28. Margaret R. Spitz, Scott M. Lippman, Hong Jiang, et al., "Mutagen Sensitivity as a Predictor of Tumor Recurrence in Patients with Cancer of the Upper Digestive Tract," *Journal of the National Cancer Institute* 90 (1998): 243–45.

29. Richard G. Martin interview; Gilbert Fletcher interview; Marsha McNeese interview. In 2007, epidemiologists also implicated HPV in the persistence of cancer of the throat, attributing it to the increasing prevalence of oral sex in the American population.

30. Richard Doll and Richard Peto, "The Causes of Cancer: Quantitative Estimates of Avoidable Risks of Cancer in the United States Today," *Journal of the National Cancer Institute* 66 (1981), 1193–1308.

31. Artemis P. Simopoulos, "Obesity and Carcinogenesis: Historical Perspective," *American Journal of Clinical Nutrition* 45 (1987): 271–76; Guy Newell, "Screening for Cancer," MDA, CAL, 1J, Cancer Prevention.

32. Guy Newell, "Nutrient Analysis System," Guy R. Newell to Frank J. Rauscher, Nov. 3, 1980, MDA, CAL, 1J, Cancer Prevention; Charles A. LeMaistre to Richard G. Mund, Aug. 25, 1981, MDA, CAL, 1D, Development, General, 80–81.

33. "National Objectives for Disease Prevention and Health Promotion for the Year 2000," Testimony Submitted by Guy R. Newell and Charles A. LeMaistre, MDA, CAL, 13, Texas Department of Health, 83–84.

34. *Messenger* 25 (Dec. 1996): 3.

35. Ellen R. Gritz interview; Charles A. LeMaistre interview.

36. Office of Vice President for Research to Division Heads, Department Chairs, and Members of the Clinical and Research Staff, Sept. 13, 1983, MDA, CAL, 1G, Grants Administration; Waun Ki Hong, J. Endicott, L. M. Itri, et al., "13*cis*-Retinoic Acid in the Treatment of Oral Leukoplakia," *New England Journal of Medicine* 315 (1986), 1501; *Messenger* 25 (January 1996): 6.

37. David Plotkin, "Good News and Bad News about Breast Cancer," *Atlantic Monthly*, June 1996, 55–56; DeGregorio and Wiebe, *Tamoxifen and Breast Cancer*, 72–83.

38. Olson, *Bathsheba's Breast*, 200.

39. *Medical Industry Today* (May 16, 2000): 1–3; *Atlanta Constitution*, May 19, 2000; *Electronic Telegraph* no. 1820 (May 19, 2000); John C. Bailar III and Elaine M. Smith, "Progress against Cancer," *New England Journal of Medicine* 314 (May 8, 1986): 1226–32; John C. Bailar III, "Rethinking the War on Cancer," *Issues in Science and Technology* (Fall 1987): 16–21; Proctor, *Cancer Wars*, 250–51; Olson, *Bathsheba's Breast*, 200–202.

40. Cairnes, *Cancer*, 159–61. Also see Russell, *Is Prevention Better than Cure?* and Patterson, *Dread Disease*, 267–71.

41. "National Objectives for Disease Prevention and Health Promotion for the Year 2000," Testimony Submitted by Guy R. Newell and Charles A. LeMaistre, MDA, CAL, 13, Texas Department of Health, 83–84.

42. Barbara J. Culliton, "Breast Cancer: Second Thoughts about Routine Mammography," *Science* 193 (Aug. 13, 1976): 555–58; Ross, *Crusade*, 102–11; Laurence and Weinhouse, *Outrageous Practices*, 102–11; G. J. Subak-Sharpe, "Is Mammography Safe? Yes, No, and Maybe," *New York Times Magazine*, Oct. 24, 1976, 42–44.

43. Steven F. Woolf, "Screening for Prostate Cancer with Prostate-Specific Antigen (PSA): An Examination of the Evidence," *New England Journal of Medicine* 333 (Nov. 23, 1995): 1401–5.

44. I. Van der Waal and N. De Vries, "Prevention of Oral Cancer," *Advanced Dentistry Research Monthly* (1995).

45. *New York Times*, Jan. 1, Apr. 7, and Apr. 30, 1992, and June 29, 1994; DeGregorio and Wiebe, *Tamoxifen and Breast Cancer*, 72–83.

46. *Cancer* 78 (1996): 2045–48; *Medical Industry Today*, May 16, 2000, 1–3; *Atlanta Constitution*, May 19, 2000; *Washington Times*, July 12, 1999.

47. "The Future of Cancer Prevention," *M. D. Anderson Backgrounder*, updated Mar. 2006; Scott M. Lippman and Bernard Levin, "Cancer Prevention: Strong Science and Real Medicine," *Journal of Clinical Oncology* 23 (Jan. 10, 2005): 249–53.

48. www.beyonddiscovery.org/content/view.page.asp?I=108.

CHAPTER 12. John Mendelsohn and the New Frontiers in Oncology, 1996–2000

1. Edward Coffey interview; Joan Coffey interview.

2. Olby, *The Path to the Double Helix*; John Mendelsohn interview.

3. John Mendelsohn interview; Edwin D. Kilbourne, "The Emergence of the Physician-Basic Scientist in America," *Daedalus* 115 (Spring 1986): 43–54.

4. *Proceedings of the National Academy of Science USA* 80 (1983): 1337–41.

5. *Cancer Research* 44 (1984): 1002–7.

6. *CancerWise* (May 2003): 1–4; John Mendelsohn interview.

7. J. Baselga, D. Tripathy, J. Mendelsohn, et al., "Phase II Study of Weekly Intravenous Recombinant Humanized Anti-185 (HER2) Monoclonal Antibody in Patients with HER2/*neu*-overexpressing Metastatic Breast Cancer," *Journal of Clinical Oncology* 14 (1996): 601–9.

8. Edward M. Copeland III, "Heroes and Friends," *Annals of Surgery* 231 (May 2000): 617–24.

9. Richard G. Martin interview; Helmuth Goepfert interview; James Bowen interview; Charles A. LeMaistre interview.

10. James Bowen interview.

11. James Bowen interview.

12. www.fda.gov/oc/voneschenbach/bio.html; *Oncologist* 5 (Feb. 2000), 84–86.

13. H. Masui, T. Kawamoto, J. D. Sato, B. Wolf, G. H. Sato, and J. Mendelsohn, "Growth Inhibition of Human Tumor Cells in Athymic Mice by Anti-EGF Receptor Monoclonal Antibodies," *Cancer Research* 58 (1998): 2825–31; *Messenger* 36 (June 1997): 2.

14. I. Craig Henderson, "Paradigmatic Shifts in the Management of Breast Cancer," *New England Journal of Medicine* 332 (Apr. 6, 1995): 707–11, 951–52.

15. George Blumenschein interview; Emil Freireich interview.

16. Gianni Bonadonna, Pinuccia Valagussa, Angela Molitieri, et al., "Adjuvant Cy-

clophosphamide, Methotrexate, and Fluorouracil in Node-Positive Breast Cancer," *New England Journal of Medicine* 332 (Apr. 6, 1995): 901–6.

17. *Messenger* 26 (Oct. 1997): 3.

18. James D. Cox interview; John Mendelsohn interview.

19. *Wall Street Journal*, Aug. 29, 2000; Micklethwait and Wooldridge, *Witch Doctors.*

20. *Messenger* 26 (Jan. 1998): 6; *Messenger* special edition 27 (Dec. 1998): 3; *Messenger* 28 (Sept. 1999): 3; *Messenger* 29 (June 2000): 3; *Messenger* 29 (Sept. 2000): 2; *Messenger* 29 (Dec. 2000): 3.

21. Elizabeth Travis interview.

22. Ludmerer, *Time to Heal*, 256–59; Janet Bickel, "Special Needs and Affinities of Women Medical Students," in *Empathic Practitioner*, ed. More and Milligan, 237–49; Morantz-Sanchez, *Sympathy and Science.*

23. Ludmerer, *Time to Heal*, 255–58.

24. Harry Robert Gibbs interview.

25. www.nih.gov/about/almanac/organization/NCCAM.htm. For a history of integrative medicine in the United States, see Whorton, *Nature Cures*; "Acupuncture OK at MDA," Minutes, Administrative Committee, May 29, 1973, MDA, RLC, Administrative Committee, Committees, MDAH, 1971.

26. www.nih.gov/about/almanac/organization/NCCAM.htm. For a history of integrative medicine in the United States, see Whorton, *Nature Cures.*

27. Lorenzo Cohen interview.

28. Armstrong, *It's Not about the Bike*, 76–80, 86, 92–95, 102–11.

29. The 2005 Bristol-Myers Squibb Tour of Hope, "Lance Armstrong Clip from Rally," Office of Public Affairs, the University of Texas M. D. Anderson Cancer Center.

30. *USA Today*, Mar. 1 and Aug. 24, 1999; *Houston Chronicle*, Aug. 20, 1999; *Jet* (Sept. 6, 1999): 51; *SI ONLINE*, Sept. 13, 1999; *Messenger* 29 (Sept. 2000): 1–6.

31. *Washington Times*, July 12, 1999; *Medical Industry Today*, May 16, 2000, 1–3; *Atlanta Constitution*, May 19, 2000; *Washington Times*, July 12, 2000; *Electronic Telegraph*, Issue 1820, (May 19, 2000); Sontag, *Illness as Metaphor*, 57–69.

CHAPTER 13: New Offensives, 2001–2007

1. *Messenger* 8 (July 1979): 8; Page Lawson interview; Steve Stuyck interview; Roy Lawson interview.

2. *New York Times*, Apr. 16, 2008; Richard Bloch interview; NCCS (National Coalition for Cancer Survivorship), Annual Reports Financial Statements, 2003, "Advocacy. Authenticity. Passion," 1–7, www.canceradvocacy.org/about/financial/; Evan Hersh interview.

3. www.canceradvocacy.org/news/press/2007/stovall-to-step-down-as-nccs-president.html; Ellen Stovall, "The March: Stand with Us and Say 'No More,'" *Oncologist* 3 (1998): 135–36, also available at www.StemCells.com/cgi/content/full/16/4/240; *Washington Post*, Sept. 26–27, 1998; www.ncsdf.org/.

4. www.cdc.gov/mmwr/preview/mmwrhtml/mm5324a3.htm; Joan Coffey interview; www.cancer.org/docroot/STT/content/STT_1x_Breast_Cancer_Facts__Figures_2003-2004.asp.

5. *Messenger* 24 (Aug. 1995): 5; Joseph T. Painter interview; Maureen Valenza interview.

6. Rush, *Letters*, 2:1104–6; Frances Cording interview; Richard Cording interview.

7. Joan Coffey interview; Edward Coffey interview.

8. Charles A. LeMaistre interview.

9. Richard G. Martin interview; Melvin Samuels interview; Raphael Pollock interview.

10. *Messenger* 30 (Mar. 2001): 7; *Messenger* 30 (Nov. 2001): 2; *Messenger* 31 (July 2002): 3; Melvin L. Samuels interview.

11. Mary Nell Jeffers Lovett interview.

12. John Mendelsohn, "Antibodies against Receptors for Growth Factors;" John Mendelsohn interview.

13. Justin Gillis, "Patients Weren't Told of Stake in Cancer Drug," *Washington Post*, June 30, 2002; Arnold S. Relman, "Dealing with Conflicts of Interest," *New England Journal of Medicine* 310 (1984): 1182–83; www.commondreams.org/headlineso2/0630–03.htm. Also see Rodwin, *Medicine, Money, and Morals*.

14. John Mendelsohn interview.

15. Ralph Freedman interview; John Mendelsohn interview; *Washington Post*, June 30, 2002.

16. *Washington Post*, June 30, 2002.

17. *Houston Chronicle*, June 2, 2003; *Barron's*, June 28, 1993, 14; *New York Times*, Jan. 24, 2002 and Jan. 19, 2005; *Wall Street Journal*, Sept. 27, 2002.

18. *New York Times*, Jan. 4, 1993; J. Moloney, "Dr. Frank J. Ruscher Jr.: An Appreciation," *Journal of the National Cancer Institute* 85 (1993): 174–75; Ralph Arlinghaus interview; S. Giralt, H. Kantarjian, and M. Tapaz, "The Natural History of Chronic Myelogenous Leukemia in the Interferon Era," *Seminars in Hematology* 32 (1995): 152–58; J. Q. Guo, Y. M. Xian, M. S. Lee, et al., "BCR-ABL Protein Expression in Peripheral Blood Cells of Chronic Myelogenous Leukemia Patients Undergoing Therapy," *Blood* 83.6 (1994): 3629–37; H. M. Kantarjian, M. Talpaz, J. Cortez, et al., "Quantitative Polymerase Chain Reaction Monitoring BCR-ABL During Therapy with Imatinib Mesylate (STI571; Gleevec) in Chronic Phase Chronic Myelogenous Leukemia," *Clinical Cancer Research* 9 (2003): 160–66.

19. John Mendelsohn interview.

20. *Houston Chronicle*, Dec. 6, 2005.

21. *OncoLog* 49 (Nov. 2004): 1–4.

22. http://gsbs.uth.tmc.edu/tutorial/fidler.html; Isaiah J. Fidler, "The Organ Microenvironment and Cancer Metastasis," *Differentiation* 70 (Dec. 2002): 498–505.

23. Frederick Becker interview; *New York Times*, May 3–7, 1998; *Washington Times*, July 12, 1999; Cooke, *Dr. Folkman's War*; Ellis quoted in Olson, *Bathsheba's Breast*, 244; Laurie Tarkan, "Scientists Begin to Grasp the Stealthy Spread of Cancer," *New York Times*, Aug. 15, 2006, D1, D.

24. *Economist* 379 (June 2006): 23–26.

25. www.mdanderson.org/departments/mccombs/display.cfm?id=d623696f-5a40–47c2–8db3a188f25170d4&method=displayfull&pn=feecb1fe-8b77–4ccb-b5392daaf1cc812d.

26. *San Francisco Chronicle*, Mar. 15, 2005; *USA Today*, Apr. 6, 2005.

27. *Texas Medical Center News*, 26 (Dec. 15, 2004): 1, 11.

28. P. Pommier, C. Ginestat, and C. Carrie, "Is Conformational Radiotherapy Progressing?" *Cancer Radiotherapy* 5 supp. (Nov. 2001): 57–67.

29. www.llu.edu/proton/physician/technical.html; James D. Cox interview.

30. James D. Cox interview.

31. Ibid., "M. D. Anderson, Partners Break Ground on Proton Therapy Center," UTMDACC, Communications Office, News Release, May 7, 2003; *Houston Chronicle*, Nov. 6, 2005.

CHAPTER 14: Tipping History?

1. Marsha Stringer interview.

2. *CancerPro* 1 (Nov.–Dec. 2004).

3. www.vioxx.com/rofecoxib/vioxx/consumer/index.jsp; John Mendelsohn interview. www.docguide.com/dg.nsf/PrintPrint/EA5634D7249C9537852568F2005733D2.

4. www.eurekalert.org/pub_releases/2005–12/uotm-ttf121605.php.

5. *Seattle Times*, Oct. 23, 2007.

6. Scott M. Lippman and Bernard Levin, "Cancer Prevention: Strong Science and Real Medicine," *Journal of Clinical Oncology* 23 (Jan. 10, 2005): 249–52.

7. *Atlanta Business Chronicle*, Sept. 17, 2007; Lovell Jones interview; *Houston Chronicle*, Dec. 3, 2007.

8. John Mendelsohn, "State of the Institution Address," University of Texas M. D. Anderson Cancer Center, Sept. 21, 2006. Mendelsohn's research interests have been recognized in recent years by many honors from peers, including the Joseph Burchenal Clinical Research Award from the American Association for Cancer Research, the David A. Karnofsky Memorial Award from the American Society of Clinical Oncology, the Bristol-Myers Squibb Freedom to Discover Award for Distinguished Achievement in Cancer Research, and the Dan David Prize in Cancer Therapy from the Israeli government. He is a member of the Institute of Medicine of the National Academy of Sciences.

9. R. Lee Clark to Theophilus Painter, May 14, 1948, MDA, RLC, Board of Regents, Misc. Correspondence, Inactive, University of Texas, 1942–51.

10. Waun Ki Hong, widely considered as a founding father of the field of chemoprevention, is the author of more than 660 articles and a recipient of the Distinguished Service Award of the American Cancer Society, the Joseph H. Burchenal Award of the American Association for Cancer Research, the Rosenthal Foundation Award, and the David A. Karnofsky Memorial Award.

11. *Messenger* 29 (June 2000): 6; *Messenger* 30 (July 2001): 6; *Messenger* 35 (May–June 2006): 2, 20–21; M. D. Anderson News Release, Jan. 22, 2002.

12. John Mendelsohn interview; Thomas D. Anderson interview; Patrick Mulvey interview.

13. John Mendelsohn interview.

14. www.cbsnews.com/stories/2006/02/09/health/main1299899.shtml; *New York Times*, Feb. 9–10, 2006.

15. John Mendelsohn interview.

16. Charles A. LeMaistre interview.

17. *Houston Chronicle*, Sept. 17, 2007.

18. www.hhs.gov/news/press/2006pres/20060627.html.

19. Bazell, *Her-2*; Olson, *Bathsheba's Breast*, 156–57.

20. *New York Times*, Dec. 14–16, 2006, and Jan. 17, 2007; *FYI: Weekly News and Notes for Patients, Families, and Visitors* 8 (Jan. 1, 2007): 1.

21. www.cbsnews.com/stories/2007/01/17/health/main2366335.shtml?source=RSSattr =HOME_236 6335; *Albuquerque Tribune,* Dec. 14, 2008;www.msnbc.msn.com/id/ 3079465.

22. www.cancer.org/aspx/blog/Comments.aspx?id=126; Emil Freireich interview; Welch, *Should I Be Tested?* 20–22, 35–37.

23. Peter A. Steck, Mark A. Pershouse, Samar A. Jasser, et al., "Identification of a Candidate Tumour Suppressor Gene, MMAC1, at Chromosome 10q23.3 That Is Mutated in Multiple Advanced Cancers," *Nature Genetics* 15 (1997): 356–62.

24. John Mendelsohn interview.

25. Douglas Hanahan and Robert Weinberg, "The Hallmarks of Cancer," *Cell* 100 (Jan. 7, 2000): 57–70.

26. *OncoLog* 49 (June 2004): 1–3.

27. John Mendelsohn interview.

28. www.mdanderson.org/sp/Care_Centers/Hematology/dIndex.cfm?pn=FA66 2522–7556–11D4–AEC300508BDCCE3A.

29. *OncoLog* 50 (Apr. 2005): 1–4.

30. Ibid., 52 (Mar. 2007): 1–2.

31. *Houston Chronicle,* June 27, 2005.

32. *Houston Chronicle,* Apr. 7, 2005.

33. www.forbes.com/forbes/2004/1101/096_print.html, Robert Langreth, "Men, Cancer, and Hope," Forbes.com, Nov. 1, 2004; John Mendelsohn interview.

34. Eric Berger, "Cancer: Looking Beyond Mutations," *Houston Chronicle,* June 27, 2005.

35. Frederick Becker interview.

BIBLIOGRAPHY

Manuscript Collections

Aikin, A. M., Jr., Papers. Center for American History, University of Texas, Austin, Texas.

Aikin, A. M., Jr., Papers. James Gilliam Gee Library. Texas A&M University—Commerce, Commerce, Texas.

Aikin, A. M., Jr., Papers. Paris Junior College, Paris, Texas.

Bates, William. Papers. John P. McGovern Historical Collection and Research Center, Houston Academy of Medicine–Texas Medical Center Library, Houston, Texas.

Bertner, Ernst W. Papers. John P. McGovern Historical Collection and Research Center, Houston Academy of Medicine–Texas Medical Center Library, Houston, Texas.

Bocker, Truman. Papers. John P. McGovern Historical Collection and Research Center, Houston Academy of Medicine–Texas Medical Center Library, Houston, Texas.

Bruche, Hilda. Papers. John P. McGovern Historical Collection and Research Center, Houston Academy of Medicine–Texas Medical Center Library, Houston, Texas.

Clark. R. Lee. Papers. John P. McGovern Historical Collection and Research Center, Houston Academy of Medicine–Texas Medical Center Library, Houston, Texas.

Clark. R. Lee. Papers. The University of Texas M. D. Anderson Cancer Center, Houston, Texas.

Cowdry, E. V. Papers. Bernard Becker Library, Washington University School of Medicine, St. Louis, Missouri.

Dmochowski, Leon. Papers. John P. McGovern Historical Collection and Research Center, Houston Academy of Medicine–Texas Medical Center Library, Houston, Texas.

Elliott, Frederick. Papers. John P. McGovern Historical Collection and Research Center, Houston Academy of Medicine–Texas Medical Center Library, Houston, Texas.

Goff, Frances. Papers. Center for American History, University of Texas, Austin, Texas.

Haas, Felix. Papers. John P. McGovern Historical Collection and Research Center, Houston Academy of Medicine–Texas Medical Center Library, Houston, Texas.

Hickey, Robert. Papers. John P. McGovern Historical Collection and Research Center, Houston Academy of Medicine–Texas Medical Center Library, Houston, Texas.

Hsu, T. C., Papers. The University of Texas M. D. Anderson Cancer Center.

Johnson, Lyndon Baines. Papers. Lyndon B. Johnson Presidential Library, Austin, Texas.

Kelsey, Mavis P. Papers. John P. McGovern Historical Collection and Research Center, Houston Academy of Medicine–Texas Medical Center Library, Houston, Texas.

Leake, Chauncey. Papers. University of California, San Francisco.

LeMaistre, Charles A. Papers. Personal collection of C. A. LeMaistre.

LeMaistre, Charles A. Papers. The University of Texas M. D. Anderson Cancer Center, Houston, Texas.

Russell, William O. Papers. University of California at Davis.
Seybold, William D. Papers. John P. McGovern Historical Collection and Research Center, Houston Academy of Medicine–Texas Medical Center Library, Houston, Texas.
Shivers, Allan. Papers. Allan Shivers Library and Museum, Woodville, Texas.
Shivers, Allan. Papers. Center for American History, University of Texas, Austin, Texas.
Sutow, Wataru W. Papers. John P. McGovern Historical Collection and Research Center, Houston Academy of Medicine–Texas Medical Center Library, Houston, Texas.
Yarborough, Ralph. Papers. Center for American History, University of Texas, Austin, Texas.

Interviews

All interviews were conducted by James S. Olson except as noted. Names of institutions from which outside interviews were obtained are

CU	Oral History Research Office, Columbia University Libraries
NLM	Oral History Collection, U.S. National Library of Medicine, Bethesda, Maryland
TAMU-C	Special Collections, James Gilliam Gee Library, Texas A&M University–Commerce
UCD	Special Collections, Peter J. Shields Library, University of California at Davis
UTMDA	University of Texas M. D. Anderson Cancer Center, Research Medical Library

Aikin, A. M. (TAMU-C)
Aikin, John
Ainsworth, Joseph T.
Aldridge, Bettie
Alexanian, Raymond
Almond, Peter
Alt, Joyce
Anderson, Hesta
Anderson, Roger W.
Anderson, Thomas D.
Anglin, Maud
Arlinghaus, Ralph B.
Arnim, Joan L.
Bachrach, David J.
Baines, Rosie L.
Baker, Dennis
Balch, Charles M.
Ballantyne, Alando
Baumgartner, Karen
Baynham-Fletcher, Laura
Becker, Frederick F.
Belcher, Betty
Berger, Vesta

Beto, George
Blanco, Armandina
Bloch, Richard
Blumenschein, George
Bodey, Gerald P.
Bowen, James M.
Bryant, Teresa
Bull, Lynn
Cato, James
Clark, Randy
Clark, R. Lee (TAMU-C)
Clayton, William (CU)
Coatney, Frances
Coe, David
Coffey, Edward
Coffey, Joan
Cohen, Lorenzo
Cole, Thomas
Cooper, Sue
Cording, Frances
Cording, Richard
Cox, James D.
Cumley, Rosemary

Danburg, Debra
DeBakey, Michael (NLM)
Delclos, Luis
Deng, Furgen
DePew, William
DeWees, Andrew
Dickson, Jayn
Dodd, Gerald D.
Dodd, Gerald D. (UTMDA)
Douglas, Paul Clayton (CU)
Dowdy, Ray
Etheredge, M. B.
Ferrin, Brent
Fletcher, Gilbert H.
Frazier, Charles
Freedman, Ralph S.
Frei, Emil (UTMDA)
Freireich, Emil J
Freireich, Emil J (UTMDA)
Frost, Dodie
Garnett, Betty
Gates, Amos
Gebert, Alvin
Gehan, Edmund A.
Gehan, Edmund A. (UTMDA)
Gibbs, Harry Robert
Goepfert, Helmuth
Goodwin, Gale
Gritz, Ellen R.
Grona, Robert
Guillory, Cassandra
Guinee, Vincent
Gutterman, Jordan
Hale, JoAnne
Harston, Clive
Henri-Foley, Daphne
Hersh, Evan M.
Hickey, Rose
Hilkemeyer, Renilda
Hill, C. Stratton
Hogstrom, Kenneth R.
Holmes, Harry D.
Hsu, T. C.
Hubona, Colleen
Hutchinson, Marvin
Ivy, A. C. (NLM)
Jackson, William T.

Jaffe, Norman
Johnson, Jimmy A.
Johnson, Pamela
Jones, Lovell A.
Jones, Martha
Kaled, Susan
Kelsey, Mavis P.
Kolocsay, Attila
Krakoff, Irwin H.
Lasker, Albert David (CU)
Lasker, Mary (CU)
Lawson, Page
Lawson, Roy
Leake, Chauncey (NLM)
LeClear, Patti Watson
Lee, Annie
LeMaistre, Charles A.
LeMaistre, Charles A. (UTMDA)
Lindberg, Robert D.
Lovett, Mary Nell Jeffers
Macdonald, Eleanor
Macdonald, Eleanor (UTMDA)
Mansell, Peter W. A.
Mansell, Peter W. A. (UTMDA)
Martin, Richard G.
Mathis, James
McKelvey, Eugene M.
McKinley, James
McNeese, Marsha D.
Mendelsohn, John
Meyer, Margaret W.
Moore, William T.
Morton, Dorothy
Mulvey, Patrick
Munger, Edward
Musgrove, John
Olson, Julia
Pagel, Walter
Painter, Joseph T.
Parker, Jack
Pollock, Raphael E.
Polunsky, Blanche
Prier, David
Puduvalli, Vijay
Ratliff, Karen
Ratliff, Sophie
Ricketts, Harry

Rios, Mary Beth
Roberts, Randy
Rockefeller, William (CU)
Rogge, Mary Lou
Rosenblum, Michael
Russell, William O. (UCD)
Samuels, Melvin L.
Satre, Thomas
Scarstad, John
Schier, Mary Jane
Schmidt, Benno (CU)
Schnipper, Hester Hill
Schover, Leslie R.
Seybold, William D.
Shivers, Allan
Shotwell, Judith
Sinkovics, Joseph G.
Sinkovics, Joseph G. (UTMDA)

Skarstad, John
Smith, Alfred R.
Smith, Samantha
Spears, Betty
Stringer, Marsha
Stuyck, Stephen C.
Tedder, Patricia
Thompson, Geneva
Tomasovic, Stephen P.
Travis, Elizabeth
Trujillo, Elizabeth
Trujillo, Luisa
Valenza, Maureen
Van Nevel, Paul
Wall, Marion (Lowrey)
Willett, Ellen
Woodard, Lois
Zimmerman, Stuart O.

Selected Secondary Sources

Altman, Roberta. *Waking Up, Fighting Back: The Politics of Breast Cancer*. Boston: Little Brown, 1996.

Armstrong, Lance. *It's Not about the Bike: My Journey Back to Life*. New York: Putnam, 2000.

Aronowitz, Robert A. *Unnatural History: Breast Cancer and American Society*. New York: Cambridge University Press, 2007.

ASCO. *40 Years of Quality Care: History of ASCO*. Alexandria, Va.: American Society of Clinical Oncology, 2004.

Batt, Sharon. *Patient No More: The Politics of Breast Cancer*. Charlottetown, Canada: Gynergy, 1994.

Bazell, Robert. *Her-2: The Making of Herceptin, a Revolutionary Treatment for Breast Cancer*. New York: Random House, 1998.

Beeman, Edward A. "Robert J. Huebner, M.D.: A Virologist's Odyssey." Office of NIH History, 2005. *http://history.nih.gov/01docs/historical/2015.htm*.

Bishop, J. Michael. *How to Win the Nobel Prize: A Life in Science*. Cambridge: Harvard University Press, 2003.

Boylan, Anne. *The Origins of Women's Activism: New York and Boston, 1797–1840*. Chapel Hill: University of North Carolina Press, 2002.

Boston Women's Health Book Collective. *Our Bodies, Ourselves*. New York: Simon and Schuster, 1971.

Böttcher, Helmuth. *Wonder Drugs: A History of Antibiotics*. Philadelphia: Lippincott, 1964.

Brandt, Alan. *No Magic Bullet: A Social History of Venereal Disease in America*. New York: Oxford University Press, 1985.

Brynner, Rock, and Trent D. Stephens. *Dark Remedy: The Impact of Thalidomide and Its Revival as a Vital Medicine*. Cambridge, Mass.: Perseus, 2001.

Burns, Chester R. *Saving Lives, Training Caregivers, Making Discoveries: A Centennial*

History of the University of Texas Medical Branch at Galveston. Austin: Texas State Historical Association, 2003.

Bush, Barbara. *Barbara Bush: A Memoir.* New York: Scribner's, 1994.

Bush, George. *All the Best, George Bush: My Life in Letters and Other Writings.* New York: Scribner's, 1999.

Cairnes, John. *Cancer: Science and Society.* San Francisco: W. H. Freeman, 1978.

Cayleff, Susan E. *Babe: The Life and Legend of Babe Didrikson Zaharias.* Urbana: University of Illinois Press, 1995.

Clark, Joseph Lynn. *Thank God, We Made It! A Family Affair with Education.* Austin: University of Texas, 1969.

Clow, Barbara. *Negotiating Disease: Power and Cancer Care, 1900–1950.* Montreal: McGill-Queens University Press, 2001.

Cohen, John. *Shots in the Dark: The Wayward Search for an AIDS Vaccine.* New York: W. W. Norton, 2001.

Committee on Diet and Health, Food and Nutrition Board, Commission on Life Sciences, National Research Council. *Diet and Health: Implications for Reducing Chronic Disease Risk.* Washington, D.C.: National Academy Press, 1982.

Cooke, Robert. *Dr. Folkman's War: Angiogenesis and the Struggle to Defeat Cancer.* New York: Random House, 2001.

Cox, Patrick. *Ralph W. Yarborough: The People's Senator.* Austin: University of Texas Press, 2001.

Creager, Angela N. H., Elizabeth Lunbeck, and Londa Schiebinger, eds. *Feminism in Twentieth-Century Science, Technology, and Medicine.* Chicago: University of Chicago Press, 2001.

Crile, George, Jr. *Cancer and Common Sense.* New York: Viking, 1955.

Dally, Ann. *Women Under the Knife: A History of Surgery.* New York: Routledge, 1991.

Davis, Loyal. *Fellowship of Surgeons: A History of the American College of Surgeons.* Chicago: American College of Surgeons, 1988.

DeGregorio, Michael, and Valerie J. Wiebe. *Tamoxifen and Breast Cancer.* New Haven: Yale University Press, 1993.

Dodd, Gerald. *A History of Diagnostic Imaging.* Houston: University of Texas M. D. Anderson Cancer Center, 2006.

Drug Treatment of Cancer Pain in a Drug-Oriented Society. Ed. C. Stratton Hill Jr. and William S. Fields. New York: Raven Press, 1987.

Duffy, John. *From Humors to Medical Science: History of American Medicine.* Urbana: University of Illinois Press, 1993.

Efron, Edith. *The Apocalyptics: Cancer and the Big Lie: How Environmental Politics Controls What We Know about Cancer.* New York: Simon and Schuster, 1984.

Epstein, Samuel. *Politics of Cancer.* Garden City, N.Y.: Anchor, 1978.

The First Twenty Years of the University of Texas M. D. Anderson Hospital and Tumor Institute. Houston: M. D. Anderson Hospital and Tumor Institute, 1964.

Fleming, Lamar, Jr. *Growth of the Business of Anderson, Clayton, and Company.* Houston: Texas Gulf Coast Historical Association, 1966.

Frei, Terry. *Horns, Hogs, and Nixon Coming: Texas vs. Arkansas in Dixie's Last Stand.* New York: Simon and Schuster, 2002.

Freireich, Emil J, and Noreen A. Lemak. *Milestones in Leukemia Research and Therapy.* Baltimore: Johns Hopkins University Press, 1991.

Fujimura, Joan H. *Crafting Science: A Sociohistory of the Quest for the Genetics of Cancer.* Cambridge: Harvard University Press, 1996.

Gallo, Robert. *Virus Hunting: AIDS, Cancer, and the Human Retrovirus: A Story of Scientific Discovery.* New York: Basic, 1991.

Gardner, Kirsten E. *Early Detection: Women, Cancer, and Awareness Campaigns in the Twentieth-Century United States.* Chapel Hill: University of North Carolina Press, 2006.

Geiger, Roger L. *To Advance Knowledge: The Growth of American Research Universities, 1900–1940.* New York: Oxford University Press, 1986.

Gimlin, Debra L. *Body Work: Beauty and Self-Image in American Culture.* Berkeley: University of California Press, 2002.

Giroud, Françoise. *Marie Curie: A Life.* New York: Holmes and Meier, 1986.

Grob, Gerald N. *The Deadly Truth: A History of Disease in America.* Cambridge: Harvard University Press, 2002.

Guither, Harold D. *Animal Rights: History and Scope of a Radical Social Movement.* Carbondale: Southern Illinois University Press, 1998.

Harvey, A. McGehee. *Science at the Bedside: Clinical Research in American Medicine.* Baltimore: Johns Hopkins University Press, 1981.

Henderson, Richard B. *Maury Maverick: A Political Biography.* Austin: University of Texas Press, 1970.

Hilts, Philip J. *Protecting America's Health: The FDA, Business, and One Hundred Years of Regulation.* New York: Knopf, 2003.

Hynes, Patricia. *The Recurring Silent Spring.* New York: Pergamon, 1989.

International Union against Cancer. Tenth International Cancer Congress. *Abstracts.* Austin: University of Texas Press, 1970.

Jones, James. *Bad Blood: The Tuskegee Syphilis Experiment.* New York: Free Press, 1993.

Kelsey, Mavis Parrott, Sr. *Doctoring in Houston: Including My Story of the Kelsey-Seybold Clinic and the Kelsey-Seybold Foundation, Memoirs from 1949 to 1996.* Houston: The Foundation, 1996.

Kenney, Martin. *Biotechnology: The University-Industrial Complex.* New Haven: Yale University Press, 1986.

King, Larry. *Confessions of a White Racist.* New York: Viking, 1971.

Kluger, Jeffrey. *Splendid Solution: Jonas Salk and the Conquest of Polio.* New York: Putnam, 2005.

Kluger, Richard. *Ashes to Ashes: America's Hundred-Year Cigarette War, the Public Health, and the Unabashed Triumph of Philip Morris.* New York: Knopf, 1996.

Laurence, Leslie, and Beth Weinhouse. *Outrageous Practices: The Alarming Truth about How Medicine Mistreats Women.* New York: Fawcett Columbine, 1994.

Lazlo, John. *The Cure of Childhood Leukemia: Into the Age of Miracles.* New Brunswick, N.J.: Rutgers University Press, 1995.

Lebineau, Jonathan. *Medical Science and Medical Industry: The Formation of the American Pharmaceutical Industry.* Baltimore: Johns Hopkins University Press, 1987.

Leopold, Ellen. *A Darker Ribbon: Breast Cancer, Women, and Their Doctors in the Twentieth Century.* Boston: Beacon Press, 1999.

Lerner, Barron H. *The Breast Cancer Wars: Hope, Fear, and the Pursuit of a Cure in Twentieth-Century America.* New York: Oxford University Press, 2001.

Ludmerer, Kenneth M. *Learning to Heal: The Development of American Medical Education.* New York: Basic, 1985.

———. *Time to Heal: American Medical Education from the Turn of the Century to the Era of Managed Care.* New York: Oxford University Press, 1999.

Macon, N. Don. *Clark and the Anderson: A Personal Profile.* Houston: Texas Medical Center, 1976.

———. *Monroe Dunaway Anderson, His Legacy: A History of the Texas Medical Center.* Houston: Texas Medical Center, 1994.

———. *Mr. John H. Freeman and Friends: A Story of the Texas Medical Center and How It Began.* Houston: Texas Medical Center, 1973.

———. *South from Flower Mountain: A Conversation with William B. Bates.* Houston: Texas Medical Center, 1975.

Marcus, Alan. *Cancer from Beef: DES, Federal Food Regulation, and Consumer Confidence.* Baltimore: Johns Hopkins University Press, 1994.

Markell, Regina Morantz. *Sympathy and Science: Women Physicians in America.* New York: Oxford University Press, 1985.

McCay, Mary A. *Rachel Carson.* New York: Twayne, 1993.

Micklethwait, John, and Adrian Wooldridge. *The Witch Doctors: Making Sense of the Management Gurus.* New York: Times Books, 1996.

Morantz-Sanchez, Regina Markell. *Sympathy and Science: Women Physicians in American Medicine.* New York: Oxford University Press, 1985.

More, Ellen Singer, and Maureen Milligan, eds. *The Empathic Practitioner: Empathy, Gender, and Medicine.* New Brunswick, N.J.: Rutgers University Press, 1994.

Morgen, Sandra. *Into Our Own Hands: The Women's Health Movement in the United States, 1969–1980.* New Brunswick, N.J.: Rutgers University Press, 2002.

Moursund, Walter H. *A History of Baylor University, College of Medicine, 1900–1953.* Houston: n.p., 1956.

Nevidjon, Brenda, ed. *Building a Legacy: Voices of Oncology Nurses.* Durham, N.C.: Duke University Press, 1995.

Nuland, Sherwin B. *Doctors: The Biography of Medicine.* New York: Knopf, 1988.

Olby, Robert C. *The Path to the Double Helix.* Seattle: University of Washington Press, 1974.

Olson, James S. *Bathsheba's Breast: Women, Cancer, and History.* Baltimore: Johns Hopkins University Press, 2002.

———. *The History of Cancer: An Annotated Bibliography.* Westport, Conn.: Greenwood, 1989.

Oncology, 1970: Being the Proceedings of the Tenth International Cancer Congress. Chicago: Year Book Medical Publishers, 1971.

Patterson, James T. *The Dread Disease: Cancer and Modern American Culture.* Cambridge: Harvard University Press, 1987.

Pernick, Martin S. *A Calculus of Suffering: Pain, Professionalism, and Anesthesia in Nineteenth-Century America.* New York: Columbia University Press, 1985.

Proctor, Robert N. *Cancer Wars: How Politics Shapes What We Know and Don't Know about Cancer.* New York: Basic, 1995.

Rather, L. J. *The Genesis of Cancer: A Study in the History of Ideas*. Baltimore: Johns Hopkins University Press, 1978.

Rettig, Richard A. *Cancer Crusade: The Story of the National Cancer Act of 1971*. Princeton, N.J.: Princeton University Press, 1977.

Reynolds, Moira Davison. *How Pasteur Changed History: The Story of Louis Pasteur and the Pasteur Institute*. Bradenton, Fla.: McGuinn and McGuire, 1994.

Roberts, Randy, and James S. Olson. *John Wayne: American*. New York: Free Press, 1995.

Rodwin, Marc A. *Medicine, Money, and Morals: Physicians' Conflicts of Interest*. New York: Oxford University Press, 1993.

Ross, Walter. *Crusade: The Official History of the American Cancer Society*. New York: Arbor House, 1987.

Rothman, David. *Strangers at the Bedside: A History of How Law and Bioethics Transformed Medical Decision Making*. New York: Basic, 1991.

Rothstein, William G. *American Medical Schools and the Practice of Medicine: A History*. New York: Oxford University Press, 1987.

Rush, Benjamin. *Letters*. 2 vols. Ed. L. H. Butterfield. Princeton, N.J.: Published for the American Philosophical Society by Princeton University Press, 1951.

Russell, Louise. *Is Prevention Better than Cure?* Washington, D.C.: Brookings Institution, 1986.

Ruzek, Sheryl Burt. *The Women's Health Movement: Feminist Alternatives to Medical Control*. New York: Praeger, 1978.

Schacter, Bernice. *The New Medicines: How Drugs Are Created, Approved, Marketed, and Sold*. Westport, Conn.: Praeger, 2006.

Schaffer, Kristen. *Daniel H. Burnham: Visionary Architect and Planner*. New York: Rizzoli, 2003.

Scully, Matthew. *Dominion: The Power of Man, the Suffering of Animals, and the Call to Mercy*. New York: St. Martin's, 2002.

Sharif, Myron. *Fury on Earth: A Biography of Wilhelm Reich*. New York: St. Martin's, 1983.

Shaughnessy, Donald. "The Story of the American Cancer Society." Ph.D. diss., Columbia University, 1957.

Sibley, Marilyn M. *The Port of Houston: A History*. Austin: University of Texas Press, 1968.

Smith, Henry Nash. *The Controversy at the University of Texas, 1939–1946*. Austin: University of Texas Students Association, 1946.

Sontag, Susan. *Illness as Metaphor*. New York: Farrar, Straus and Giroux, 1978.

Spangenburg, Ray, and Diane K. Moser. *Disease Fighters since 1950*. New York: Facts on File, 1996.

Stabiner, Karen. *To Dance with the Devil: The New War on Breast Cancer*. New York: Delacorte, 1997.

Starr, Paul. *The Social Transformation of American Medicine: The Rise of a Sovereign Profession and the Making of a Vast Industry*. New York: Basic, 1982.

Steinmark, Freddie. *I Play to Win*. Boston: Little, Brown, 1971.

Swaim, Joan Hewatt. *Walking TCU: A Historic Perspective*. Fort Worth: Texas Christian University Press, 1992.

Taylor, Harvey Grant. *Remembrances and Reflections*. Houston: University of Texas Health Science Center at Houston, 1991.

Tobey, James. *Cancer: What Everyone Should Know about It.* New York: Knopf, 1932.

Troetel, Barbara R. "Three-Part Disharmony: The Transformation of the Food and Drug Administration." Ph.D. diss., City University of New York, 1996.

Van Eys, Jan, and James M. Bowen. *The Common Bond: The University of Texas System Cancer Code of Ethics.* Springfield, Ill.: Thomas, 1986.

Walsh, Mary Roth. *"Doctors Wanted: No Women Need Apply": Sexual Barriers in the Medical Profession, 1945–1975.* New Haven: Yale University Press, 1977.

Weinberg, Robert A. *Racing to the Beginning of the Road: The Search for the Origin of Cancer.* New York: Harmony, 1996.

Welch, H. Gilbert. *Should I Be Tested for Cancer? Maybe Not and Here's Why.* Berkeley: University of California Press, 2005.

Wheatley, Steven C. *The Politics of Philanthropy: Abraham Flexner and Medical Education.* Madison: University of Wisconsin Press, 1988.

Whorton, James C. *Nature Cures: The History of Alternative Medicine in America.* New York: Oxford University Press, 2000.

Yafa, Stephen. *Big Cotton: How a Humble Fiber Created Fortunes, Wrecked Civilizations, and Put America on the Map.* New York: Viking, 2004.

INDEX

Page numbers preceded by a "G" refer to illustrations in the gallery.